5G CORE NETWORKS

5G CORE NETWORKS
Powering Digitalization

STEFAN ROMMER

PETER HEDMAN

MAGNUS OLSSON

LARS FRID

SHABNAM SULTANA

CATHERINE MULLIGAN

ELSEVIER

ACADEMIC PRESS
An imprint of Elsevier

Academic Press is an imprint of Elsevier
125 London Wall, London EC2Y 5AS, United Kingdom
525 B Street, Suite 1650, San Diego, CA 92101, United States
50 Hampshire Street, 5th Floor, Cambridge, MA 02139, United States
The Boulevard, Langford Lane, Kidlington, Oxford OX5 1GB, United Kingdom

Notices
Knowledge and best practice in this field are constantly changing. As new research and experience broaden our understanding, changes in research methods, professional practices, or medical treatment may become necessary.

Practitioners and researchers must always rely on their own experience and knowledge in evaluating and using any information, methods, compounds, or experiments described herein. In using such information or methods they should be mindful of their own safety and the safety of others, including parties for whom they have a professional responsibility.

To the fullest extent of the law, neither the Publisher nor the authors, contributors, or editors, assume any liability for any injury and/or damage to persons or property as a matter of products liability, negligence or otherwise, or from any use or operation of any methods, products, instructions, or ideas contained in the material herein.

Library of Congress Cataloging-in-Publication Data
A catalog record for this book is available from the Library of Congress

British Library Cataloguing-in-Publication Data
A catalogue record for this book is available from the British Library

ISBN 978-0-08-103009-7

For information on all Academic Press publications
visit our website at https://www.elsevier.com/books-and-journals

Publisher: Mara Conner
Acquisition Editor: Tim Pitts
Editorial Project Manager: Isabella C. Silva
Production Project Manager: Anitha Sivaraj
Cover Designer: Greg Harris

Typeset by SPi Global, India

Contents

Foreword   *ix*
Acknowledgments   *xi*

1. Introduction   1

   1.1 5G—A new era of connectivity   1
   1.2 A step change   1
   1.3 A new context for operators   2
   1.4 The road to 5G network deployments   2
   1.5 3GPP release 15 and 16   2
   1.6 Core requirements   4
   1.7 New service grades   4
   1.8 Structure of this book   5

2. Drivers for 5G   7

   2.1 Introduction   7
   2.2 New use cases   7
   2.3 New technologies   9

3. Architecture overview   15

   3.1 Introduction   15
   3.2 Two perspectives on 5G Core   19
   3.3 Service-based architecture (SBA)   22
   3.4 The core of the core   26
   3.5 Connecting the core network to mobile devices and radio networks   28
   3.6 Mobility and data connectivity   30
   3.7 Policy control and charging   35
   3.8 5GC interworking with EPC   37
   3.9 Voice services   41
   3.10 Messaging services   44
   3.11 Exposure of network information   46
   3.12 Device positioning services   48
   3.13 Network analytics   49
   3.14 Public warning system   50
   3.15 Support for devices connected over non-3GPP access networks   52
   3.16 Network slicing   54
   3.17 Roaming   55

3.18 Storage of data   59
3.19 5G radio networks   59

4. EPC for 5G   **73**
4.1 Introduction   73
4.2 Key EPC functions   77
4.3 (Enhanced) Dedicated Core Networks ((e)DECOR)   84
4.4 Control and User Plane Separation (CUPS)   89

5. Key concepts   **105**
5.1 Architecture modeling   105
5.2 Service Based Architecture   105
5.3 Identifiers   107

6. Session management   **111**
6.1 PDU Session concepts   111
6.2 PDU Session types   114
6.3 User plane handling   121
6.4 Mechanisms to provide efficient user plane connectivity   126
6.5 Edge computing   132
6.6 Session authentication and authorization   134
6.7 Local Area Data Network   135

7. Mobility Management   **137**
7.1 Introduction   137
7.2 Establishing connectivity   138
7.3 Reachability   144
7.4 Additional MM related concepts   146
7.5 N2 management   150
7.6 Control of overload   157
7.7 Non-3GPP aspects   161
7.8 Interworking with EPC   162

8. Security   **171**
8.1 Introduction   171
8.2 Security requirements and security services of the 5G system   172
8.3 Network access security   176
8.4 Network domain security   192
8.5 User domain security   198
8.6 Lawful intercept   198

9. Quality-of-Service   **203**

    9.1 Introduction   203
    9.2 Flow based QoS framework   205
    9.3 Signaling of QoS   207
    9.4 Reflective QoS   210
    9.5 QoS parameters and characteristics   213

10. Policy control and charging   **217**

    10.1 Introduction   217
    10.2 Overview of policy and charging control   217
    10.3 Access and mobility related policy control   222
    10.4 UE policy control   224
    10.5 Management of Packet Flow Descriptions   227
    10.6 Network status analytics   228
    10.7 Negotiation for future background data transfer 228
    10.8 Session Management related policy and charging control 229
    10.9 Additional session related policy control features   237
    10.10 Charging   242

11. Network slicing   **247**

    11.1 Introduction   247
    11.2 Management and orchestration   249
    11.3 Network Slice selection framework   251

12. Dual connectivity   **265**

    12.1 Introduction   265
    12.2 Multi-RAT Dual Connectivity overall architecture 268
    12.3 MR-DC: UE and RAN perspective   272
    12.4 MR-DC: Subscription, QoS flows and E-RABs, MR-DC bearers 274
    12.5 Managing secondary RAN node handling for mobility and session
             management   278
    12.6 Security   282
    12.7 Reporting User Data Volume traversing via SN   283

13. Network functions and services   **287**

    13.1 5G core network functions   287
    13.2 Services and service operations   293

14. Protocols   **337**

    14.1 Introduction   337
    14.2 5G non-access stratum (5G NAS)   337

14.3 NG application protocol (NGAP)   343
14.4 Hypertext transfer protocol (HTTP)   347
14.5 Transport layer security (TLS)   360
14.6 Packet forwarding control protocol (PFCP)   363
14.7 GPRS tunneling protocol for the User Plane (GTP-U)   378
14.8 Extensible Authentication Protocol (EAP)   379
14.9 IP security (IPSec)   382
14.10 Stream Control Transmission Protocol (SCTP)   387
14.11 Generic routing encapsulation (GRE)   392

15. Selected call flows   **395**
15.1 Introduction   395
15.2 Registration and deregistration   396
15.3 Service Request   400
15.4 UE Configuration Update   404
15.5 PDU Session Establishment   407
15.6 Inter-NG-RAN handover   409
15.7 EPS interworking with N26   416
15.8 EPS fallback   422
15.9 Procedures for untrusted non-3GPP access   424

16. Architecture extensions and vertical industries   **431**
16.1 Overview   431
16.2 Architecture enhancements and extensions   431
16.3 New feature capabilities   437

17. Future outlook   **465**

References   *467*
Abbreviations   *471*
Index   *475*

Forewords

It is an extremely exciting time for the telecommunications industry with the roll-out of 5G. Previous generations of network technologies were to a greater or lesser extent about increasing speeds and accessibility for end-user consumers. More than just improving bandwidth and reducing latency, 5G is enabling truly disruptive solutions to emerge across all manner of industries. Globally, we can see numerous 5G networks and trials are being installed—ready to deliver on the promise of the 5G system. At Telstra, we are delivering Australia's first 5G network that will transform the way we live and work through providing more capacity, faster speeds and lower latency. Rollouts have commenced in all major cities across Australia and the rest of the country is not be far behind.

In addition to providing consumers high-speed downloads and uploads for streaming and sharing on social media, we are forging new partnerships to illustrate the potential of the technology across a wide variety of industries. One example is our collaboration with the Commonwealth Bank of Australia that brings together technology providers and the financial services sector to explore 5G Edge computing use cases and network architectures. Through this, we are exploring what the future of banking will look like over 5G—showcasing what the bank of the future might look like and exploring how 5G edge compute can reduce network infrastructure requirements at individual bank branches.

This is just the beginning for our engagement in 5G, with many other industrial solutions in development that will bring a step-change improvement to the lives of people in Australia. We're exploring IoT, automotive and drone safety in 5G networks with a variety of corporate partners, including Thales to imagine a safe and secure ecosystem for the management of Australia's low-altitude airspace. These are merely the tip of the iceberg of services that will be enabled by the transition to and we're proud to be leading the transition in Australia.

The changes enabled by 5G are due to new radio and different spectrum. At the same time, however, there have been fundamental shifts in the core network specifications. For 5G Core, 3GPP has set out to define a new 5G core that fully integrates web protocols and is adapted to cloud native environments.

This book therefore comes at an important time—as 5G starts initial deployments. It provides a much needed and accessible description of the 3GPP standards that make up the 5G core. The topics in this book are key to understanding how the core network is constructed for 5G and will enable readers to quickly learn the inner workings of this crucial evolution of the core network. The team has been deeply involved in the development of the 3GPP specifications and are uniquely placed to explain these concepts.

Håkan Eriksson
Group CTO, Telstra, Melbourne, VIC, Australia

5G is happening. In one way it is the obvious thing; after 4G, 5G must follow, with higher bandwidth, lower latency and new frequency bands. In other ways it is less obvious holding the promise of those enhancements not just improving current use cases, but fundamentally changing things by passing thresholds for example for latency that enables totally new use cases. With the possibility of more radical changes in the network design of the core network comes the possibility of enabling the internet of things in a different way. It is still not obvious to me how this will play out, and it reminds me a bit of when the industry developed 3G in the 1990s and we used to spend time over coffee debating what on earth people would use the mind blowing 3G bit pipe for. We extrapolated some of the fixed internet use cases and anticipated video calls as the expected prolongation of voice. We totally missed out on things like social media, open source development of apps and everything going mobile that later came with the birth of the smart phone and the true network improvements that 4G in the end delivered.

In a similar way I believe we have only scratched the surface of what connecting other things than mobile phones will bring. In automotive for instance while connectivity has been around for quite some time, the development is moving from connectivity being a high end option in the past, to connectivity being standard and the fundamental enabler of continuously improving cars over time with software. As all cars become connected all the time we are able to do some rather obvious things like upgrading the software continuously over the air, but also dynamically making use of the connectivity to enhance features in the car. And while cars for a foreseeable future will have mission critical software installed in on board computers, the 5G connectivity holds the promise of a network with such high performance, low latency and low cost that it can allow us to dynamically enhance the on-board capabilities of vehicles with off-board processing and connection to surrounding systems, enabling everything from true real time updated maps to enhancements of advanced driver support and autonomous drive functions. This development is also the foundation of personalizing cars and many other things, meaning users bring their digital world with them wherever they go, making them the center of all eco systems rather than having to move between different eco systems each centered round different things in their life.

Earlier in my career, the previous edition of this book was my go-to book for mobile core technologies. I look forward to the new edition being the same for 5G.

Ödgärd Andersson
Chief Digital Officer, Volvo Car Corporation, Gothenburg, Sweden

Acknowledgments

A work of this nature is not possible without others' support.

The authors would like to gratefully acknowledge the contribution of many of our colleagues at Ericsson, in particular David Allan, Aldo Bolle, Åke Busin, Torbjörn Cagenius, Qian Chen, George Foti, Jesus De Gregorio, Magnus Hallenstål, Maurizio Iovieno, Ralf Keller, Vesa Lehtovirta, Alessandro Mordacci, Stefan Parkvall, Anders Ryde, Alexander Vesely, Mikael Wass, and Frank Yong Yang.

We would also like to thank our families. Writing this book would not have been possible without their generosity and support throughout the process.

CHAPTER 1

Introduction

1.1 5G—A new era of connectivity

The telecommunications industry has embarked on a dramatic transition and one that—if successful—will see it redefine its role in industry and society. 5G, while often portrayed as a tool for higher speeds or critical to the development of the so called Industry 4.0, represents a foundational shift for wireless communications—one that places it directly at the center of a truly digital economy. This overhaul of communications is therefore unlike any that have gone before it—it is not the same as the move from 2G to 3G or 3G to 4G—it is a step change that the industry may not see again for quite some time.

The 5G architecture itself consists of two parts—the new Radio Network (NG-RAN) supporting the New Radio (NR), and the 5G Core Network (5GC). Both have changed considerably compared to previous generations of technology. This book focuses on 5GC, providing short forays into NR where it aids understanding of the interactions towards the core network. A detailed description of NR is, however, beyond the scope of this book and interested readers are directed to Dahlman et al. (2018).

1.2 A step change

The first broad scale adoption of mobile technologies started with GSM (2G)—released in 1991, which focused on calls and text messaging. WCDMA (3G), released in 1999 gave consumers the ability to browse the internet and use feature phones. It was not until the introduction of LTE (4G)—in 2008, however, that we saw the broad adoption of Mobile Broad Band (MBB) and the uptake of video and data traffic on the all-IP network including the development of 'apps' on smartphones. Each generation saw a large increase in bandwidth and speeds provided with end-user consumers as the core focus. 5G is unlike the previous generation of networks; it represents a shift from operators having end-users as customers to over time having industries as their main customers. This represents not just a technology shift, but a business model shift unlike any previously as well. New players may very well enter the market because of the disruptive capabilities of 5G.

5G is a more ambitious approach to network architectures—not only incorporating requirements from the telecommunications industry but other industries and at the same time including cloud-native and web scale technologies such as HTTP. It is quite simply a new approach to developing architecture and delivering services on a global scale.

5G Core Networks
https://doi.org/10.1016/B978-0-08-103009-7.00001-6
1

1.3 A new context for operators

Broken up into building blocks covering access, transport, cloud, network applications and management (including orchestration and automation), 5G systems aim to provide a higher level of abstraction designed to simplify network management and operations. In addition, new services will need to be rapidly implemented on the network as new business models emerge that demand operators move to programmable, software-based networks that deliver services on-demand and in an 'as a Service' manner. Throughout this book, we illustrate where the technology itself overlaps with some of these new business models providing a unique insight into how some of those decisions have been made. In addition, where previously human customers were making requests of the networks, with 5G there is an increased level of non-human, i.e., machine and software, requests that means the entire way services are developed and delivered needs to change.

1.4 The road to 5G network deployments

The initial work on defining the requirements and vision on 5G networks was carried out in ITU-R in 2012. ITU formally refers to this as IMT-2020. A good reference is Dahlman et al. (2018). This was followed by multiple more detailed studies in ITU-R itself, as well as in industry fora and research projects around the world.

The initial work to develop the 5G specifications to meet the ITU-R IMT-2020 requirements was done in 2014, picking up speed in 2015 and 2016. Trials of 5G systems have been in place in several countries, with commercial rollouts planned for most markets around 2020. Outlining the core network evolution in an easy to use and accessible manner so that engineers and other interested parties can understand the changes brought about by 5G is therefore the core reason for us writing this book.

Several early commercial 5G systems became available already from late 2018 and early 2019. Some initial 5G network deployments include:

- Verizon and AT&T have both launched USA's first 5G services during 2018 and 2019
- Telstra has rolled out multiple 5G areas across Australia during 2018 and 2019
- Services targeting enterprise use cases launched by all three Korean operators by the end of 2018
- Early eMBB services were launched in Korea, the U.S., Switzerland and the U.K. in the first half of 2019

1.5 3GPP release 15 and 16

5G Core is described in a set of specifications developed by the 3rd Generation Partnership Project (3GPP) and captured in Release 15 (Rel-15) and subsequent releases. Rel-15 was the first full set of 5G standards and was released in several steps between June 2018 and early 2019. Rel-16 is planned to be released early 2020 and planning of work has commenced on Release-17 with an aim to have specifications ready in 2021 or 2022.

Rel-15 contained e.g.:

- Architecture for Non-Stand Alone (NSA), i.e., New Radio (NR) used with the LTE and EPC infrastructure Core Network
- Architecture for Stand-Alone (SA), i.e., NR is connected to the 5G Core Network (5GC)
- 5GC using a Service-Based Architecture (SBA)
- Support of virtualized deployment
- Network functionalities to provide registration, deregistration, authorization, mobility and security
- Data communication with IP, Ethernet and Unstructured data
- Support of concurrent local and central access to a data network
- Support for Edge Computing
- Network Slicing
- Unified access control
- Converged architecture to support non-3GPP access
- Policy framework and QoS support
- Network capability exposure
- Multi-Operator Core Network, i.e., sharing same NG-RAN by multiple core networks
- Support of specific services such as SMS, IMS, Location Services for emergency services
- Public Warning System (PWS)
- Multimedia Priority Services (MPS)
- Mission Critical Services (MCS)
- PS Data Off
- Interworking between the 5GS and 4G

Rel-16 is set to contain several additions, many specifically aimed at different industry verticals:

- V2X
- Access Traffic Steering, Switch and Splitting support in the 5G system architecture (ATSSS)
- Cellular IoT support and evolution for the 5G System (5G_CIoT)
- Enablers for Network Automation for 5G (eNA)
- Enhancing Topology of SMF and UPF in 5G Networks (ETSUN)
- Enhancement to the 5GC Location Services (5G_eLCS)
- Enhanced IMS to 5GC Integration (eIMS5G_SBA)
- 5GS Enhanced support of Vertical and LAN Services—5G-LAN aspects
- 5GS Enhanced support of Vertical and LAN Services—TSN aspects

- 5GS Enhanced support of Vertical and LAN Services—non-public network aspects
- System enhancements for Provision of Access to Restricted Local Operator Services by Unauthenticated UEs (PARLOS) NOT FOR 5G
- Enhancements to the Service-Based 5G System Architecture (5G_eSBA)
- Enhancement of URLLC supporting in 5GC (5G_URLLC)
- User Data Interworking and Coexistence (UDICOM)
- Optimizations on UE radio capability signaling (RACS)
- Wireline support (5WWC)

1.6 Core requirements

The 5GC has been designed to implicitly and explicitly support several architectural principles:
- Support for a service-based architecture for modularized network services
- Consistent user experience between 3GPP and non-3GPP access networks
- Harmonization of identity, authentication, QoS, policy and charging paradigms
- Adaption to cloud native and web scale technologies
- Edge Computing and nomadic/fixed access; bring computing power closer to the point where sensor data from remote, wireless devices would be collected, eliminating the latency incurred by public cloud-based applications
- Improved quality of service, and extend that quality over a broader geographic area
- Machine-to-machine communications services that could bring low-latency connectivity to devices such as self-driving cars and machine assembly robots;

The architectural impacts of these are described more fully in Chapter 3.

1.7 New service grades

5G allows for three service grades that may be tuned to the special requirements of their customers' business models:
- Enhanced Mobile Broadband (eMBB) aims to service more densely populated metropolitan centers with downlink speeds approaching 1 Gbps (gigabits-per-second) indoors, and 300 Mbps (megabits-per-second) outdoors.
- Massive Machine Type Communications (mMTC) enables machine-to-machine (M2M) and Internet of Things (IoT) applications that a new wave of wireless customers may come to expect from their network, without imposing burdens on the other classes of service
- Ultra-Reliable and Low Latency Communications (URLLC) would address critical needs communications where bandwidth is not quite as important as speed—specifically, an end-to-end latency of 1 ms or less.

1.8 Structure of this book

This book is roughly divided into four separate parts.

1.8.1 Part one: Introduction, architecture and scope of book

Chapters 2–4 provide an introductory overview and scope of the book. This includes the key technologies used within 5GC and a high-level architectural introduction. Chapter 3 forms the basis of understanding for the rest of the book. Chapter 4 meanwhile illustrates EPC for 5G—more details of this is beyond the scope of this book, but interested readers are referred to 3GPP TS 23.401.

1.8.2 Part two: Core concepts of 5GC

Chapters 5–12, meanwhile provide a comprehensive overview of all the core concepts of 5GC that readers require to understand the entirety of the system. This includes modeling, session management, mobility, security, QoS, charging, network slicing and dual connectivity solutions. These concepts form a fundamental base for the remaining chapters.

1.8.3 Part three: 5GC nuts and bolts

Chapters 13–15 provide the in-depth knowledge required for all practitioners in the 5GC space, going into detail of how the core concepts in part two fit together and work as a unified whole to deliver the 5G Core Network. Readers are presented with deep dive into Network functions, reference points, protocols and call flows. After reading part 3, readers will be ready to work with 5GC.

1.8.4 Part four: Release 16 and beyond

Chapters 16 and 17 conclude the book with a description of architecture extensions in Release 16 and some overview of the support for vertical industries. The book concludes with a future vision for the development of 5GC going forward.

CHAPTER 2

Drivers for 5G

2.1 Introduction

The requirements on mobile and other types of communications networks have been growing significantly over the past decade. From humble beginnings just providing phone calls and text messages, these networks are now expected to form the underlying infrastructure for a truly digital economy—enabling new means of operation as the world transitions from 20th century operating models into ones that are designed for the challenges of the 21st. The drivers for 5G are far more, therefore, than merely the drive for a new core network but rather the result of intersecting requirements and demands—namely

(1) Business case demands from a broader set of economic actors, including industrial companies driving new use cases,
(2) New technologies for delivering core network components creating expectations of more efficient and flexible operations, and
(3) Shifts in how business, society and environmental needs are balanced to deliver services in a new way.

2.2 New use cases

Previous versions of mobile technologies illustrated the potential of these technologies to deliver innovative, previously un-thought of services to a global subscriber base. These have driven ideas and expectations about what the next generation of mobile technologies could bring—creating a broad ranging set of market expectations on what value 5G technologies will bring to different industries and areas of society. The possibilities for both significant cost savings and new revenue enablers has therefore created a large interest in 5G across multiple industries, not only among traditional mobile service providers and users.

For services that already are offered using 4G or older technologies, such as mobile broadband services, 5G is providing both an enhanced user experience and a more cost-efficient solution. The enhanced user experience is mainly experienced as overall higher data rates—not so much about higher peak data rates, but more about providing an increased average data rate across the network. Users of mobile broadband services will therefore experience a higher quality of service.

5G Core Networks
https://doi.org/10.1016/B978-0-08-103009-7.00002-8

Also related to the consumer segment, there are expectations that the low latency of 5G radio access would nicely suit time-sensitive services such as mobile gaming. While the full business case to design infrastructure to cater for mobile gaming or other low latency-sensitive services remains to be developed, the types of possibility that 5G enables are one of the core drivers for its implementation.

From the service provider side, a major challenge is the ever-increasing data volumes in the networks, and 5G comes with the promises of being able to offer capacity expansion more cost efficiently than if the expansion is done with existing 4G/LTE technologies.

On the network operations side, meanwhile, expectations are that the new 5G network architecture would give additional benefits in terms of increased support for automation of various operational processes. This could be for example network capacity scaling, software upgrades, automatic testing, and usage of analytics to optimize network performance. Also, the possibility to deploy new software and new services easier and at lower initial cost is imperative for many operators.

While 3GPP is actively working on enablers for automation and for cloud deployment, it must also be acknowledged that some of the possible gains in this area are coming from implementation decisions by the companies designing the infrastructure software. Not everything is subject to standardization or is even possible to standardize.

5G is not just about mobile networks either—fixed wireless access solutions are receiving an increased interest with the emergence of 5G solutions. The market for connecting residential homes and enterprises with high capacity broadband solutions is growing significantly globally, and with 5G technologies there is a new option on the table for service providers that provides high speeds without the costs of implementing fixed infrastructure. It can be assumed that for some geographical areas, delivering broadband services over the air using 5G access technologies is among the best and most cost-efficient solutions. This adds to the interest for 5G among some service providers.

One of the initial key drivers for the new 5G Core architecture and the associated principles for access-technology independence was converging the operations for various types of technologies. This would mean that a service provider that offers both mobile and fixed services to its customers could in the future utilize a single operational team, a uniform set of infrastructure solutions, and identical operational processes across the different service offerings. If this happens this would mean that the concept of "fixed-mobile convergence" would finally be realized, a wish since long from large service providers with significant fixed service business and extensive cost for their operations across mobile and fixed services.

When looking beyond the enhancement of today's services from a user experience, capacity optimization or operational efficiency perspective, a whole new area of use cases are creating drivers for 5G technologies.

This is coming from the collective set of use cases that can be applied to "industry digitalization" meaning that the special characteristics of 5G technologies in terms of very low latency, very high data capacity and very high reliability can be utilized to optimize existing industrial processes or solutions, or even realize completely new ones. Many new business opportunities can be envisioned here and has been outlined by many, for example, Ericsson and Arthur D. Little (A.D. Little, 2017). The wide range of industry sectors that are being targeted and explored include for example industrial manufacturing, public safety, energy production and distribution, automotive and transport and healthcare.

This could, for example, mean utilizing the massive capacity scalability targeted with 5G to support data collection from large numbers of sensors and devices in order to perform advanced data analytics on different IoT and CPS solutions. It could also mean utilizing the very high reliability or low latency of 5G to design more flexible and robust industry communication solutions, for example for real time control of robots in a variety of different industrial manufacturing and other systems. Another potential use case area is to enhance industrial processes using AR/VR technologies to support operational personnel in trouble-shooting, general maintenance or to safely perform operations in dangerous environments.

While it can be assumed that all use cases will not be commercially or technically viable, the sheer range of use cases being explored will mean that 5G can be expected to play a significant role in general industry digitalization for the years to come. This is one of the main drivers for why the global community across multiple industry sectors is increasingly looking at 5G as a key component for their future business operations.

2.3 New technologies

Many new technologies have driven the development of 5G, in this section we very briefly discuss the main ones:
(1) Virtualization,
(2) Cloud native,
(3) Containers,
(4) Microservices, and
(5) Automation

2.3.1 Virtualization

Traditionally Mobile core network element functional designs are distributed applications which scale horizontally and run on dedicated hardware such as processor blades in a chassis. The network element architecture is distributed internally onto specific types of blades that perform specific tasks. For example, blades that execute software that is responsible for overall management of the network element versus blades that perform the actual work of managing mobile core subscribers. Scale is achieved primarily by internal horizontal scaling of working blades.

The first major step of virtualization was to migrate those application-specific blades to virtualized resources such as virtual machines (VMs) and later containers. ETSI NFV (Network Function Virtualisation) and OPNFV was created to facilitate and drive virtualization of the telecoms networks by harmonizing the approach across operators. The network element could then be realized as an application that is distributed among several virtual hosts. Because the application was no longer constrained by the resources and capacity of a physical chassis, this step allows much greater flexibility of deployment and for harmonization of the installed hardware. For example, the operator can deploy much larger (or even much smaller) instances of the network element. This first step was also mainly for proving that a virtualized host environment could scale appropriately to meet the subscriber and capacity demands of today's mobile core. However, most applications in this phase are like a 2-Tier application design wherein the second (Logic) tier the application itself was tightly coupled to state storage it required. The storage design to maintain state was ported from physical systems where individual blades had their own memories.

The next step in the mobile core architecture evolution is to a cloud-native design to take advantage of the flexibility offered in using cloud technology and capabilities. In this step, the mobile core network element design that was tightly integrated together in pre-defined units and ratios is now decoupled both logically and physically to provide greater flexibility and independent scalability. For example, this step sees further separation of control plane and user plane of a network function. Also, in this cloud evolution, mobile core functions begin to implement the network architecture of web applications.

2.3.2 Cloud native

Cloud Native architectures have gained a lot of interest over the past years and service operators attempt to emulate the efficiencies captured by so-called hyperscalers (e.g., Facebook, Google, Amazon) has led to a much heightened interest in this area. Simply put, the architectures and technologies (service-based interfaces, microservices, containers, etc.) used in web-scale applications bring benefits to networking infrastructure in elasticity, robustness and deployment flexibility. Cloud-native applications and infrastructure should not be viewed as another level of complexity on top of a cloud transformation that still is not fully up and running; rather, it should be viewed as a natural evolution of the cloud transformation that is already in progress in the telecom industry today.

A cloud-native strategy therefore allows service providers to accelerate both the development and deployment of new services by enabling practices such as DevOps, while the ability to rapidly scale up or scale down services allows for resource utilization to be optimized in real-time, in response to traffic spikes and one-time events.

There are several cloud-native design principles that hold for all installations, including:

- *Infrastructure Agnostic:* Cloud-native applications are independent and agnostic of any underlying infrastructure and resources.
- *Software decomposition and life cycle management:* Software is decomposed into smaller, more manageable pieces, utilizing microservice architectures. Each piece can be individually deployed, scaled, and upgraded using a CaaS (Container as a Service) environment.
- *Resiliency:* In legacy applications, the MTBF (Mean Time Between Failures) of hardware has been the base metric for resiliency. In the cloud, we instead rely on distribution and independence of software components that utilize auto-scaling and healing. This means that failures within an application should cause only temporary capacity loss and never escalate to a full restart and loss of service.
- *State-optimized design:* How we manage state depends on the type of state/data and the context of the state. Therefore, there is no "one size fits all" way of handling state and data, but there should be a balance between performance, resiliency, and flexibility.
- *Orchestration and automation:* A huge benefit of cloud-native applications is increased automation through, for example, a Kubernetes-based CaaS layer. A CaaS enables auto-scaling of microservices, auto-healing of failing containers, and software upgrades including canary testing (small-scale testing) before larger deployments.

2.3.3 Containers

Virtualization has revolutionized IT infrastructure and enabled tech vendors to offer diverse IT-based services to consumers. From a simplistic perspective, system-level virtualization allows instances of an Operating System (OS) to run simultaneously on a single-server on top of something called a hypervisor. A hypervisor is a piece of computer software that creates and runs virtual machines. System-level virtualization allows multiple instances of OS on a single server on top of a hypervisor.

Containers on the other hand are isolated from each other and share OS kernels among all containers. Containers are widely used in sectors where there is a need to optimize hardware resources to run multiple applications, and to improve flexibility and productivity. In addition, the eco systems and tooling for container based environment, e.g., Kubernetes are rapidly expanding.

Containers are especially useful for telecommunications applications

- Where low-latency, resilience and portability are key requirements—e.g., in Edge Computing environments.

- For implementing short-lived services, i.e., for highly agile application deployments.
- In machine learning or artificial intelligence when it is useful to split a problem up into a small set of tasks—it is expected therefore that containers will assist to some extent with automation.

2.3.4 Microservices

Microservices are an architectural and organizational approach to software development where rather than be developed in a monolithic fashion, software is composed of small independent services that communicate over well-defined APIs. It is often considered a variant of the service-oriented architecture approach. The overall aim with microservices architectures is to make applications easier to scale and faster to develop, enabling innovation and accelerating time-to-market for new features. They also, however, come with some increased complexity including management, orchestration and create new data management methods.

Microservice disaggregation has several benefits:
- Microservice instances have a much smaller scope of functionality and therefore changes can be developed more quickly.
- An individual feature is expected to apply to a small set of microservices rather than to the entire packet and 5GC function.
- Microservice instances can be added/removed on demand to increase/decrease the scalability of their functions.
- Microservices can have independent software upgrade cycles.

Therefore, rather than deploying replicated pre-packaged instances of functionality, with microservices the operator can deploy functionality on demand at the scale required. This approach further enhances the efficiency of resources utilization. It also greatly simplifies deployment of new functionality because the operator can add features/perform upgrades on a set of microservices without impacting adjacent services.

2.3.5 Automation

One of the main drivers for the evolution of the core network is the vision to deliver networks that take advantage of automation technologies. Across the wider ICT domain, Machine Learning, Artificial Intelligence and Automation are driving greater efficiencies in how systems are built and operated. Within the 3GPP domains, automation within Release 15 and Release 16 refer mainly to Self-Organising Networks (SON), which provide Self-Configuration, Self-Optimisation and Self-Healing. These three concepts hold the promise of greater reliability for end-users and less downtime for service providers. These technologies minimize lifecycle costs of mobile networks through eliminating manual configuration of network elements as well as dynamic optimization and troubleshooting.

Operators using SON for LTE have reported Accelerated rollout times, simplified network upgrades, fewer dropped calls, improved call setup success rates, higher end-user throughput, alleviation of congestion during special events, increased subscriber satisfaction, and loyalty, and operational efficiencies – such as energy and cost savings and freeing up radio engineers from repetitive manual tasks (SNS Telecom and IT, 2018).

5G holds unique challenges, however, which makes automation of configuration, optimization and healing a core part of any service providers network. The drivers for this include the complexity of having multiple radio networks running and connecting to different cores simultaneously, the breadth of infrastructure rollouts required and the introduction of concepts such as network slicing, dynamic spectrum management, predictive resource allocation and the automation of the deployment of virtualization resources outlined above.

In addition, we expect that Machine Learning and Artificial Intelligence will become further integrated across all aspects of the mobile systems in the coming years.

CHAPTER 3

Architecture overview

3.1 Introduction

3.1.1 Balancing evolution and disruption

Work on designing and specifying a Core network for 5G was done in parallel with and in close cooperation with the teams designing the 5G radio network.

One key principle with the design of the 3GPP 5G Core architecture was not providing backwards compatibility for the previous generations of radio access networks, i.e., GSM, WCDMA and LTE. Previously, when new access network generations were developed, each one had a different functional split between the core network and the radio network, as well as new protocols for how to connect the radio and core networks. For example, when GPRS packet data services for GSM (2G) was designed back in the mid 90's, it included a Frame Relay-based interface (Gb) between radio and core. WCDMA (3G), designed a couple of years later came with an ATM-influenced interface (Iu) for connecting radio and core. Finally, when LTE (4G) was designed around 2007–2008, it brought the new IP-based S1 interface for connecting radio and core networks. In addition, the different methods for addressing battery savings and scheduling on devices meant that each new generation came with similar—but still slightly different—functionality and used different data communication protocols for the networking layer. Over time this has created complexity in network architecture, as most service providers have deployed a combination of 2G, 3G and 4G on different frequency bands to provide as good coverage and capacity as possible for a heterogenous fleet of devices.

The 5G Core, however, brought a mindset shift aiming to define an "access-independent" interface to be used with any relevant access technology as well as technologies not specified by 3GPP such as fixed access. It is also, therefore, intended to be as future-proof as possible. The 5G Core architecture does not include support for interfaces or protocols towards legacy radio access networks (S1 for LTE, Iu-PS for WCDMA and Gb for GSM/GPRS). It instead comes with a new set of interfaces defined for the interaction between radio networks and the core network. These interfaces are referred to as N2 and N3 for the signaling and user data parts respectively. The N2/N3 protocols are based on the S1 protocols defined by 3GPP for 4G LTE (S1-AP and GTP-U), but efforts have been made to generalize them in the 5G System with the intention to make them as generic and future proof as possible. N2/N3 are described in Section 3.5.

15

While GSM and WCDMA access technologies were not discussed much during the 3GPP work to define the 5G Core architecture, LTE was. This is because LTE is the most important mobile radio access technology globally and will likely remain so for a long time. Because of this, efforts were made to define how to connect LTE access to the new 5G architecture. Backwards compatibility for devices and LTE radio access was not addressed, but the LTE specifications were complemented to make it a second access technology supporting the same architecture and (i.e., the same N2/N3 interfaces) protocols as NR.

Essentially this means that any access network that supports N2/N3 could be connected to the new 5G Core architecture. In the context of the new architecture, 3GPP has so far specified such support for LTE, NR, and combinations of LTE and NR.

3.1.2 3GPP architecture options

The outcome of the 3GPP work on the 5G network architecture was a number of architecture options, based on 3GPP making three important decisions:
- To specify LTE support for the new 5G architecture
- To specify support for combinations of LTE and NR access
- To specify an alternative 5G architecture based on an evolution of LTE/EPC

We will discuss each of these below. The key document for the technical study on the 5G Network Architecture in 3GPP is the technical report 3GPP TR 23.799.

The fact that LTE access support is specified for the new 5G architecture means that an LTE access network in practice has two ways of connecting with a core network, potentially simultaneously and selected on a per device basis:
- Using S1 connectivity to an EPC core network
- Using N2/N3 to a 5GC core network

Note that it is not only the network interface and associated logic that needs to change when migrating from S1 to N2/N3 but connecting LTE to 5G Core also requires a new Quality-of-Service concept that impacts the radio scheduler.

While this is within the scope of 3GPP specifications in Release-15, it remains to be seen if any LTE networks will actually be converted to connect to the 5GC core network, or if service providers will instead rely on maintaining the S1 connection to EPC combined with interworking between EPC and 5GC, a solution we will describe further in Section 3.8.

When defining the 5G radio access network specifications, two variants of combining LTE and the new 5G radio access technology (NR) were discussed. Each one relies on the assumption that one of the technologies will have a larger geographical coverage and therefore be used for all signaling between devices and the network, while the other radio technology would be used to boost user traffic capacity inside geographical areas where both access technologies are present.

3.1.2.1 The non stand-alone (NSA) architecture

In conjunction with extending the new 5G architecture to not only include NR access but also LTE access, a parallel track was started in the 3GPP Release 15 work. This was driven by a widely established view in the telecom industry that there was a need for a more rapid and less disruptive way to launch early 5G services. Instead of relying on a new 5G architecture for radio and core networks, therefore, a solution was developed that maximizes the reuse of the 4G architecture. In practice it relies on LTE radio access for all signaling between the devices and the network, and on an EPC network enhanced with a few selected features to support 5G. The NR radio access is only used for user data transmission, and only when the device is in coverage. See Fig. 3.1.

One drawback with this architecture is that NR can only be deployed where there is already LTE coverage. This is reflected in the name of the solution—the NR Non-Stand-Alone (NSA) architecture. Another drawback is that the available network features are limited to what is supported by LTE/EPC. The main differences in terms of capabilities are in the areas of Network slicing, Quality-of-Service handling, Edge computing flexibility and overall core network extensibility/flexibility for integrating towards applications in an IT-like environment. These will be discussed in subsequent chapters.

In summary, there are four ways that LTE and/or NR can be deployed:
- Only LTE for all signaling and data traffic
- Only NR for all signaling and data traffic
- A combination of LTE and NR where LTE has the larger coverage and is used for signaling while both LTE and NR are used for data traffic
- A combination of LTE and NR where NR has the larger coverage and is used for signaling while both LTE and NR are used for data traffic

Add two possible core networks—EPC and 5GC—and you therefore get $4 \times 2 = 8$ possible network architectures.

In order to create a common terminology around different variants of deploying radio access technologies, the concept of "options 1–8" was proposed during the initial technical work with the 5G architecture (3GPP SP-160455, 2016). These are illustrated in Fig. 3.2.

It was decided at an early stage that options 6 and 8 should not be progressed further as they assumed connecting NR access directly to EPC, something that would impose too

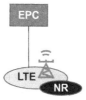

Fig. 3.1 The non stand-alone architecture.

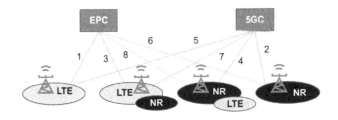

Access network:	LTE only	NR only	LTE with NR for data only	NR with LTE for data only
EPC core network	Option 1 (=4G)	Option 6 (disregarded)	Option 3	Option 8 (disregarded)
5GC core network	Option 5	Option 2	Option 7	Option 4

Fig. 3.2 The possible combinations of 5G radio and core networks.

many limitations on NR in order to provide for backwards-compatible with EPC functionality. Since option 1 referred to the existing 4G architecture, this meant that the technical work proceeded on options 2, 3, 4, 5, and 7. Out of these, priority in the specification work was given to the two variants that were assumed to have the largest market value—option 3 and option 2.

Irrespective of the decisions to limit the number of options, this is an area where it may be argued that 3GPP has created too much flexibility for its own good as so many variants may increase cost and complexity across the industry ecosystem for radio networks and devices. The full impact of this remains to be seen.

From a 5G Core network perspective, the four combinations of radio access technologies (options 2, 4, 5 and 7) all use more or less the same interface, protocols and logic. This is the first attempt to create an access independent interface between the core network and whatever access technology that is used.

Option 3 is the popular name for Non Stand-Alone, or NSA, architecture described above. It was the first 5G network architecture to enter commercial services as it allows for expanding from the existing 4G LTE/EPC architecture, facilitating a smooth introduction of 5G, even if it is mainly addressing existing mobile broadband services.

The formal name of the NSA radio network solution is EN-DC, short for "E-UTRAN-NR Dual Connectivity". We describe the key EPC features to support 5G NSA in Chapter 4, and the new radio architecture concept in Chapter 12. We then dedicate the rest of this book to the new technologies and concepts defined for the 5G Core architecture as defined in 3GPP Release 15. The two key reference documents for this are the 3GPP specifications 3GPP TS 23.501 and 3GPP TS 23.502.

The rest of this chapter provides readers with a high-level description of the 5G Core architecture and introduces the key components and functionality. In subsequent sections, we describe the details and logic of each network function as well as the protocols specified for different parts of the network architecture.

We start off describing the most fundamental aspects of 5G Core and the most important features in Sections 3.2–3.10.

In addition to this there are more capabilities defined that can optionally be used to support more advanced use cases. These are described in Sections 3.11–3.18.

Concluding this chapter is a brief overview of the 5G Radio technology and network architecture in Section 3.19.

A final note—in this chapter we refer to the mobile device connecting to the network simply as a "device", while in subsequent chapters this is often referred to as a "UE", the abbreviation for the 3GPP term "User Equipment". The same goes for the radio base station that is later referred to using the 3GPP term "gNB" for NR.

3.2 Two perspectives on 5G Core

When comparing to the existing EPC architecture, the 5GC architecture is simultaneously very similar and very different.

The user data processing parts, as well as the integration with 3GPP radio access networks, are quite similar between the new 5GC and the traditional EPC network architecture, originally defined for 4G/LTE. The part of the network that contains signaling-only functionality, is on the other hand very different.

Another difference between the EPC architecture used for 4G and 5G NSA is that the architecture of 5G Core can be visualized and described in two different ways.

The first visualization shows the way different network functions are connected. The major difference compared to previous 3GPP architectures in this visualization is the concept of Service-Based interfaces. It means that the network functions that include logic and functionality for processing of signaling flows are not interconnected through point-to-point interfaces but instead exposing and making available services to the other network functions. For each interaction between network functions, one of these acts as a "Service Consumer", and the other as a "Service Producer". We will describe this concept in more detail in Section 3.3.

This representation of the architecture is shown in Fig. 3.3.

At a first glance, this architecture may look quite complex to the reader, and we will therefore describe the functionality and key features of the different parts of the architecture step by step below.

Firstly, however, let's look at the other visualization of the architecture that illustrates how network functions interact with other network functions, represented by traditional point-to-point interfaces. Showing these interfaces can be useful to illustrate which of the network functions that utilize, or consume, the services of which other network functions. Even if all the network functions in theory could be connected in a full connectivity mesh, the actual call flows define which service combinations that apply in real operations. And these combinations are visualized as logical interfaces, or more correctly—*reference points*, in the view shown in Fig. 3.4. That is the main value of the point-to-point representation.

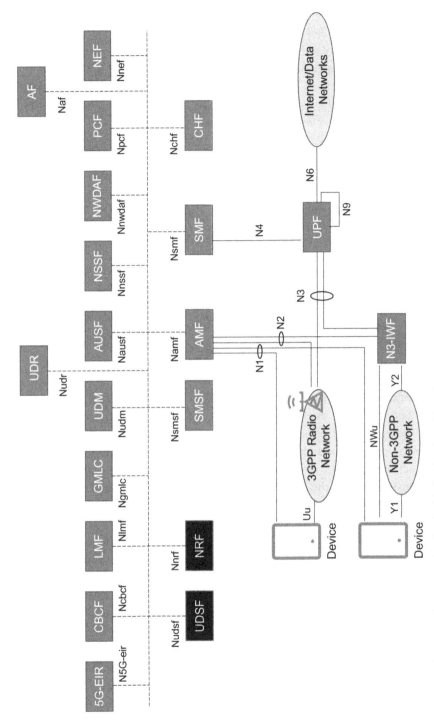

Fig. 3.3 5G Core architecture visualized with Service-Based interfaces.

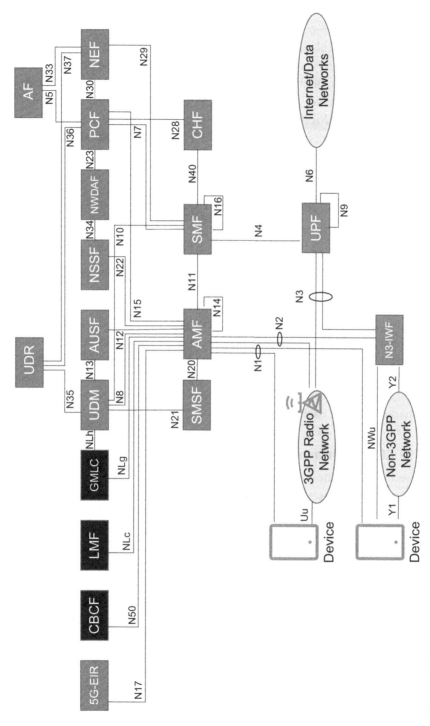

Fig. 3.4 The 5G Core architecture visualized with point-to-point interfaces.

Another difference between the two ways of representing the architecture, apart from illustrating how different Network Functions interconnect is that some Network Functions are only applicable in one of the representations.

In the Service-Based representation in Fig. 3.3, the two Network Functions NRF and UDSF are visible (highlighted in black). They are described below, but for now it can be noted that they are only applicable to the Service-Based representation of the architecture view. UDSF has a point-to-point interface name assigned (N18) but it is less useful to illustrate as it can connect to any other Network Function. See Section 3.18 for more details on UDSF.

3.3 Service-based architecture (SBA)

3.3.1 The concept of services

A major difference in 5G Core compared to previous generations of traditional network architectures represented by "nodes" or "network elements" connected by interfaces, is the usage of service-based interactions between Network Functions.

This means that each Network Function offers one or more services to other Network Functions in the network. In the 5GC architecture, these services are made available over Network Function interfaces connected to the common Service-Based Architecture (SBA). In practice this means that functionality supported in a specific Network Function is made available and accessible over an API (Application Programming Interface). It shall be noted that this architecture applies to signaling functionality only, not to the transfer of user data.

3.3.2 HTTP REST interfaces

The communication method defined for 5G Core relies on the widely used "HTTP REST paradigm" that are a set of rules or guidelines that define how web communication technologies access services from distributed applications using APIs. "REST" is short for "Representational State Transfer" and defines a set of design rules for how to implement the communication between different software modules in a networked architecture. This is the standard way of designing IT networking applications today, and it has been selected by 3GPP as a means of allowing for tighter integration between the mobile networks and surrounding IT systems, as well as for allowing for shorter and simplified service development efforts. The expectation is that the network capabilities shall be easier to extend when using the relatively light-weight Service Based Interface (SBI) concept, than if using a more traditional point-to-point architecture that relies on detailed and extensive protocol specification efforts.

Using SBI and APIs can also be seen as a logical choice by 3GPP when specifying the 5G Core Network, as the 5GC software applications that implement the Network

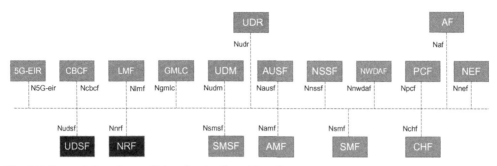

Fig. 3.5 Network functions utilizing Service-Based interfaces.

Functions are assumed to be executing in an IT-like or even shared IT environment, typically in a cloud data center. A harmonization of both software technologies and IT architecture across the mobile network solution and supporting IT applications is to some extent possible with this approach.

Fig. 3.5 shows 3GPP network functions utilizing HTTP REST for service-based communication. They are logically interconnected to a common networking infrastructure.

HTTP REST uses message syntax from the widely used HTTP web protocol, and relies on the concept of Resource Modeling, which means that a distributed software application can be addressed through Uniform Resource Identifiers (URIs), in practice a web address pointing at a resource or set of resources. On top of this, a very simple set of commands, standard HTTP "methods", are being used. The most important ones are listed below.

GET—this is used to fetch data from a server. It shall not change any data.
POST—this is used to send data to a server.
PUT—this is also used to send data to a server, but it replaces existing data.
DELETE—this is used to remove data from a server.

An important aspect of REST is that all communication must include the full set of information needed for a specific processing action. It must not rely on previous messages, and hence it can be considered as stateless. Utilizing this principle for software design allows for excellent scalability and distribution capabilities for the system. More details on the HTTP protocol are available in Chapter 13.

3.3.3 Service registration and discovery

When two Network Functions communicate over the 3GPP SBA architecture, they take on two different roles. The Network Function that sends the request has the role of a **Service Consumer**, while the Network Function that offers a service and triggers some action based on the request has the role of a **Service Producer**. Upon completion of the requested action, the Service Producer responds back to the Service Consumer.

So far so good, but a critical part of this concept is the mechanism for how the Service Consumer can locate and contact a Service Producer that can provide the requested service. The solution is based on the concept of *Service Discovery*.

Service Discovery relies on that a well-known function in the network keeps track of all available Service Producers and what services they offer. This is achieved through that each Service Producer, for example a 3GPP Network Function like the PCF, registers that its services are available. In the 5GC architecture, this registration is done to a dedicated Network Function that is called the Network Repository Function (NRF). This concept allows the NRF to keep track of all available services of all Network Functions in the network. It also means that each individual Network Function needs to be provisioned or configured with the address of one or more NRFs, but it does not need, and shall not have, addresses to all other Network Functions configured.

Let's look at a practical example involving three actual Network Functions—PCF, AMF and NRF. The detailed roles and key functionality of AMF and PCF will be more extensively explained below, so for now, just assume they are any Network Functions that need to interact as part of a specific call flow.

It starts with PCF doing a **Service Registration**.

During the actual registration, the PCF acts as a Service Consumer, and the NRF as a Service Producer, basically offering the service of "Network Resource Registration" to the PCF.

Fig. 3.6 illustrates the initial part of the call flow. The PCF registers with the NRF using an HTTP PUT message that includes information about the PCF such as available services, network address and identity. The NRF verifies that the request is valid, stores the data associated with the PCF registration, and acknowledges the PCF registration with a response back to the PCF. Now the PCF services are available to other Network Functions through querying the NRF.

In the next phase, another Network Function like the AMF wants to utilize the services of a PCF. This is achieved through first querying the NRF for a list of PCFs offering these services. This phase is called the **Service Discovery**. In this case the AMF is the Service Consumer and the NRF is the Service Producer. See Fig. 3.7.

The AMF sends a query to the NRF, stating what sort of Network Function it is asking for, and what services this NF shall support to be of interest. This is done using

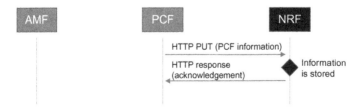

Fig. 3.6 First part of the call flow—Service Registration.

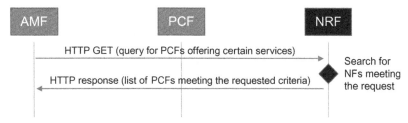

Fig. 3.7 Second part of the call flow—Service Discovery.

an HTTP GET message. The NRF filters out all Network Functions that are registered and are providing the requested services, and then responds back to the AMF.

When this step is finalized, the AMF can make a selection of a PCF that fulfills the service requirements, and then contact the selected PCF with a **Service Request**. In this step, the AMF is again the Service Consumer, while the PCF is the Service Producer. This is done using an HTTP POST message.

Note that the Service Request referred to here is not be mixed up with the Service Request a mobile device sends to the network when it is to move from idle to connected mode.

Upon reception of this service request, the PCF determines the applicable policy that is requested by the AMF and responds back to the with an HTTP response (Fig. 3.8).

The call flow including all three steps is shown in Fig. 3.9.

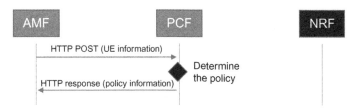

Fig. 3.8 Third part of the call flow—Service Request.

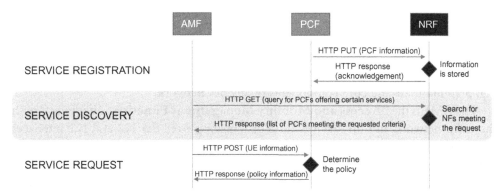

Fig. 3.9 Consolidated call flow.

Note that these three parts do not usually happen in direct sequence. A Network Function typically registers with the NRF when it is put into service, while the service discovery and service requests may for example take place when a device connects to the network.

The rest of the call flow and the subsequent interaction between the Network Functions is beyond the scope of this chapter, but the concept remains the same through each step, and for all other call flows between Network Functions interacting with HTTP within the Service-based architecture. One Network Function acts as the Service Producer, another one as the Service Consumer. And all communication is done using the HTTP protocol.

There is another way of interaction between a Service Producer and one or many Service Consumers. This is based on that the fact that one or several Network Functions can *subscribe* to a service from another Network Function. The Network Function acting as Service Producer then sends notifications to all the Service Consumers when some specific criteria are met, for example when certain information has been changed. The concept of Subscribe and Notify removes the need for Service Consumers to frequently request information from the Service Producer, instead allowing them to wait for the Service Producer to notify when something has happened.

3.4 The core of the core

Having described the new mechanisms for how different Network Functions communicate, let's now return to the functional view of the network.

The core functionality of the network architecture includes functionality for establishing sessions in a secure way and to forward user data to and from mobile devices providing data connectivity. This is the part of the network that cannot be excluded from any 5G Core deployment. In addition to Radio Network and the NRF described in Section 3.3, it includes the following six Network Functions:

- AMF
- SMF
- UPF
- AUSF
- UDM
- UDR

Fig. 3.10 illustrates the core components of any 5G network.

The AMF is the "**Access and Mobility Management Function**". It interacts with the radio network and the devices through signaling over the N2 and N1 interfaces respectively. Connections towards all other Network Functions are managed via service-based interfaces. The AMF is involved in most of the signaling call flows in a 5G network. It supports encrypted signaling connections towards devices, allowing these to register, be authenticated, and move between different radio cells in the network. The AMF also supports reaching and activating devices that are in idle mode.

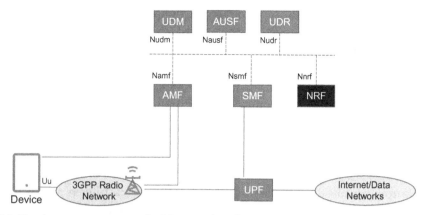

Fig. 3.10 Mandatory components of a 5G network architecture.

One difference to the EPC architecture is that the AMF (as opposed to the MME) does not handle session management. Instead the AMF forwards all session management-related signaling messages between the devices and the SMF Network Function. Another difference is that the AMF (as opposed to the MME) does not perform device authentication itself, instead the AMF orders this as a service from the AUSF Network Function.

The AMF functionality is described in more detail in Chapter 7.

The SMF is the "**Session Management Function**", meaning as the name suggests that the SMF manages the end user (or actually device) sessions. This includes establishment, modification and release of individual sessions, and allocation of IP addresses per session. The SMF indirectly communicates with the end user devices through that the AMF forwards session-related messages between the devices and the SMFs.

The SMF interacts with other Network Functions through producing and consuming services over its service-based interface, but also selects and controls the different UPF Network Functions in the network over the N4 network interface. This control includes configuration of the traffic steering and traffic enforcement in the UPF for individual sessions.

In addition to this, the SMF has a key role for all charging-related functionality in the network. It collects its own charging data, and controls the charging functionality in the UPF. The SMF supports both offline and online charging functionality. Furthermore, the SMF interacts with the PCF Network Function for Policy Control of user sessions.

The SMF functionality is described in more detail in Chapter 6.

The "**User Plane Function**" (UPF) has as the main task to process and forward user data. The functionality of the UPF is controlled from the SMF. It connects with external IP networks and acts as a stable IP anchor point for the devices towards external networks, hiding the mobility. This means that IP packets with a destination address belonging to a specific device is always routable from the Internet to the specific UPF that is serving this device even as the device is moving around in the network.

The UPF performs various types of processing of the forwarded data. It generates traffic usage reports to the SMF, which the SMF then includes in charging reports to other Network Functions. The UPF can also apply "packet inspection", analyzing the content of the user data packets for usage either as input to policy decisions, or as basis for the traffic usage reporting.

It also executes on various network or user policies, for example enforcing gating, redirection of traffic, or applying different data rate limitations.

When a device is in idle state and not immediately reachable from the network, any traffic sent towards this device is buffered by the UPF which triggers a page from the network to force the device back to go back to connected state and receive its data.

The UPF can also apply Quality-of-Service (QoS) marking of packets towards the radio network or towards external networks. This can be used by the transport network to handle each packet with the right priority in case of congestion in the network.

The UPF functionality is described in more detail in Chapter 6.

The UDM is the "**Unified Data Management Function**". It acts as a front-end for the user subscription data stored in the UDR (more on that further down) and executes several functions on request from the AMF.

The UDM generates the authentication data used to authenticate attaching devices. It also authorizes access for specific users based on subscription data. This could for example mean applying different access rules for roaming subscribers and home subscribers.

In case there are more than one instance of AMF and SMF in the network, the UDM keeps track of which instance that is serving a specific device.

The UDR—the **"Unified Data Repository"**—is the database where various types of data is stored. Important data is of course the subscription data and data defining various types of network or user policies. Usage of UDR to store and access data is offered as services to other network functions, specifically UDM, PCF and NEF.

The functionality of the **"Authentication Server Function"** (AUSF) is quite limited but very important. It provides the service of authenticating a specific device, in that process utilizing the authentication credentials created by the UDM. In addition, the AUSF provides services for generating cryptographical material to allow for secure updates of roaming information and other parameters in the device.

3.5 Connecting the core network to mobile devices and radio networks

The description above outlines the key parts of the core network architecture. The connections to the radio network and the devices are shown in Fig. 3.11.

N2 is a key reference point in the 5G network architecture. All signaling between the radio networks and the core network (fronted by the AMF) is carried across this reference

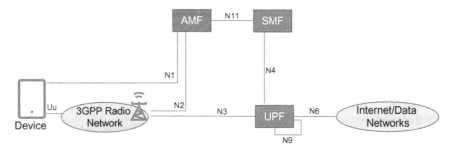

Fig. 3.11 Connecting 5G RAN and 5G Core.

point. It should be noted that there is a naming inconsistency in the 3GPP set of Release-15 specifications here, as the specifications developed by the RAN working groups use the term "NG-C interface" while the Architecture and Core teams use the term "N2 reference point" in their specifications. In this book we consistently use N2.

The signaling carried across N2 is based on the NG-AP protocol. There are multiple types of signaling procedures supported over N2.

- Procedures supporting management of N2 such as configuration of the interface itself. One gNB (the 5G radio base station) can be connected to multiple AMFs, for load sharing, resiliency and network slicing purposes.
- Procedures related to signaling for a specific UE/device. Each UE/device is always only associated with a single AMF (except for some special cases related to simultaneous roaming in 3GPP and non-3GPP networks). This signaling can be divided into three different types of procedures:
  - Signaling related to forwarding of messages between the device and the core network. This is based on the NAS protocol, short for "Non-Access Stratum". In the 5GC architecture individual NAS messages are either managed by the AMF or the SMF. The SMF manages NAS messages related to Session Management, and in that case, messages are passed between SMF and device with the help of the AMF, as there is no direct connection between the SMF and the radio network. The AMF manages all other NAS messages without involving the SMF. It shall be noted that NAS can also be used to transparently carry some messages to and from other Network Functions. This is described in more detail in Chapter 14.
  - Signaling related to modification of the stored data for a specific device, the "UE context"
  - Signaling related to management of events like handovers between radio cells or access networks and paging of devices that are in idle mode.

One difference to the EPC architecture is that for 5GC, also the reference point between the device and the Core Network (the AMF) has its own name. This is called N1, over which the NAS messages described above are carried. It means in practice that the NAS messages are carried transparently over the air interface Uu and the RAN-Core interface N2. But from a logical perspective, N1 is shown as its own reference point in the architecture.

The NAS messages that relate to the AMF functionality are of course managed by the AMF, while the NAS messages that relate to SMF functionality are forwarded by the AMF to the applicable SMF over the logical N11 interface, after that the AMF has performed basic NAS message processing for e g security. In practice N11 is realized through utilizing the services available over the Namf and Nsmf interfaces in the service-based architecture.

It should be noted that one device is always served by one single AMF, but the same device can be utilizing data sessions managed by more than one SMF. This gives additional flexibility compared to the EPC architecture, and for example allows for simultaneous connections to multiple logical networks with different treatment, policies and rules being applied for the routing of user data. Note again as mentioned above, there is actually a case when the device is simultaneously served by two AMFs, but that is a very specific case related to non–3GPP access, and is out of scope for this chapter.

Fig. 3.12 shows the case when one device is served by a single AMF but have sessions established with two SMFs, each with its own UPF. This concept is further described in Section 3.16 in the context of Network slicing.

3.6 Mobility and data connectivity

As described above, user data is handled in the UPF, the "User Plane Function", in the Core Network. Data is transported between the radio access network and the UPF over the N3 reference point. Data is tunneled across N3, meaning that IP routing is done on the tunnel header IP address instead of the end user IP address. This allows for maintaining a stable IP anchor point even though the device is moving in the network. It also

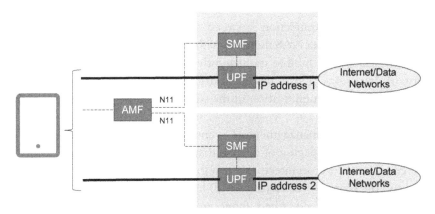

Fig. 3.12 Multiple service connections with individual SMFs and UPFs.

allows the same mechanisms and transport routing to be used independently of the type of data that is carried. Besides IP packets, the 5G architecture specifications also include support for Ethernet frames and so called "unstructured data".

The concept is very similar to how it is done in the EPC architecture, where the corresponding interface or reference point is called S1-U. N3 includes a new way of managing the Quality-of-Service of specific data flows and how to map data flows to tunnels.

On the other side, the UPF connects to external data networks. Here IP packets are in general routed based on the actual IP address of the device, meaning that the traffic is not tunneled. The reference point is called N6 and corresponds to SGi in the EPC architecture. For Ethernet sessions, N6 is a layer 2 link instead of a routable IP network. There are also possibilities to tunnel end user data over N6 using virtual private networking, creating secure tunnels for example for enterprise connections.

Fig. 3.13 shows the interfaces related to user data processing and transport in the 5G architecture.

The SMF controls the behavior of the UPF. This is done through signaling over the N4 reference point. As described above, there could be several SMF/UPF pairs simultaneously managing traffic for one and the same device.

The control of the UPF is done per end user data session, where the SMF can create, update and remove session information in the UPF. In addition, some functions are defined also for individual data flows.

Some of the key features in the SMF related to UPF control include
• Controlling traffic detection rules to be used in the UPF
• Controlling packet forwarding rules to be used in the UPF
• Controlling usage reporting rules to support policy and charging functionality in SMF. The UPF then reports usage based on these rules towards the SMF. Reporting can be done both for the total traffic associated with a data session, as well as for individual traffic flows
• Providing Quality-of-Service parameter values to the UPF for QoS enforcement of data flows, for example limitations of the available data rate

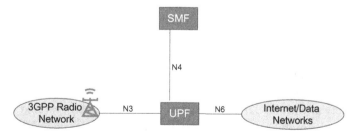

Fig. 3.13 The user plane connection to radio networks and external data networks.

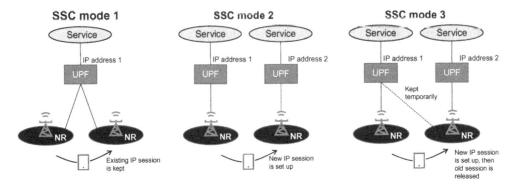

Fig. 3.14 Session and Service Continuity modes 1, 2 and 3.

The 5G Core architecture includes more extensive support and flexibility for different levels of data mobility compared to the EPC architecture specified for 4G and inherited for use with 5G NSA.

One fundamental concept is the three "Session and Service Continuity" modes, abbreviated to SSC modes 1, 2 and 3. They indicate different ways of dealing with existing data sessions when the device moves across the network. This allows for a more flexible selection between prioritizing a stable mobility anchor point or prioritizing low user data delays. The SSC modes require corresponding support from the device, otherwise they will not work.

Fig. 3.14 is an overview of the three SSC modes.

SSC mode 1 means that the IP address is maintained regardless of movements in the network. The same IP anchor point (UPF) is accessible and can be used across the network.

SSC mode 2 means the opposite of SSC mode 1. The network will release and trigger the device to reestablish new sessions as the device moves around in the network. The network decides to release the session based on operator policies, for example based on a request from an application function in the network. When the device requests a new session, the network can select a new UPF which is more suitable to the service, for example a UPF that is located closer to where the device is currently located. As opposed to SSC mode 1, SSC mode 2 means a short interruption of the service, which may or may not be acceptable depending on what end user service that is being targeted.

SSC mode 3 is a bit more advanced, as it tries to combine some benefits of both options 1 and 2. It allows for the same low delays as SSC mode 2 through triggering release and reestablishment of IP sessions using new UPFs, but allows for a continuous service availability as with SSC mode 1, albeit likely with a delay that may not fully meet the needs during the mobility phase. This is done through first establishing the new session and connection to the new UPF before releasing the session and connection anchored in the old UPF. This puts additional requirements on the device as it needs to maintain two sessions and two IP addresses for the same service for a limited time.

The selection of a suitable SSC mode is preferably be done based on needs of the service itself. One example is if a service requires very low network delays while being implemented in a network covering a large geographical area. A large coverage area means that the IP anchor point for SSC mode 1 would need to be centralized to a location where all radio network base stations can be reached, and preferably with a not too long and decently uniform delay. But having the IP anchor point (the UPF) in this location may not meet the delay requirements, which instead calls for IP anchor points closer to the access, to reduce the delay coming from the transport networks connecting different cities or even different parts of the country. So, the SSC mode 2 may in this case be needed to meet the delay requirements, with the drawback that the IP address and the location of the IP anchor point and the application server where the service is executing needs to be changed as the device moves in the network.

The selection of SSC mode for a session is done by the SMF, based on a combination of allowed SSC modes in the subscription data, and the request from the device. The SSC mode does not change once a session is established.

One limitation that should be noted is that while SSC modes 1 and 2 can be used for both IP and Ethernet type sessions, SSC mode 3 only works for IP.

Somewhat related to the ability to access local services using SSC mode 2 or 3, the concept of "Local Area Data Network", abbreviated LADN, supports restricting access to certain services to only be possible in certain geographical areas, defined as several Tracking Areas. From a simplistic perspective, a tracking area can be viewed as a collection of radio cells that combined together cover a larger geographical area. A mobile network normally contains many tracking areas that each contain many cells.

Using the LADN mechanisms, the operator can define some services to be available only in certain geographical areas. For a device to access such a service, the subscription used for the device need to include the corresponding LADN service support. Further details on SSC modes and LADN are available in Chapter 6.

A special feature of the 5G Core UPF is that two UPFs can be deployed in series, then connected via an interface referred to as N9. There are three main use cases for this:
1. Network-wide mobility
2. Break-out of selected data flows
3. Roaming with home routing
We cover the first two cases below. The third case will be described in Section 3.17.

To provide full mobility with a stable IP anchor across the full network, there may be a need to connect two UPFs. If this is required or not depends on the operator network configuration, specifically how the transport network between the base stations in the radio network and the various core network sites has been designed.

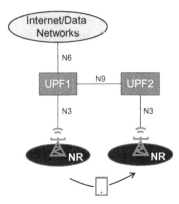

Fig. 3.15 IP mobility when interconnecting two UPFs.

An example is shown in Fig. 3.15. Assume a device attaches over a radio cell in the left-hand NR coverage area. The UPF1 will be selected and be serving as the IP anchor for the device, connecting to the Internet or another data network that offers some end user services, e.g., IMS.

The device then moves to another part of the network, where the radio base stations cannot connect to UPF1 due to limitations in the transport network configuration. The SMF then allocates UPF2 to serve as a termination point for the new N3 interface and connect back to UPF1. Through this method, there is no change to the IP address of the device or the point of interconnect.

The second case is a new concept in 5GC compared to EPC, the ability to apply classification and traffic management in the UPF to selectively send IP packets to different IP interfaces. The typical use case for this scenario is to allow for some traffic to be terminated at—or close to—the edge of the network, for example, to secure the lowest possible data plane latency, or to protect sensitive data from being intercepted in the more centralized parts of the network. The architecture therefore involves two UPFs again, connected in series. See Fig. 3.16.

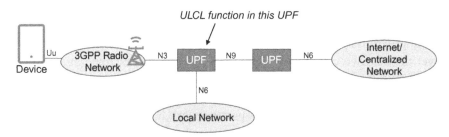

Fig. 3.16 Break out of selected data flows using Uplink Classifier.

This concept relies on a new mechanism in the UPF called the "Uplink Classifier" (ULCL), that filters out IP packets coming uplink from the device and that match certain classification criteria, and sends these packets to a separate IP interface connected to a local network. This interface happens to also be called N6. For this to work, the ULCL function must be applied in the UPF closest to the access network.

Packets that do not meet the selection criteria are sent to the centralized UPF over the N9 interface.

In the downlink, data flowing towards the device from both the centralized UPF and the local UPF is combined into one single data stream in the UPF closest to the access network.

The ULCL function is controlled by network rules provided by the SMF that manages the specific IP session. The SMF decides based on policies to either include or not include an ULCL function and an extra UPF in the data path for a given IP session. Signaling for ULCL is handled over the N4 reference point between SMF and UPF(s).

This functionality is completely transparent to the devices and the devices are therefore not aware of if ULCL and local breakout of traffic is applied in the network or not.

There is also another solution for providing breakout of selected data flows. This relies on IPv6 being used with multi-homing. More information on both ULCL and IPv6 multi-homing is provided in Chapter 6.

3.7 Policy control and charging

The 5GC architecture includes extensive functionality for policy-based control of traffic and user services. This functionality is similar to what is supported in the 3GPP EPC architecture but includes additional new functionality that we describe below. Policies can be viewed as rules for how users and data sessions and data flows shall be controlled or managed, including what services are allowed or disallowed, how charging shall be done, what quality-of-service that applies etc.

Policies can be applied with different levels of granularity, e.g.,
- Policy rules that apply to all users in the network
- Policy rules that apply to all services for a specific user
- Policy rules that apply to specific data sessions or data flows for a given user

A center-piece of the 5GC Policy Control architecture is naturally the Policy Control Function (PCF), which interacts with several other Network Functions as shown in Fig. 3.17. It should be noted that we have not as yet introduced some of these Network Functions in this book. These are therefore illustrated in black and are further described in later chapters.

The main document describing the 5G Policy Architecture is the 3GPP Technical Specification 23.503.

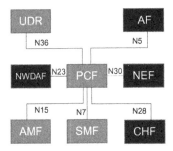

Fig. 3.17 PCF connections to other Network Functions.

On a high level, the functionality of the PCF belongs to one of two main areas:
- Policy control related to data sessions
- Policy control not related to data sessions

Policy control not related to data sessions, refers to how service providers can control the way a specific user can access the network, for example through restricting the geographical area within which the user can attach or move with retained connectivity. It can also be used to define which radio access technologies that a user can utilize. It builds on interaction between the PCF, the UDR and the AMF.

An even more advanced policy control functionality is the ability to provide user-specific information which the AMF can convey to the radio access network to control mobility schemes between radio access types or even frequency bands. This concept uses a parameter called RFSP Index, but more detailed explanations of the radio-related functionality is beyond the scope of this chapter. Readers interested in more details on 5G Radio Access technologies beyond what we cover in this book is suggested to look at for example (Dahlman et al., 2018).

A new feature in 5G is the possibility for the PCF to provide some policies to the device via the AMF. This feature is similar to the Access Network Discovery and Selection functionality (ANDSF) used for 4G, with the difference that ANDSF policies were carried to the device embedded in user data packets while for 5G it is carried in normal NAS signaling messages. Two types of rules can be provided from PCF to the device:
- UE Route Selection Policy (URSP) information. These rules indicate to the device how application traffic shall be sent over the network, e.g., which data session, slice, SSC mode etc. shall be used for a certain application. When an application starts in the device, the URSP can be used to determine if an existing session can be used or a new session is needed with an appropriate SSC mode, etc.
- Access Network Discovery and Selection Policy (ANDSP) Information. These rules are applicable for non-3GPP network access, which we will describe in Section 3.15. This information is used to guide the device in determining which Wi-Fi networks to select, e.g., what WLAN SSIDs to prioritize.

Policy Control for data sessions is similar as for EPC. It is a key part of the concept called PCC (Policy and Charging Control) which exists already for EPC, and which has been extended to cover also 5GC.

The PCC concept is designed to enable flow-based charging, including for example online credit control, as well as policy control which includes support for service authorization and QoS management. Charging and policy control functions rely on that all IP flows are classified in the UPF using unique packet filters that are defined from the SMF, and that operate in real time on the IP data flows flowing through the UPF. Policy Control for data sessions is also supported for Ethernet sessions.

The PCF can interact with external applications over the N5 or Rx interfaces. Rx is inherited from the EPC architecture, where Rx is terminated by the PCRF. Rx is Diameter-based also for 5GC Rel-15 and can be used by external application servers that do not support service-based interaction with PCF to control policies for specific network services. In EPC, Rx is for example used by the P-CSCF for controlling the bearers related to voice services carried over LTE (VoLTE).

Charging support in the 5GC architecture is provided through the Charging Function (CHF), which interacts with PCF and SMF to provide support for charging services.

More details on Policy Control and Charging can be found in Chapter 10.

3.8 5GC interworking with EPC

When radio access technologies that connect with the 5G Core Network instead of to an existing EPC network start to be deployed, initial geographical coverage will be quite limited for two main reasons. Firstly, it takes time to build coverage, so in the initial phases of a service launch the coverage of the new radio technologies can be assumed to be quite spotty.

Secondly, new radio technologies such as 3GPP NR are in many cases deployed on higher frequency bands than existing radio technologies, both due to the fact that new spectrum made available for mobile services are typically in higher frequency bands than existing bands, but also as these higher frequency bands provide superior network capacity as there is more spectrum available. However, as the ability to cover a given geographical area with a given base station output power quickly decreases with higher frequencies, coverage will be limited. Simply put, the gain in increased data capacity achieved with moving to the higher spectrum is balanced against much less coverage.

For users that want wide-area mobility while retaining stable IP addresses, the solution is to rely on other radio access technologies when out of coverage of the new access technology served by 5GC. A stable IP anchor point is kept while the access network being used at a specific location and time is changing depending on mobility patterns,

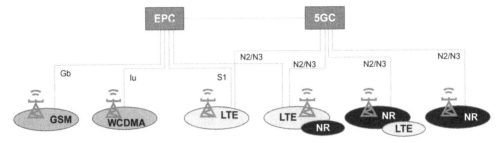

Fig. 3.18 Options for connecting 2G/3G/4G/5G radio networks to EPC and/or 5G Core.

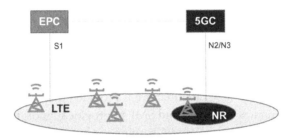

Fig. 3.19 Simplified architecture for interworking between EPC and 5GC.

coverage of different technologies, and operator policies. This requires connectivity between some of the EPC and 5GC network elements.

In theory any combination of the technologies shown in Fig. 3.18 could apply, but it is unlikely that a typical service provider deployment would have all of the variants we have illustrated.

Let us simplify the architecture a bit so we can focus on the 4G–5G interworking case, illustrated in Fig. 3.19.

Here we assume that LTE has overlapping coverage with NR and that users would best be served by NR when in coverage of both technologies. Devices that are in areas covered by NR access are served by 5GC, but they will need to be served by LTE and hence EPC when they are or move outside of the NR coverage area.

3.8.1 Interworking using the N26 interface

The concept relies on that IP sessions for 5G-capable devices are always anchored in the 5GC architecture. The detailed architecture is shown in Fig. 3.20.

Readers should note several important aspects:

- The SMF and UPF need to support EPC PGW logic and functionality across the S5-C and S5-U interfaces. This means that the EPC SGW is unaffected.

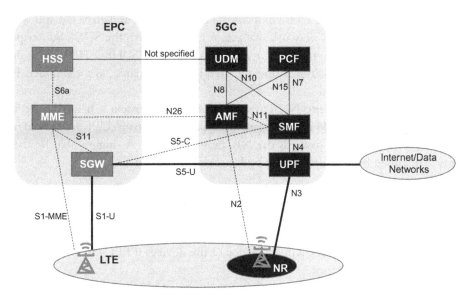

Fig. 3.20 Detailed architecture for interworking between EPC and 5GC.

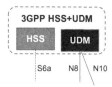

Fig. 3.21 Functionality of User Data Management solution.

- The PCF need to support the necessary policy data parameters across the N7 interface in order to support the PGW-C functionality of the SMF.
- The functionality supported over the N26 interface is in practice a subset of the functionality specified in EPC for the S10 inter-MME communication interface. This approach minimizes the impact on the EPC MME.
- Existing 4G users of course do not need to be migrated to the 5GC architecture. They should be completely unaffected by the introduction of 5G. This is done through maintaining existing EPC PGW and PCRF functionality for the 4G users (not shown in the picture).
- In the first generation of 5G specifications, 3GPP did not specify how the HSS and UDM interact or are implemented. The standard simply outlines a combined HSS + UDM, leaving the solution to the infrastructure vendors. In effect, this means implementing a combined solution that supports both HSS and UDM functionality, or providing an interface in between these. This interface was later specified as part of 3GPP Release-16 specifications. See Fig. 3.21.

So 5G users that connect over LTE will be served by an MME and an SGW and will be allocated an SMF in the 5GC network that acts as PGW towards the SGW. Subscription data for the user will be provided to the MME from the HSS. How HSS gets hold of subscription data for the 5G user is not specified in the standards, as a combined HSS + UDM is all that is outlined.

When a 5G user moves into NR coverage, the session context is handed over from MME to AMF across the N26 interface, and user data tunnels from UPF are moved from S5-U towards SGW to N3 towards the NR radio network. A single stable IP anchor point is maintained by the UPF across both access technologies.

3.8.2 Interworking without an N26 interface

It is possible to provide mobility support across EPC and 5GC without the AMF-MME connection N26. The network will announce to the devices if EPC-5GC interworking using N26 is supported or not.

Without an N26 interconnection, the MME and AMF cannot exchange session information. Instead, information on which SMF (with PGW functionality) to use as an anchor point for the device is maintained in the HSS and UDM. Remember that these need to be combined or interconnected. In addition, the device will indicate to the MME or AMF, depending on in which direction it is moving, when it attaches to the network if an existing session exists or not.

There are two variants of interworking without the N26 interface:
- Single mode registration
- Dual mode registration

While a solution without N26 may simplify the network setup somewhat, it has the drawback of a longer interruption time of the data session during the mobility phase at least if using Single mode registration. This may be acceptable to some applications, but it is not good enough for example for voice or other real time critical applications.

The main difference from the Single mode registration variant is that the Dual Mode registration variant allows the device to register in the target access network before releasing the connection in the network it is leaving. This can reduce the service interruption time during the mobility phase; the main drawback is a more complex solution. Dual registration means that the device may be simultaneously registered in EPC and 5GC and have multiple radio signaling connections active simultaneously. This means a more complex device, and support for Dual mode registration is also optional, while Single mode registration support is mandatory in the device.

It shall be noted that the 3GPP specifications Release 15 and 16 do not include support for 5GC/NR interworking with GSM and WCDMA access networks. This would involve the SGSN and would require additional specification of parameter support over N7 between SMF and PCF and over either Gn between SMF and SGSN or S5 between SMF and SGW.

3.9 Voice services

3.9.1 Overview of 5G voice

Except for dedicated systems for connectivity between non-humans, for example industrial applications interconnecting sensors and processing logic, it can be assumed that most networks that provides data connectivity to mobile devices also need to support voice and messaging services. This is the case also for the 3GPP 5G network architecture. As both LTE and NR are packet-only access technologies, the voice and messaging solutions designed for these access networks rely on IP-based communications. The exception is the first voice solution for LTE, referred to as CS fallback, which triggers the device to move to GSM or WCDMA in order to use circuit-switched connections for the voice call. This was designed to allow for voice support in 4G networks which did not yet support voice-over-LTE, referred to as VoLTE.

The 3GPP-specified voice and multimedia services are based on the IMS solution. IMS is short for IP Multimedia Subsystem and is the technology that is used for the widely available Voice-over-LTE (VoLTE) services that are now supported in many 4G networks world-wide.

There are two main options for how to realize support for voice services in a 5G-capable device

- EPS fallback
- Voice-over-NR

Both rely on usage of IMS for providing the service logic and handle the control signaling with the devices using the SIP protocol. The difference is mainly how the access networks are used—if the voice calls are carried over the 5G/NR access network connected to 5GC or over an 4G/LTE access network connected to EPC.

3.9.2 EPS fallback

EPS is short for "Evolved Packet System" and is the formal 3GPP term used for a 4G network including the LTE radio access and the EPC core network. EPS fallback is realized through forcing the device to use LTE access once it needs to make or receive a call. A prerequisite for this to work is that there is enough LTE coverage everywhere where this is NR coverage, and that VoLTE is enabled as a service in the network.

EPS fallback is used when a device is within NR access coverage and is attached to 5GC. When a call is to be made from the device, or a call is incoming to the device, the network triggers the device to change radio access network from NR to LTE *before* the call is established. This benefits from the existence of the N26 interface between MME and AMF as that reduces the call setup time significantly. N26 has however no impact on an ongoing voice call, as that is not carried over NR access at all. Fig. 3.22 shows the EPS fallback architecture.

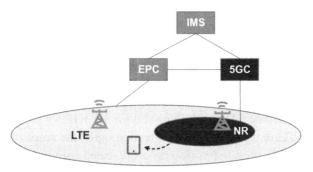

Fig. 3.22 EPS fallback.

Once the device is connected to LTE, the call is established and served as an ordinary VoLTE call utilizing LTE/EPC/IMS. All data sessions ongoing on 5G/NR are also moved to 4G/LTE for the duration of the call. When the call is terminated, the device should preferably move back to the NR access if still in coverage.

EPS Fallback can typically be assumed to be used in the early phases of 5G deployment, when NR radio access networks may not yet have the full support for voice and multimedia bearers or may not yet have been tuned or configured for such services.

3.9.3 Voice-over-NR

Voice-over-NR is different in that it does not trigger a radio access change of a device that is within NR radio coverage. Instead, all SIP signaling and establishment of the IMS bearers is done over NR access, utilizing the 5GC network interconnected with the IMS domain. This requires support from the NR radio access network for voice and multimedia traffic.

Also, when Voice-over-NR is used, mobility to 4G/LTE may be needed. The difference to EPS Fallback is that mobility to 4G/LTE is optional and used only *after* establishing the call, for handing over ongoing calls to LTE in case the device is moving and losing NR coverage. Fig. 3.23 shows the architecture for Voice-over-NR.

Note again the difference between the two solutions, despite that the two Figs. 3.22 and 3.23 may look very similar. With EPS Fallback, devices are always moved to 4G/LTE in order to establish voice calls. With Voice-over-NR, calls can instead be established over 5G/NR and are only moved to 4G/LTE if NR coverage is lost due to movements of the device during the call.

When a handover of an ongoing call from NR to LTE is to take place, it is very important that the handover time is as short as possible, as this is in the middle of an ongoing call. There is no real option than to utilize an inter-radio handover and the N26 interface to minimize the interruption time.

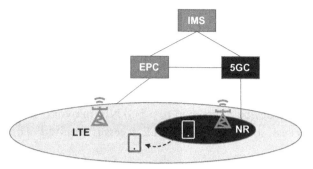

Fig. 3.23 Voice-over-NR.

When serving voice over NR and 5GC, regulatory requirements in the country where the network is deployed may require support for Emergency Calls. This brings additional requirements to the NR/5GC network compared to the case when only normal calls may be served over NR/5GC and Emergency Calls instead may be handled either over VoLTE or over Circuit-switched access networks like GSM or WCDMA, based on an access selection by the device.

One additional network functionality is typically needed when voice services are to be deployed in a network. This covers the scenario that more than one PCF is being used in the network. The PCF that shall serve a specific data session is selected by the SMF, typically after querying the NRF. This means that different data sessions may be served by different PCFs (as well as different SMFs). For an external application to locate the specific PCF that is serving a specific data session, it will query a separate Network Function that has as its role to maintain records of which sessions that are served by which PCFs. This Network Function is called the Binding Support Function (BSF) and connects to the 5G Core SBA architecture. The BSF offers services to PCFs to register and deregister information about data sessions, and services to other applications to query the BSF which PCF that serves a specific data session. For voice services the Application Function (AF) is the IMS P-CSCF. The concept is not very complicated but is a very important function in order to connect the IMS and the 5GC domains when multiple Network Functions are deployed. The BSF is defined with a service-based interface called Nbsf. Fig. 3.24 illustrates the BSF concept.

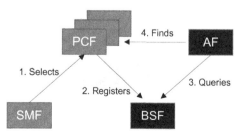

Fig. 3.24 Scaling the Voice-over-NR capacity with BSF and multiple PCFs.

3.10 Messaging services

3.10.1 Overview of messaging services

The Short Message Service (SMS) is used both for end user messages, but also for messages sent between the network and the device without human interaction.

Just as for voice, there are two main options for how to support Short Message Services (SMS) when the device is attached to a 5G/NR network. Both variants are supported over NR access, so no fallback to LTE access is required. The two options are:
- SMS over IP
- SMS over NAS

3.10.2 SMS over IP

For this solution to work, there needs to be an IMS system connected to the 5GC system, just as for the voice solutions described above. See Fig. 3.25. The functionality of the IMS system connected to 5GC is basically identical with the IMS functionality for 4G/EPC as the service principles are the same in both cases.

It is beyond the scope of this book to discuss the IMS principles in any detail, but the basic architecture is shown in Fig. 3.26 and the functionality works like this.

First the IMS-capable device connects to the network, is authenticated and establishes a data session, similar to if it was to connect to the Internet. Over this data connection, the device communicates using the IETF SIP protocol with the IMS node called the Call Session Control Function (CSCF). There are actually three CSCFs in an IMS network, the P-, the I- and the S-CSCFs, but in order to simplify the description, we treat it is as one here.

The SIP signaling between the CSCF and the device is used to control the IMS session and the services being used. For IMS services such as voice, the actual media session (the voice call) is handled through the SBG function in IMS, and the CSCF interacts with the PCF in the 5GC network to establish the required Quality-of-Service support for the voice session. But this is not needed when IMS is used only for the SMS service.

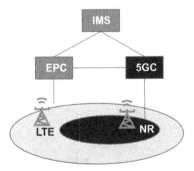

Fig. 3.25 IMS simultaneously connected to EPC and 5GC.

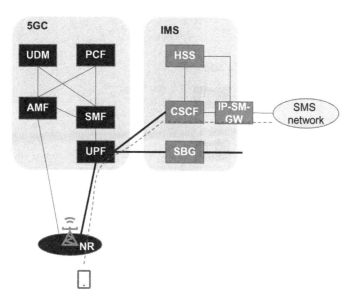

Fig. 3.26 Architecture when interconnecting IMS and 5GC for SMS over IP.

SMS-over-IP instead relies on that the short messages to and from the device are encapsulated inside SIP messages, the IMS control signaling, carried transparently through the access network and the 5GC UPF function. The short messages are then forwarded between the CSCF and the SMS system via a gateway function that 3GPP calls IP-SM-GW ("IP Short Message Gateway").

3.10.3 SMS over NAS

The other solution for supporting SMS in a 5GC network does not rely on the existence of an IMS system. This may be relevant specifically for devices that are not smartphones but instead appliances like sensors, routers or other industry devices. These may rely on short messages to communicate, e.g. to boot/restart or to perform software upgrades. And they typically do not come with an IMS software stack. Also, smartphones that can do IMS signaling can rely on SMS over NAS for allowing the operator to provide low level device configuration or provisioning data over the air interface.

The architecture is shown in Fig. 3.27.

Like the SMS over IP solution, the SMS over NAS solution relies on encapsulating SMS messages within control signaling for efficiency reasons. Here it is however the NAS control signaling between the AMF and the device that is used to carry the SMS messages, not the SIP signaling between the CSCF and the device. Hence the name "SMS over NAS". The AMF that terminates the NAS signaling to and from the device forwards the decapsulated SMS messages to and from the **Short Message Service Function** (SMSF) in the 5G Core architecture. The SMSF is an optional Network Function that is only

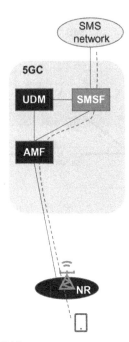

Fig. 3.27 Architecture for SMS over NAS.

deployed for enabling SMS over NAS services. Just as other Control Plane network functions in the 5G architecture, the SMSF exposes a service-based interface relying on HTTP communication, called Nsmsf. It connects with the AMF and the UDM Network Functions within the 5G Core architecture. The SMSF checks subscription data through interacting with the UDM, generates charging records and forwards SMS messages between the AMF and the SMS networks external to the 5G Core architecture.

3.11 Exposure of network information

The Network Exposure Function (NEF) supports interaction with external applications. It exposes selected network capabilities that can be used in various ways by these applications. This is of interest for opening up new business opportunities for service providers through enabling more advanced services to be offered by third party application providers. One key functionality supported by the NEF is to allow external applications to trigger devices to perform specific actions related to the application, including connecting to the NEF or the application itself. Besides this, three different types of capability exposure are specified in Release-15, visualized in Figs. 3.28–3.30. Additional use cases are being defined as part of later 3GPP releases.

• Monitoring—allows external applications to monitor some of the network events associated with a specific mobile device, for example if the device is reachable, what

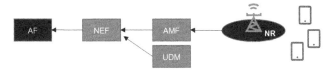

Fig. 3.28 Architecture for exposure of network event monitoring.

Fig. 3.29 Architecture for data provisioning from external applications.

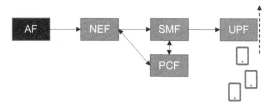

Fig. 3.30 Architecture for policy control from external applications.

the location is, and if it is roaming or not. The NEF retrieves this information from the UDM and the AMF.

- Provisioning—allows external applications to provision information into the system that would apply to selected devices. So far, the parameters defined for external provisioning relates to expected device mobility patterns. This can be used by the AMF to instruct the radio network how to tune certain settings, including how to minimize state changes for the devices, to optimize the overall network signaling capacity.

- Policy and Charging Control—allows external applications to control various aspects of data sessions. One example is the ability to influence traffic routing, for example to influence when and how local breakout and routing should be applied for certain devices. Also, QoS and charging policies can be controlled and enforced from external applications via the NEF, providing information about the data traffic. The NEF interacts with the PCF for this, which in turn determines the Quality-of-Service and charging information based on the application information provided by AF/NEF. Other mechanisms include the support for negotiation about the transfer policies for future background data transfer, and NEF support for allowing an external application to define the templates used in the UPF to detect that certain traffic is related to specific external application traffic. The NEF interacts with the SMF to achieve this, and the SMF forwards these templates to the UPF. This functionality is visualized in Fig. 3.30.

It goes without saying that for all use cases, authentication of the external applications that want to interact with the NEF is very important. This to protect both the network and the devices from malicious interference or unauthorized information gathering.

3.12 Device positioning services

To determine and keep track of the geographical position of a specific device may be important for two reasons:
- For emergency situations where the user of a device quickly needs to be accurately located for safety or medical reasons
- To provide position information to external applications, for example to be used to provide location-based information.

In the first versions of 3GPP specifications, location services are optional and only defined for regulatory use case such as positioning for emergency calls. This can be expected to change in later releases of the standard.

It is of course important to recall that most devices today have built-in capabilities to determine their own location, either using GPS satellite signals or through other means such as identifying the identities of nearby WiFi networks, and then comparing this information with what is stored in an external database. Communication is performed in the data session connecting the device with the Internet. The significant majority of today's smartphones include a GPS antenna and receiver.

However, positioning methods relying on the device capabilities may not be possible in all situations. Devices that are not smartphones do not always come with GPS or WiFi capabilities. This is the case for many low-cost devices for machine-type applications, for example sensors. In addition, the GPS antenna in a device may be turned off, or satellite coverage may be blocked for instance when being indoors. Furthermore, there may be regulatory requirements to be able to locate a user device, independent of a device's capability to locate itself.

The 5G network architecture comes with the capability to locate a device using network capabilities. There are two specific Network Functions in the 5G Core architecture that are key to providing these capabilities—the Location Management Function (LMF) and the Gateway Mobile Location Center (GMLC). The positioning functionality also relies on support by the AMF and the radio network.

An external application requesting the position of a device will contact the GMLC, which first authorizes the request. If the positioning occurs during an emergency session the AMF serving the specific device may have informed the GMLC that it has established a session with the device. The GMLC will query the AMF about the position of the device. In case the GMLC does not know which AMF that is serving the device, it will first query the UDM to find out which AMF that is the right one. The AMF will in its turn query the LMF about the position, and the LMF will return data about the position

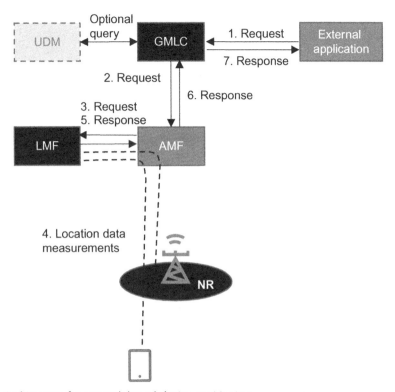

Fig. 3.31 Architecture for network-based device positioning.

to the AMF. The AMF will forward this to GMLC which forwards to the external application (Fig. 3.31).

The LMF finds and calculates the position of the device based on interaction with the radio network and/or the device itself. It sends a request to the radio network for location information. Depending on capabilities, the network can either send for example the identity of the current cell, or make measurements to locate the device. All communication between the LMF and the radio network is carried via the AMF, then conveyed over the N2 interface to the radio network. The communication between the device and the LMF is also done via the AMF but is carried transparently over the radio network within the NAS protocol used between the core network and the device.

3.13 Network analytics

The Network Data Analytics Function (NWDAF) was a late addition to the first version of the 5G specifications and is an optional component in the 5GC network architecture. NWDAF collects various type of network and subscriber data, applies an "analysis" to

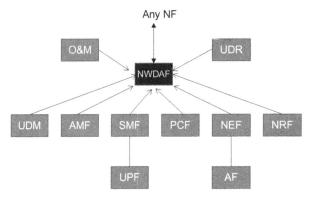

Fig. 3.32 NWDAF interactions with other Network Functions.

this data, and offers the results to other Network Functions through a service-based interface.

NWDAF was only very briefly outlined in the 3GPP Release-15 specifications, while the more detailed functionality is being defined as part of Release-16 work.

The NWDAF collects data from several other Network Functions (NFs) using event exposure services offered by these NFs. It also collects data from the O&M system as well as subscriber-related data from the UDR.

The services offered by NWDAF can in theory be consumed by any other NF, and even by an external application (AF), then via the NEF. Primary consumers of the NWDAF services are NSSF and PCF, which is visible in the architecture in Fig. 3.32.

The analysis done by the NWDAF on the collected data could for example be a summary of statistical/historical data, or an attempt to predict future data values. The analytics data could be used by other NFs to apply certain actions in the network, like selection of a specific slice or modification of QoS for a service.

More information on NWDAF can be found in 3GPP TS 23.288.

3.14 Public warning system

The ability to quickly send important messages to many or all users in a network in more or less real time is a very important capability of mobile networks, and even a legal requirement in many countries. Typical use cases for a Public Warning System (PWS) include sending warning messages related to natural disasters such as earthquakes, tsunamis or severe storms, or ongoing criminal actions like child abductions or terrorist actions. It can also be used for example to transmit road traffic conditions.

Public Warning System support in the 5G architecture relies on the usage of the concept called Cell Broadcasting—the ability to trigger the radio network to transmit one

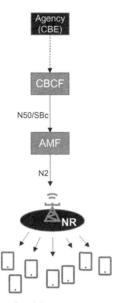

Fig. 3.33 Architecture for broadcasting of public warning messages.

single short message to multiple devices in the network simultaneously. Messages are broadcast either within the whole network or within certain geographical areas, down to the size of a single radio cell. This is controlled from the network on request from the originator of the message.

Cell Broadcast capabilities are nothing unique to 5G, instead it is the same concept that has been used for 2G, 3G and 4G. The 5G implementation details are, however, a bit different as the Network Functions and protocols are new. Fig. 3.33 shows the PWS architecture when 5G Core and the new 5G Radio is used.

Messages are originated within and sent from what 3GPP refers to as a Cell Broadcast Entity (CBE), without further specifying what it is or how it interconnects with the Cell Broadcast Center (CBC) in the mobile network. The CBE can typically belong to an organization controlled by the national authorities in the country where the network is deployed, dealing with public safety or crime prevention or both.

The Cell Broadcast Center controls the transmission area, duration and how frequent re-transmissions shall be made for a given message.

The request is sent from the CBC to the applicable AMFs, over either the N50 or SBc interface. The reason why there are two options is that 3GPP has specified two implementation variants for this interconnection. Essentially—if the CBC has a service-based interface, service-based interfaces are used, including Namf in AMF. As shown in Fig. 3.33, this interaction using service-based interfaces is visualized as N50 in the logical point-to-point architecture. And then CBC is strictly called CBCF as in Fig. 3.33.

If the CBC only has a traditional Diameter interface used for Cell Broadcasting over 4G/LTE, SBc is used. It is beyond the scope of this book to further outline the implementation options. The functionality and the information sent is in any case the same—the CBC sends the message to the applicable AMFs, together with information on in which geographical area the message shall be broadcasted.

The AMF receives the message transmission request and subsequently sends a corresponding request to all radio base stations within the requested geographical area over the N2 interface. As opposed to normal messaging between the Core Network and the devices like SMS, no interaction between AMF and the devices take place in the case of Cell Broadcasting, instead the message is sent from the AMF to the devices via the radio network. The AMF also reports back to the CBC if the transmission was successful or not, using reports from the radio network.

More information on Cell Broadcasting can be found in 3GPP TS 23.041.

3.15 Support for devices connected over non-3GPP access networks

The 5G Core architecture includes support for devices connecting over what is referred to as "non-3GPP access technologies". In most cases this means that the device is connected over a WiFi access network instead of over a 3GPP radio network, but in theory it could be any type of access network that is supported by the device and offers IP connectivity.

When EPC was defined for 4G, it included multiple variants of non-3GPP access support, relying on either "trusted" or "untrusted" access, and "network-based mobility" or "host-based mobility". It is outside the scope of this book to describe the EPC architecture in any detail, so interested readers are referred to for example (Olsson et al., 2012).

For the Release 15 5G Core architecture, fewer options are defined for supporting non-3GPP access than for the EPC architecture. Firstly, it is assumed that there are only "untrusted" non-3GPP. "Untrusted" in this context simply means that the operator of the 3GPP-defined mobile network does not trust the security of the non-3GPP access network. This is obvious since public or private WiFi networks typically use password-based access authorization methods and sometimes lack payload encryption, which is not acceptable for permitting access to mobile network infrastructure and services. In Release 16, 3GPP included support for trusted non-3GPP access networks as well as wireline access.

The 5G Core architecture includes the Non-3GPP InterWorking Function (N3IWF) that acts as a Gateway to the mobile network and the connection point for devices that gain access over a non-3GPP access network. Note that since this is about untrusted access, the architecture does not specify how a non-3GPP access network is connected to the 5GC architecture, it instead specifies how *devices* utilizing *any* untrusted non-3GPP access connect to the 5G Core using the N3IWF Network Function.

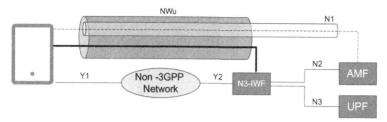

Fig. 3.34 Connections between the device and the core network for the non-3GPP access case.

Traffic to and from these devices could in theory be routed between the untrusted access network and the mobile network across the public Internet.

Fig. 3.34 illustrates the setup.

The device connects to the non-3GPP network, is authorized, granted access and given an IP address. This connection is referred to as Y1 in the picture but how and when this is done is out of the operator's control nor is it specified by 3GPP. Y1 is typically a WiFi air interface. The device selects an N3IWF and connects to this N3IWF using the IP access service offered by the non-3GPP network. This connection is referred to as Y2 and is also not specified by 3GPP. As explained above, Y2 may very well be the public Internet. Then a secure and encrypted IPsec tunnel is established between the device and the N3IWF through which both signaling and data traffic between the device and the mobile network can be forwarded. This tunnel is referred to as NWu.

The N3IWF selects an AMF, which in almost all cases shall be the same AMF as is already serving the device over a 3GPP access if this is the case. An N2 interface is established between N3IWF and the selected AMF. Then an N1 interface carrying NAS signaling is established between the device and the AMF. This is a difference to the EPC architecture. In the 5GC architecture, devices connecting over non-3GPP access networks are managed in more or less the same way as if they are connecting over a 3GPP access networks. NAS signaling no longer only apply to 3GPP access networks as in the EPC architecture. So, the NAS signaling is carried over the N1 interfaces, across NWu and N2, between device and AMF and NWu is the tunnel on top of Y1, the non-3GPP access network, and Y2.

Once a UPF is selected, an N3 interface between N3IWF and UPF is established for data transmission. Data is carried over NWu between the device and the N3IWF, and then across N3 between N3IWF and UPF.

Fig. 3.35 shows a slightly more complete picture of a device simultaneously connected to 3GPP and non-3GPP access networks. For simplicity, not all details within the core network are shown.

It should be noted that a device can be simultaneously registered over both 3GPP and non-3GPP access, and a session can be moved between these access networks while

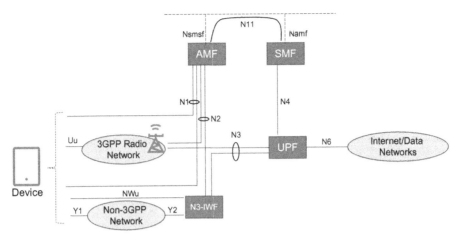

Fig. 3.35 Simultaneous access to 3GPP and non-3GPP access technologies.

maintaining a stable anchor. A device can also simultaneously have two sessions active over 3GPP and non–3GPP access respectively.

In a Release-15, one single session cannot simultaneously be active over both 3GPP and non-3GPP access. This is however possible in Release-16 compliant solutions, where the specifications require that both the network and devices support this (see Chapter 16, Section 16.3.8 for further details).

Another architecture for support of non-3GPP access is based on the ePDG as specified for the EPS architecture. In such instances, the ePDG connects to the SMF/UPF as if it was a 4G PGW in the EPC architecture. This architecture is not further described in this book. The interested reader is referred to (Olsson et al., 2012) for more information.

3.16 Network slicing

Efficient support of network slicing has been one of the main drivers when designing the 5G network architecture. There is no exact definition of network slicing across the industry, but the overall idea is to separate traffic into multiple logical networks that all execute on and share a common physical infrastructure. The reasons for this separation can for example be to address security concerns, to optimize the configuration and the network topology differently for different services, or to enable a differentiation between operator service offerings.

In the 3GPP specifications, a network slice consists of a radio network and a core network. Some parts of the network resources can be shared across multiple network slices, while some parts can be unique to a single slice. The 5G slicing concept also involves optional resource partitioning in the radio network per slice.

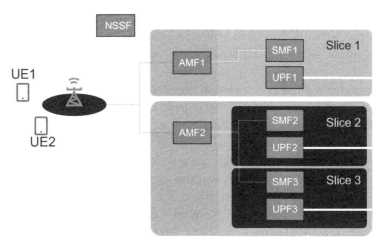

Fig. 3.36 Network slicing simplified.

The new 5G Core architecture also allows one single device to connect to more than one slice simultaneously, a feature which was not supported in the EPC architecture defined for 4G.

A specific network slice is identified by a parameter called S-NSSAI, short for "Single Network Slice Selection Assistance Information" and consisting of two sub parameters—the Slice/Service Type (SST) and the optional Slice Differentiator (SD). SD is used to differentiate between multiple slices of the same type, hence having the same SST.

The radio network serving the device will use one or more S-NSSAI values requested by the device to do the initial selection of AMF.

The selected AMF will either decide to serve the specific device or make a new slice selection itself, or it may use the Network Slice Selection Function (NSSF) for this. The NSSF has as its single role to support the selection of network slices based on a combination of S-NSSAI values defined for the network, requested by the device and allowed in the subscription.

In the simplified Fig. 3.36, device 1 (UE1) connects to slice 1 which consists of a dedicated AMF, SMF and UPF. Device 2 (UE2) simultaneously connects to slice 2 and slice 3, each of these containing an SMF and an UPF but both being served by a common AMF2. Again, note that UE is the 3GPP term used to denote a device.

A detailed description of slice selection can be found in Chapter 11.

3.17 Roaming

The 5G specifications naturally include support for connecting networks from two operators to support subscriber roaming. Compared to the non-roaming case described so far, the network architecture now becomes somewhat more complex. Some Network

Functions remain in the network where the user is attaching (the visited network, VPLMN), some Network Functions will exist in the network where the user is a subscriber (the home network, HPLMN), and some Network Functions become duplicated.

In order to achieve a secure connection between the VPLMN and the HPLMN, the "Security Edge Protection Proxy" (SEPP) is used. The SEPP is not a Network Function that produces or consumes services, instead it acts as a service relay between the consumer and the producer when these two Network Functions are in different networks. Besides protecting the communication through using message filtering and applying roaming policies, the SEPP concept also means that the topology of a network is hidden to the other party. This applies in both directions.

Fig. 3.37 shows the distribution of Network Functions in the visited and the home networks respectively. As illustrated, the NRF, NEF and PCF are duplicated and exist in both networks. The figure also shows the connection between the SEPP in the HPLMN and the SEPP in the VPLMN, referred to as N32 and used to carrying all signaling traffic between the two networks. The figure also shows that the NRFs in the respective network are interconnected via the N27 reference point, overlaid onto N32.

Besides N27, the rest of the roaming interfaces overlaid on N32 are more easily visualized in the simplified point-to-point representation in Fig. 3.38. It illustrates that not only N27 is carried over N32, but also four other reference points interconnecting PCF, AMF and SMF in VPLMN with PCF, UDM and AUSF in the HPLMN.

The basic concept for all roaming configurations means that all authentication and handling of subscription data is done in the home network. So, the AMF and SMF

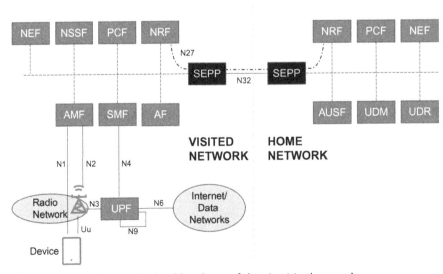

Fig. 3.37 Roaming architecture for local break-out of data in visited network.

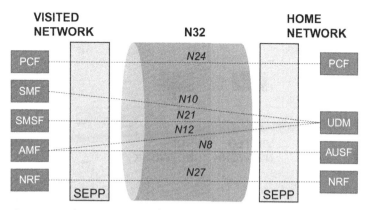

Fig. 3.38 Signaling across the roaming interface for the local breakout case.

serving the roaming devices in the visited network need to connect to UDM and AUSF in the home network. This is done across the N8, N10 and N12 reference points.

NRFs in the visited and home networks need to interconnect to support Service Discovery between Network Functions located on different sides of the network–network boundary. This is handled over N27.

N21 is an optional reference point, only applicable if SMS-over-NAS is used. If SMS-over-IP is the method for sending and receiving short messages, N21 is not applicable.

N24 is also an optional reference point and is used by the PCF in the visited network to connect to the PCF in the home network for policy-related signaling.

Roaming may not look all the complex so far, but here it shall be noted that Figs. 3.37 and 3.38 only visualize one out of two possible roaming scenarios—the so called "Local Breakout Scenario" where data traffic is routed to and from Internet or service networks directly from the visited network. There is also a "Home routed" scenario where the user data traffic is routed from the VPLMN to the HPLMN before routed to the Internet or service networks. This means that a slightly different network configuration is needed.

Compared to the local breakout configuration, the home routed configuration means that more functionality is executing in the home network. The consequence of this is also a more complex roaming interface with three additional reference points, while one is no longer applicable (Fig. 3.39).

The full set of reference points used across the roaming interface is shown in Fig. 3.40.

In addition to the signaling between visited and home networks, this case also includes the actual user data traversing the network-to-network boundary, being sent to the home network for processing and service access. This gives the home network operator more control over the services offered to its own subscribers, but it also makes the roaming setup a bit more complex. It requires that the transport network connecting the operators is dimensioned to carry all the data traffic from the users, not only the applicable signaling traffic.

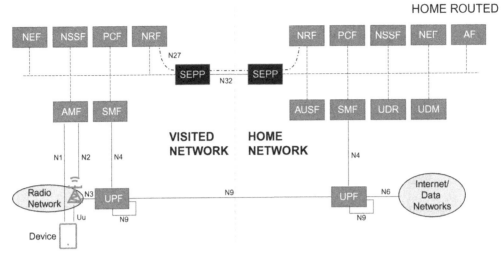

Fig. 3.39 Roaming architecture for routing of data to home network.

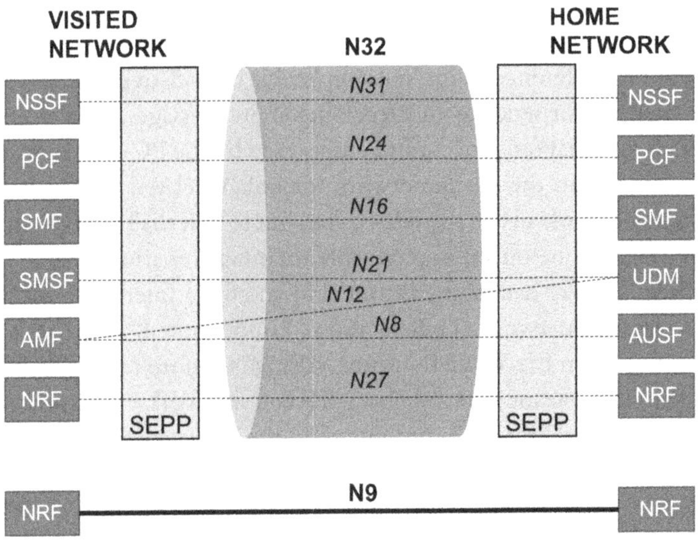

Fig. 3.40 Signaling across the roaming interface for the home routed case.

N9 is used for carrying the user data between the networks, as the device is in the visited network but the connection to the Internet or other IP services are in the home network in this network configuration. N9 is hence a roaming interface only for the home routed scenario.

N16 is used by the SMF in the visited network to connect to the SMF in the home network for session management signaling.

N31 is optional and only applies if network slicing is used in the network. N31 is then used by the NSSF in the visited network to retrieve network slicing information from the NSSF in the home network.

One interface that is no longer used across the roaming interface is N10 between SMF and UDM. N10 only exists in the home network in the home routed roaming network configuration.

3.18 Storage of data

The UDSF Network Function is a bit special compared to most other Network Functions. When specifying this, it can be argued that 3GPP moved somewhat away from the fundamental guiding principles of only specifying logical network functionality, not implementation methods. The UDSF can be viewed as being in a gray zone here, as it is specified as a generic database component in the architecture, allowing for "any" Network Functions to store and retrieve "any" of its data using the UDSF. This data is called "unstructured" by 3GPP, basically meaning unspecified. And this means that it is implementation-specific per vendor. Several Network Functions may share one single UDSF, or they may use separate UDSFs.

The UDSF is naturally an optional component in the architecture. If it exists in a specific network implementation, it also only serves Network Functions in the same network, never Network Functions across a roaming interface.

The UDSF is providing services to other NFs over the Nudsf reference point. When one or more other Network Functions uses the UDSF for data storage, they connect over N18 in the point-to-point representation of the network architecture. See Fig. 3.41.

3.19 5G radio networks

3.19.1 Overview

Even if this a book about 5G Core Networks, it is beneficial for the reader to have a basic understanding also of the fundamental architecture and concepts of the 5G Radio Networks. Defining 5G Radio and Core networks has been a combined effort in the industry that started with the work on 3GPP specifications Release-15.

The new 5G radio technology defined by 3GPP is simply named "New Radio" and abbreviated to NR.

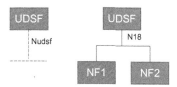

Fig. 3.41 UDSF interfaces.

3.19.2 Mobile network fundamentals

The radio network part of mobile networks (cellular networks) consists of several radio base stations, each serving wireless transmission and reception of digital information in one or several "cells", where a cell in this context refers to a smaller part of the overall geographical area that the network serves. Traditionally, a typical deployment case is that one base station serves three cells through careful antenna configurations and planning for how to utilize the available radio spectrum. See Fig. 3.42. Note that the 3GPP specifications however do not put limitations on the number of cells served by one base station.

The size and the outline of the cell are controlled by a few factors, including base station and terminal power levels, antenna configurations and frequency bands. Radio signals using lower frequencies normally propagate over longer distances than radio signals using higher frequencies if the same power level is used. The radio wave propagation environment also has a significant effect on the cell size; there is a large difference depending on whether there are lots of buildings, mountains, hills, or forests in the area, compared to a surrounding area that is fairly flat and mostly uninhabited.

A fundamental ability of a cellular network is to allow the usage of the same frequency in multiple cells. This means that the total capacity of the network is greatly increased compared to the case where different frequencies would be needed for every site. The most intuitive way of allowing this frequency reuse is to make sure that base stations supporting cells using exactly the same subset of the available frequencies are geographically located sufficiently far apart to avoid radio signals from interfering with each other. This was also the solution used in GSM, the first generation of digital systems (2G). However, all subsequent generations of mobile network technologies have functionality that allow adjacent cells to use the same frequency sets. This is achieved with advanced signal processing that targets to minimize interference from unwanted signals transmitted from neighboring cells.

Base stations are located at sites that are carefully selected to optimize the overall capacity and coverage of the mobile services. This means that in areas where many users are present, for example in a city center, the capacity needs are met through locating the base station sites more closely to each other and hence allowing more (but smaller) cells, while in the countryside, where not so many users are present, the cells are normally made larger to cover a large area with as few base stations as possible.

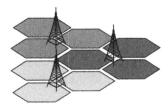

Fig. 3.42 The concept of a cellular network.

All generations of digital mobile systems defined by 3GPP since the 1990's, ranging from GSM (2G), over WCDMA (3G) and LTE (4G), to NR (5G) support the basic concepts of digital transmissions to many devices in a cellular system, but each technology generation does this with different technical solutions, resulting in differences in capabilities and service characteristics.

It should be noted that the cellular concept can be enhanced beyond the traditional three-sector cells through the optional usage of multi-beaming, something we describe in Section 3.19.5.

3.19.3 5G targets

In order to address the expectations and needs identified in the market and industry for existing and new use cases, a number of concrete targets on service characteristics were defined to serve as design goals for the 5G specification work.

At a high level, 5G technologies are designed to meet the requirements of a wide range of different use cases:

- Requirements for mobile broadband services are mainly set to address the needs for efficiently handling very large and growing data volumes in the network through (1) optimizing network capacity, and (2) providing an enhanced user experience in larger parts of the network.
- On the other hand, use cases targeting large number of small or cheap devices supporting Internet-of-Things applications have different needs. Here requirements include for example high energy efficiency to optimize the battery life for these devices, and high connection density to be able to serve large numbers of devices even in a limited geographical area.
- Finally, for business-critical industry applications, some of the most important requirements are very low latency and very high reliability.

Service requirements for 5G networks began to be formulated by multiple industry fora and regulators across the world from approximately 2015. These were summarized by the International Telecommunication Union (ITU) in the report ITU-R TR M.2410-0 (2017) as requirements on an "IMT-2020 network", where IMT-2020 is the formal ITU term used for 5G networks. The requirements have served as input to the corresponding technical study in 3GPP, out of which requirements were formulated in the technical report 3GPP TR 38.913.

A high-level summary of some of the most important 5G service requirements is shown in the table in Fig. 3.43.

As the requirements are use-case dependent and quite diverse, it meant that the NR radio technology needed to be designed in a flexible way, so that a wide range of use cases could be efficiently supported.

Another important requirement is that the NR radio shall be possible to deploy in a very wide range of frequency bands, ranging from 450 MHz up to above 52 GHz. This is a range that no previous radio access technology (2G, 3G or 4G) has supported.

Peak data rates	Up to 20 Gbit/s DL, up to 10 Gbit/s UL
Average experienced data rates	Up to 100 Mbit/s DL, up to 50 Mbit/s UL
Spectral efficiency	Up to 30 bits/s Hz DL, up to 15 bits/s/Hz UL
Connection density	Up to 1million devices/km2
Device battery life	More than 10 years
Mobility	Up to 500 km/h
User data latency	1ms for industry use cases, 4 ms for Mobile Broadband
Reliability	At least 99.999%

Fig. 3.43 5G service requirements.

The frequency range is divided into two parts:
- FR1—Frequency Range 1, ranging from 450 MHz to 6 GHz and typically referred to as "Mid/low band"
- FR2—Frequency Range 2, ranging from 24 GHz to 52 GHz and typically referred to as "High band" or "millimeter wave" (mmwave)

Fig. 3.44 shows the supported frequency bands in FR1, information extracted from the 3GPP TS 38.101-1. As can be seen, there is a wide range of frequency bands supported for NR usage.

And in Fig. 3.45 is the much shorter list of supported frequency bands in FR2, information extracted from the 3GPP TS 38.101-2.

As can be seen, NR supports both TDD and FDD duplex modes.

TDD is short for "Time-Division Duplex" and means that both the device and the base station use the same frequencies when transmitting, but they are synchronized to use different time slots to avoid interference. This is typically configured with a static capacity split between DL and UL traffic, but can optionally be dynamically adjusted in dedicated cells where this helps optimizing the performance.

FDD is short for "Frequency-Division Duplex" and means that the device and the base station use different frequencies for their respective transmissions. FDD is only supported on Mid/low bands, not on high bands for which TDD is always used. This is a consequence of the regulatory situation, the rules that are to be followed by the spectrum license holders. The lower bands are historically paired, meaning one band for uplink and another one for downlink. The higher bands are normally always unpaired, calling for the usage of TDD as a duplex scheme.

SUL and SDL are short for "Supplementary Uplink" and "Supplementary Downlink", and are bands used to complement other bands to improve the total capacity and/or coverage of the system.

It is beyond the scope of this book to discuss all detailed requirements on the radio network. Information on these requirements can be found in a few 3GPP specifications, out of which 3GPP TS 22.261 provides an overview and links to other relevant documents.

FREQUENCY BAND	UPLINK	DOWNLINK	DUPLEX MODE
n1	1920-1980 MHz	2110-2170 MHz	FDD
n2	1850-1910 MHz	1930-1990 MHz	FDD
n3	1710-1785 MHz	1805-1880 MHz	FDD
n5	824-849 MHz	869-894 MHz	FDD
n7	2500-2570 MHz	2620-2690 MHz	FDD
n8	880-915 MHz	925-960 MHz	FDD
n12	699-716 MHz	729-746 MHz	FDD
n20	832-862 MHz	791-821 MHz	FDD
n25	1850-1915 MHz	1930-1995 MHz	FDD
n28	703-748 MHz	758-803 MHz	FDD
n34	2010-2025 MHz	2010-2025 MHz	TDD
n38	2570-2620 MHz	2570-2620 MHz	TDD
n39	1880-1920 MHz	1880-1920 MHz	TDD
n40	2300-2400 MHz	2300-2400 MHz	TDD
n41	2496-2690 MHz	2496-2690 MHz	TDD
n50	1432-1517 MHz	1432-1517 MHz	TDD
n51	1427-1432 MHz	1427-1432 MHz	TDD
n66	1710-1780 MHz	2110-2200 MHz	FDD
n70	1695-1710 MHz	1995-2020 MHz	FDD
n71	663-698 MHz	617-652 MHz	FDD
n74	1427-1470 MHz	1475-1518 MHz	FDD
n75		1432-1517 MHz	SDL
n76		1427-1432 MHz	SDL
n77	3.3-4.2 GHz	3.3-4.2 GHz	TDD
n78	3.3-3.8 GHz	3.3-3.8 GHz	TDD
n79	4.4-5.0 GHz	4.4-5.0 GHz	TDD
n80	1710-1785 MHz		SUL
n81	880-915 MHz		SUL
n82	832-862 MHz		SUL
n83	703-748 MHz		SUL
n84	1920-1980 MHz		SUL
n86	1710-1780 MHz		SUL

Fig. 3.44 Supported NR frequency bands in frequency range 1.

FREQUENCY BAND	UPLINK	DOWNLINK	DUPLEX MODE
n257	26.5-29.5 GHz	26.5-29.5 GHz	TDD
n258	24.25-27.5 GHz	24.25-27.5 GHz	TDD
n260	37-40 GHz	37-40 GHz	TDD
n261	27.5-28.35 GHz	27.5-28.35 GHz	TDD

Fig. 3.45 Supported NR frequency bands in frequency range 2.

3.19.4 NR radio channel concepts

NR is designed to meet this wide range of requirements through inclusion of several key technology concepts. It builds on some of the technology concepts used in LTE but takes these further.

The modulation technology used with NR is OFDM. OFDM is the same technology as is used for LTE, but then only in the downlink direction.

OFDM is a very flexible modulation technology which is well suited to meet the wide range of requirements set for 5G. The basic concept of OFDM is that the total available radio spectrum is subdivided into several subchannels, each carrying one sub carrier. The available capacity made available for each device (resulting from usage of selected subcarriers) can be controlled in both the time and frequency domains at the same time. An example is shown in Fig. 3.46 where three devices A, B and C are flexibly allocated capacity based on needs and available channels. The allocation in the frequency dimension changes for every time slot in that more or fewer sub carriers can be used for the individual device. Note that the figure is simplified. In practice the number of sub carriers may be up to 3300, out of which allocations of one or more bundles of 12 sub carriers are done per device and time slot.

OFDM also has the benefit of being very robust against multipath fading, i.e., the variations in signal strength that are typical for mobile communications and are caused by the signal between transmitter and receiver propagating over multiple paths at the same time. Reflections of the radio waves in various objects mean that multiple copies of the signal arrive at the receiving antenna, since these are not synchronized in time due to slightly different propagation distances. See Fig. 3.47.

Deployment of NR in a wide range of different frequency ranges is made possible through a very flexible structure of the physical layer. As described above, the NR radio

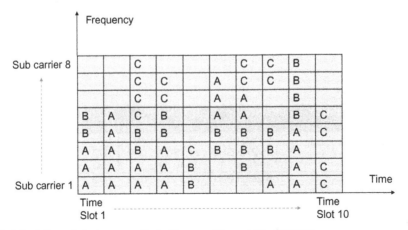

Fig. 3.46 Scheduling of device capacity in Time and Frequency domains.

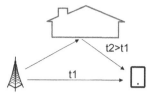

Fig. 3.47 Multipath propagation.

carrier consists, just as for LTE, of a few "sub carriers". In LTE, the sub carrier spacing is fixed to 15 kHz, while in NR there are several options.

The smallest NR sub carrier spacing is 15 kHz just as for LTE, facilitating for having both LTE and NR transmissions sharing the same radio channel. In addition to 15 kHz, several other options for wider sub carriers are defined. Another difference is that in LTE, the maximum carrier bandwidth is 20 MHz while in NR, the total bandwidth of a carrier can be as wide as 400 MHz. To relax the requirements on the devices, every NR device does not need to support the full bandwidth of an NR radio carrier, as opposed to when LTE is used when all devices need to support the full bandwidth of the carrier.

The table in Fig. 3.48 shows the options defined for LTE (as a reference) and NR.

Several NR carriers can then be combined to utilize spectrum with even higher bandwidths, a concept called Carrier Aggregation. This is supported also with LTE.

When looking into the details of the different logical radio channels and the transmission schemes for control information as well as user date, it can be noted that NR is designed with more flexibility than LTE, using a concept referred to as "Ultra Lean design". The purpose here is to have maximum flexibility for future evolution, to minimize interference and to minimize energy consumption. Examples of this include transmitting broadcast information less often, not using the full channel, and to only send reference information on demand. Also, the timing of when to send certain control information is not fixed as for LTE but can instead be sent more flexibly to optimize the overall resource usage.

	Sub carrier spacing	Max aggregated bandwidth	Applicable spectrum
LTE	15 kHz	20 MHz	All LTE bands
NR	15 kHz	50 MHz	Mid/low band (FR1)
NR	30 kHz	100 MHz	Mid/low band (FR1)
NR	60 kHz	200 MHz	All NR bands (FR1+FR2)
NR	120 kHz	400 MHz	High band (FR2)
NR (signalling only)	240 kHz	400 MHz	High band (FR2)

Fig. 3.48 Sub carrier and bandwidth options for NR and LTE.

NR also allows for low latency in that data can be sent when available, not only when a dedicated time slot is available. This is one contributing factor that allows for low latency transmissions over NR.

It is far beyond the scope of this book to describe the details of the NR radio interface. Again, the interested reader is referred to (Dahlman et al., 2018).

3.19.5 Advanced antenna techniques

In order to meet some of the requirements on very high capacity and high data rates for 5G services, there is a need to utilize two technical concepts referred to as MIMO and Beamforming. These technologies are also possible to deploy in LTE networks, but NR has more extensive functionality, including support for handling devices that are in idle mode. This means that signaling during cell search and for access requests can utilize beamforming and MIMO.

Beamforming means that the clear majority of the energy transmitted from the sender is directed towards the intended receiver, instead of being spread over the full cell. The receiver also mainly listens to the radio signals coming in the direction of the transmitter. This improves the signal-to-noise ratio, which is crucial to achieve a higher data throughput. It should be noted that in a typical deployment, the support for beamforming in the receiving direction is more common in the base station than in the device.

Multi-beam techniques mean that there are multiple antenna beams, each covering a smaller part of the cell. These beams are dynamically controllable and steerable, which is used to maximize the performance through optimizing the radio link characteristics for each connection to a device.

Fig. 3.49 illustrates the single beam and multibeam concepts.

MIMO is short for "Multiple-Input-Multiple-Output" and is a technique where the same content is simultaneously transmitted on the same frequency but over more than one propagation path, either using multiple antennas or by using beamforming techniques. The receiver is combining or selecting the best of the different signals it receives to increase the overall received signal strength. 5G radio systems typically combine these two techniques.

Single-User MIMO (SU-MIMO) means transmitting two or more copies of the same data stream in slightly different directions using beamforming, as it can be assumed that the radio signals will experience some energy loss as it passes through various types of

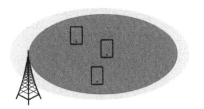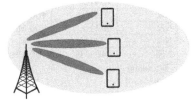

Fig. 3.49 Single beam and multibeam.

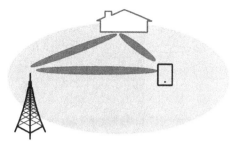

Fig. 3.50 Single user MIMO.

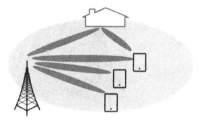

Fig. 3.51 Multi-user MIMO.

material like glass, wood etc. Signals will be reflected against for example cars and buildings located between the transmitter and the receiver. Combining multiple signals in the receiver will therefore achieve a higher aggregated signal-to-noise ratio and hence a higher data throughput. See Fig. 3.50.

When using Multi-user MIMO (MU-MIMO), the purpose is not to optimize the performance for a single user, but rather to achieve a high aggregated throughput for several users. This is required when the load is high on the network and there is a need to optimize the overall capacity usage. Beamforming is then used to simultaneously communicate with two or more users on the same frequencies. The users then need to be in different parts of the cell to allow different radio beams to be used. See Fig. 3.51.

The allocation of MIMO layers and direction of the beams are continuously adapted to the situation and the usage in the cell. This cannot be static as the radio channels constantly change as devices and other objects such as cars move in the cell. To achieve this the base station and devices make frequent estimations of the radio channel characteristics, and the base stations will then use that information to control the usage of MIMO and beamforming. It is however beyond the scope of this book to describe NR channel estimation procedures in more detail.

3.19.6 NR radio network architecture

The radio network architecture as defined by 3GPP consists of multiple radio base stations, connected both to the core network and to each other. Fig. 3.52 illustrates the architecture.

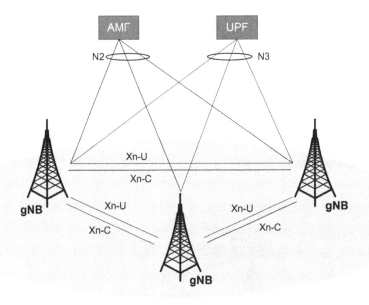

Fig. 3.52 5G radio network architecture.

3GPP defines a logical entity or node called "gNB". This is the logical functionality associated with the radio network functionality and is typically called a radio base station when it is implemented as a deployable network product. In fact, "gNB" is only used when referring to an NR base station connected to the 5G Core Network (meaning options 2 or 4 as described in Section 3.1.2). The term "ng-eNB" is used when instead referring to an LTE base station (meaning options 5 or 7). As this chapter is focusing on the NR radio access technology, we only use gNB when referring to the logical functionality of a radio base station in the subsequent text, but the same network architecture applies to both types of access networks. Even if the formal name is "gNB", we use the term "radio base station" below.

The base stations are interconnected via the Xn interface, which consists of a signaling part, Xn-C, and a data transfer part. Xn-U. All base stations connect to one or more AMF and UPF in the Core Network. In the specifications developed by the 3GPP working groups for radio technologies, the interfaces between radio and core networks are referred to as NG-C and NG-U, but in this book, we use the names defined in the 3GPP Core Network specifications—N2 and N3—as this makes the architecture description consistent with the rest of the book.

More details on the 3GPP 5G radio network architecture can be found in 3GPP TS 38.300.

User data is transferred between the base stations and between base stations and UPF using IP networking. The full protocol stack is shown in Fig. 3.53 and is the same for both Xn-U and N3 interfaces.

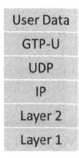

Fig. 3.53 The 5G radio network protocol stack for user data transfer.

The user data, typically IP packets, is encapsulated and transported using the 3GPP GTP-U protocol. GTP-U is well proven as it is used in all previous generations of mobile network systems and provides a reliable data communication service. GTP-U is carried over a standard UDP/IP stack, executing on top of available networking layer 2 protocols, typically Ethernet. 3GPP does however not define the lower level details of the IP networking solution.

The signaling between base stations within the radio network and between the base stations in the radio network and the AMFs in the core network also relies on IP transport, but the upper layers of the protocol stack are different. See Fig. 3.54.

Both stacks rely on the usage of SCTP instead of UDP as is the case for the user data transfer. SCTP is an IETF protocol that provides guaranteed delivery of messages as well as improved security compared to the standard TCP protocol.

The functionality supported over NG-AP includes for example mobility signaling, carrying of NAS messages between devices and the core network, and signaling for paging of devices that are in idle mode and hence are not registered in any specific base station.

The functionality of Xn-AP mainly includes mobility signaling and functions related to Dual Connectivity. The latter concept allows for combining two radio access technologies for offering enhanced service capabilities and characteristics, for example combining using NR on one frequency band and LTE on another band. Dual Connectivity is further explained in Chapter 12.

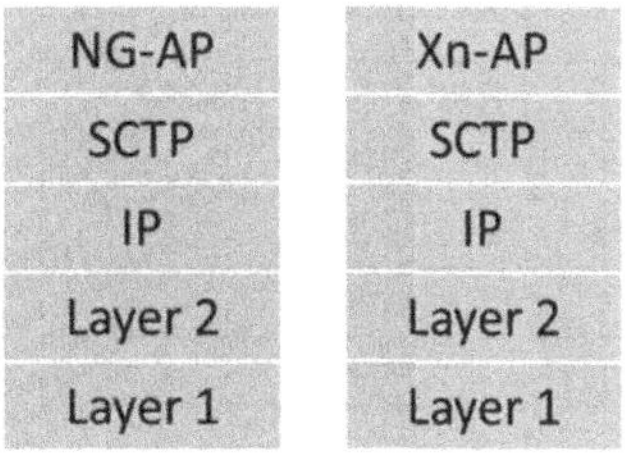

Fig. 3.54 The 5G radio network protocol stack for network-internal signaling.

3.19.7 The NR air interface

The NR air interface between the devices and the base stations is built on a protocol stack as shown in Fig. 3.55.

PHY is the "Physical layer", the lowest layer in the protocol stack and consisting of the actual radio transmission over the radio channels using the OFDM modulation scheme and the TDD/FDD multiplexing concepts. The basic service of the PHY layer is to provide transport of data bits between devices and radio base stations. These transport services are then utilized by the MAC layer protocols.

MAC is the "Medium Access Control layer" which provides transport of both signaling information and user data. The MAC layer is logically subdivided into several logical channels used for various purposes, for example access requests, information broadcasting and data transfers. The MAC layer provides support for multiplexing of data from multiple of these logical channels onto a single transport service from the Physical Layer. We will however not describe the set of logical channels in this book. Instead the interested reader is referred to a textbook on NR radio concepts, for example (Dahlman et al., 2018).

The MAC layer provides prioritization between different data flows using dynamic scheduling of data transmissions, and also applies some error correction and triggering of re-transmissions of incompletely received data packets based on reporting from the receiver.

RLC is the "Radio Link Control" protocol layer and is the layer that can provide a fully reliable transport service for selected transmissions. It supports transmission of signaling information or user data using any of three modes:

- Transparent mode (TM) which basically only provides buffering of packets in a send buffer. No feedback is received if the packets were received or not
- Unacknowledged mode (UM) which is like TM but also provides the possibility to segment the packets before transmission, and then reassembly them on the receiving side
- Acknowledged mode (AM) that includes the receiver providing feedback if the packets are correctly received or not and triggering a retransmission if needed.

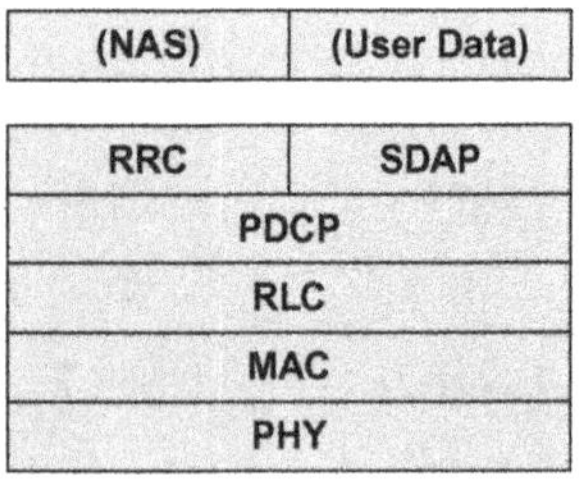

(NAS)	(User Data)
RRC	SDAP
PDCP	
RLC	
MAC	
PHY	

Fig. 3.55 The 5G NR air interface protocol stack.

On top of RLC, the "Packet Data Convergence Protocol" (**PDCP**) layer is used. It provides encryption of both user data and signaling information, as well as optional header compression of user data to enhance the channel efficiency. PDCP also handles reordering of packets that may arrive in the wrong order, based on marking the packets with sequence numbers.

On top of PDCP, the stack is different for user data and for signaling. The highest layer of signaling over the air interface is the "Radio Resource Control" (**RRC**) protocol layer. It supports various functions related to the highest level of signaling procedures between the network and the devices. This includes broadcasting of system information, delivery of encryption keys, mobility signaling, management of radio bearers, and paging of terminals who are in idle mode. RRC also transparently carries the NAS signaling between the Core Network and the devices, referred to as N1 in the 5G Core Architecture.

For user data, Service Data Adaption Protocol (**SDAP**) is used to carry packets. The main function of SDAP is to map downlink packets that are marked with different QoS classes towards the correct radio bearer, to ensure proper Quality-of-Service handling. In the other direction, SDAP secures a correct Quality-of-Service marking of packets received from the devices, before transmitting these over the N3 interface towards the UPF in the Core Network.

3.19.8 Base station internal architecture

One difference between NR and LTE is that for NR three internal interfaces to the gNB are specified by 3GPP.

These are called E1, F1-C and F1-U. The internal architecture of the gNB is shown in Fig. 3.56, and the figure also outlines which protocols in the air interface protocol stack execute in which part of the gNB.

CU is short for the Central Unit and is further separated into a Control Plane that manages signaling protocols, and a User Plane that manages user data transmissions.

In the CU-CP, the upper layer signaling protocols are used—RRC and parts of PDCP.

In the CU-UP, the upper layer user data protocols are used—SDAP and parts of PDCP.

DU—short for—Distributed Unit, is typically installed close to the antenna to minimize transmission energy losses in the cable to the antenna. The size of the loss is dependent on the radio frequency used.

In the DU, lower layer protocols are used, supporting transmissions of both user data and signaling information. These protocols are PHY, MAC and RLC.

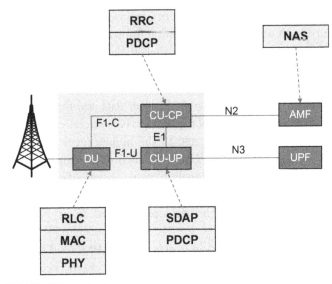

Fig. 3.56 The 3GPP NR gNB architecture.

This separation of DU and CU functionalities inside the gNB supports a modular and flexible architecture that at the same time allows for distribution of low layer functionality and centralization of the upper protocol layers, potentially executing in a data center environment using cloud technologies. Besides allowing for a geographical separation, it also provides completely independent scaling domains for the different parts of the radio base station.

CHAPTER 4

EPC for 5G

4.1 Introduction

In Chapter 3, we introduced the concept of the non-Stand-Alone Architecture (NSA) also called EN-DC in 3GPP architecture specifications. In this architecture the EPS feature known as Dual Connectivity is used to connect 5G DC-capable devices via the 5G NR Radio NR with EPC. In this chapter we will briefly describe EPC and outline how it can be used to in context of 5G. Further information on EPC architecture, functions, and features are available in Olsson et al. (2012). The important concept of Dual Connectivity is further described in Chapter 12.

The key baseline functions for the EPC-based system include support of multiple 3GPP RATs (i.e., GERAN, UTRAN, and E-UTRAN), support for non-3GPP accesses such as W-LAN, and support of Fixed wireline access. All integrated with functions as Mobility management, Session management, Network sharing, Control and User plane separation, Policy control and Charging, Subscription management, and Security. Over the years, EPC has grown with additional features such as Machine Type Communications and Cellular Internet of Things (MTC and CIoT), support for Proximity Services with Device to Device communication and Vehicle to Anything communications support (V2X), Dedicated Core Network selection (DECOR) and Control and User Plane Separation for the GWs (CUPS). DECOR and CUPS are two key enablers for the base core network architecture that enhances EPC for 5G based on EN-DC due to the flexibility and versatility they provide for the operators for deployment of differentiated core networks towards specific targeted users. Figs. 4.1 and 4.2 illustrate the key EPS architecture and the simplified architecture of EPC for 5G, respectively.

As the radio network increases its throughput and bandwidth capacity for 4G and enhanced 4G Radio, operators seek more flexibility and different grades of requirements from the user plane functions provided by the GWs. Basic EPC provided separation of control and user plane to some extent, in particular by separating the session management, user plane functions, and external data connectivity into separate GWs but these GWs (e.g., Serving GW and PDN GW) still hold session management control plane functions. CUPS, as further explained in Section 4.4, enables the SGW and PGW functions to be separated as control and user plane components. The CUPS work was driven by operator requirements to scale control and user plane functions independently of one another and the ability to deploy user plane functions in a flexible manner independently of control plane functions.

5G Core Networks
https://doi.org/10.1016/B978-0-08-103009-7.00004-1

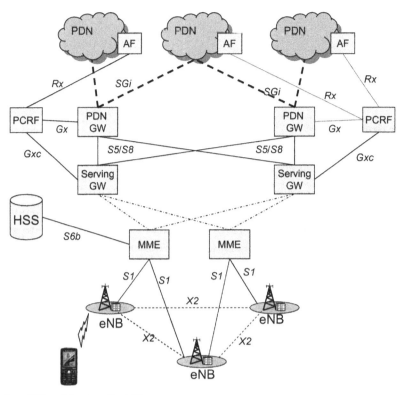

Fig. 4.1 Core EPS architecture for LTE.

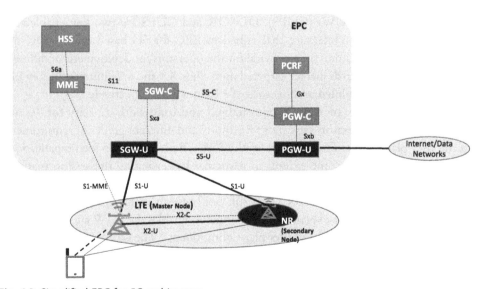

Fig. 4.2 Simplified EPC for 5G architecture.

The result enables the separation of the SGW and PGW (as well as TDF) control and user plane functions and the flexibility to have a single control plane function to control multiple user plane functions. This ability to scale the control and user plane independently allows increased user plane capacity in the network without affecting control plane components.

DEdicated CORE networks (DECOR and enhanced DECOR), meanwhile, enables operators to partition their core networks into separate dedicated core networks with potentially dedicated MME, SGW, and PGWs used for specific purposes such as dedicated core for CIoT and MBB. Together with the Dual connectivity function in the Radio Access network (for more details, see Chapter 12), where RAN can boost the throughput of the UE by adding a secondary RAT using NR 5G Radio for the UE, an operator is able to create the early 5G system using EPC. These combined features (i.e., DC, (e)DECOR, CUPS) in EPC with NR as Secondary RAT is being hailed as EPC for 5G as illustrated in Fig. 4.2.

One key aspect for the two features in EPS (DECOR and CUPS) is that both features were developed to minimize UE impacts (or have no UE impacts) and as CUPS functions developed it did not impact existing peripheral nodes such as the MME and PCRF. We discuss the details of these two features in Sections 4.3 and 4.4, respectively. In contrast, for DC to work, it requires support from the UE to simultaneously connect to the two RATs (LTE and NR), and optionally support in MME and GWs to enable additional DC related functions—this is described in Chapter 12.

Let us consider an example deployment use case where an operator plans to deploy NB-IoT and MBB. Some MBB users have IMS services, while others are using data services with high data volume requirements. An operator may decide to separate its core network components using DECOR principles into two core networks, one for NB-IoT and one for MBB. Within the MBB part of the core network, the operator additionally decides to deploy User Plane GWs for high data volume for the MBB APN and use another set of User Plane GWs for IMS services, both being controlled by a single Control Plane function. The operator may also decide to deploy DC to boost the radio. The combination of this functionality provides an EPC for 5G enabling early NR deployment, which also continues to support all 4G EPS features without any additional impacts to existing installations.

All these features together have impacts on the core network nodes selection functions (i.e., MME, SGW, PGW) and for DC, selection of the Serving and PDN GW can be further enhanced to serve dedicated UEs with DC capability and greater need to have a GW with larger capacity and throughput to support increased data traffic. This is also known as EN-DC in the 3GPP system and is further described in Chapter 12.

Without (e)DECOR and CUPS features, an example of basic selection function or these entities is illustrated in Fig. 4.3.

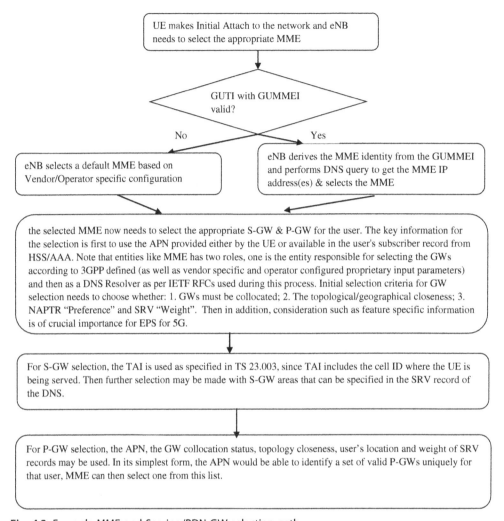

Fig. 4.3 Example MME and Serving/PDN GW selection path.

As we describe later, enabling flexible selection of the user plane GWs combined with dedicated core networks within a single PLMN enables some of the key core 5G enablers such as slicing, complete User and Control Plane separation, as well as enhanced benefit from dual connectivity of different RAT types. From an operator's perspective, the combined (e)DECOR, CUPS with DC allows separating the end to end system from 4G EPS and provides their early 5G subscribers differentiated experience. With some simple use of information in the UEs, such as, knowing when DC has been activated, it is possible to display an indication in the UE when the user is in such networks. From an end user perspective, early adopters of 5G may enjoy an enhanced experience and can look forward to even better service experience with a full 5G system.

4.2 Key EPC functions

In this section we present a high-level overview of the key EPC functions to facilitate reader's understanding of the relevant aspects associated with EPC for 5G. More details on EPC are available in Olsson et al. (2012). The key entities are HSS, MME, Serving GW, PDN GW, PCRF. Fig. 4.4 illustrates a simplified architecture for EPS, including only the key components relevant for EPC and specifically for EPC for 5G.

The 3GPP Radio access networks are GERAN, UTRAN, and E-UTRAN, with our focus being on E-UTRAN (LTE) access only. MME is the Mobility Management Entity, responsible for connectivity of the control plane signaling with the eNB, the E-UTRAN component responsible for UE connectivity. This node is also responsible for NAS termination, Registration and Tracking Area management, Paging and Authentication, and Authorization with support from HSS (and AUC) for the users and the UEs connecting to the EPS.

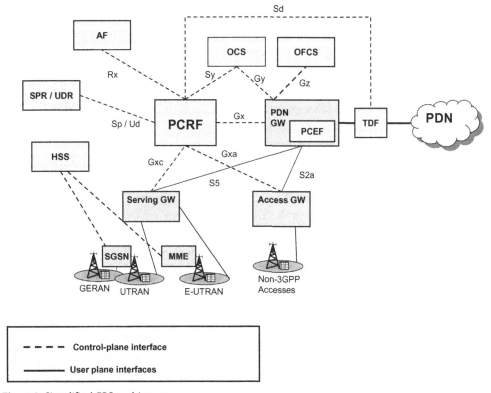

Fig. 4.4 Simplified EPS architecture.

The Serving GW (S-GW) is the user plane termination point from eNB and provides connectivity towards the PDN GW which is the anchor point towards a Packet Data Network (PDN). Each UE is normally served by a single S-GW.

The PDN GW (P-GW) is the anchor point for a specific UE accessing certain PDN(s), which is represented by the SGi interface. P-GW provides support for all packet data related enforcement functions related to policies and QoS.

The Policy and Charging Rules Function (PCRF) is the central policy handling entity in EPS. This node is responsible for functions such as QoS control, bearer binding, IP-CAN gating, and policy-related rules instructions towards the PCEF.

The Policy and Charging Enforcement Function (PCEF) is the entity part of the P-GW function responsible for enforcing policies and rules installed by PCRF.

4.2.1 Subscription and mobility management

In mobile networks there are many functions and processes that require subscription-related information. The most obvious example of user subscription data that is used in LTE/EPC networks may be the user identity and security credentials that are required when an end-user device connects to an LTE/EPC network and performs authentication. The user identity (IMSI) and the security keys are stored in the USIM card in the device and the same information is also stored for each user in the operator's core network, in the Home Subscriber Server (HSS). The subscriber data and functionality of the HSS are used for many functions in 3GPP networks.

The functionality of the HSS includes:

User security support: The HSS supports authentication and security procedures for network access by providing credentials and keys towards network entities such as SGSN, MME, and 3GPP AAA Server.

Mobility management: The HSS supports user mobility by, for example, storing information about what MME is currently serving the user.

User identification handling: The HSS provides the appropriate relations among all the identifiers uniquely determining the user in the system.

Access authorization: The HSS authorizes the user for mobile access when requested by the MME, or 3GPP AAA Server (for PS access), by checking that the user is allowed to roam to a particular visited network.

Service authorization support: The HSS provides basic authorization for mobile terminated call/session establishment and service invocation.

Service provision support: The HSS provides access to the service profile data for use within the CS domain, PS domain, and/or IMS. For the PS domain, the HSS provides the APN profiles that include what APNs the user is authorized to use. The HSS also communicates with IMS entities to support Application Services. An operator may need to have more than one HSS if the number of subscribers is too large to be handled by a single HSS. In order to support user identity to HSS resolution in such a case, Diameter agents can be deployed. In EPS, data relating to the subscriber may be managed by different network entities such as HLR/HSS and SPR. UDC has been introduced in EPC to provide convergence of user data in order to enable smoother management and deployment of new services and networks. UDC concept supports a layered architecture, keeping the actual data separate from the application logic in the 3GPP system. It does so by storing user data in a logically unique user data repository and allowing access to this data from EPC and service layer entities.

4.2.2 Mobility management

Mobility management in LTE/EPC involves keeping track of the UE as it may move among cells, via what are known as Tracking Areas. UEs are tracked while connected to the core network by the MME, which also where needed and possible, change the serving MME along the way through Tracking Area update and handover procedures. As part of the mobility procedure, the UE initially attaches to the network via a registration procedure.

In EPS the registration areas are called Tracking Areas (TAs). In order to distribute the registration update signaling, the concept of tracking area lists was introduced in EPS. The concept allows a UE to belong to a list of different TAs. Different UEs can be allocated to different lists of tracking areas. If the UE moves within its list of allocated TAs, it does not have to perform a tracking area update. By allocating different lists of tracking areas to different UEs, the operator can give UEs different registration area borders and so reduce peaks in registration update signaling, for example when a train passes a TA border.

A summary of the idle mobility procedure in EPS is:
- A TA consists of a set of cells.
- The registration area in EPS is a list of one or more TAs.
- The UE performs TA Update when moving outside its TA list.
- The UE also performs TA Update when the periodic TA Update timer expires.

When the UE reselects a new cell and realizes that the broadcast TA ID is not in their list of TAs, the UE initiates a TAU procedure to the network.

1. The first action is to send a TA update message to the MME.
2. Upon receipt of the TA message from the UE, the MME checks if a context for that particular UE is available; if not it checks the UE's temporary identity to determine which node keeps the UE context. Once this is determined the MME asks the old MME for the UE context.
3. The old MME transfers the UE context to the new MME.
4. New MME interacts with HSS to complete the procedure, including updating HSS with new MME information.

5. MME confirms to the UE about successful Tracking Area Update procedure.
Paging is used to search for Idle UEs and to trigger establishing a signaling connection.
Paging is, for example, triggered by downlink packets arriving to the Serving GW. When
the Serving GW receives a downlink, packet destined for an Idle UE, it does not have an
eNodeB address to which it can send the packet. The Serving GW instead informs the
MME that a downlink packet has arrived. The MME knows in which TA the UE is
roaming and it sends a paging request to the eNodeBs within the TA lists. Upon receipt
of the paging message, the UE responds to the MME and the bearers are activated so that
the downlink packet may be forwarded to the UE.

4.2.3 Session management

Session management is how a 3GPP system provides connectivity between the UEs and
the service network the UEs are trying to communicate with. In EPS, this connectivity is
achieved via establishing one or more PDN connections, which connects a UE through
the RAN to the PDN GW, which is the 3GPP entry/exit point towards the external
networks outside the EPC.

Providing PDN connectivity is not just about getting an IP address; it is also about
transporting the IP packets between the UE and the PDN in such a way that the user
is provided with a good experience of the service being accessed. Depending on whether
the service is a voice call using Voice-over-IP, a video streaming service, a file download,
a chat application, etc., the QoS requirements for the IP packet transport are different.
The services have different requirements on bit rates, delay, jitter, etc. Furthermore, since
radio and transport network resources are limited, and many users may share the same
available bandwidth, efficient mechanisms must be available to partition the available
(radio) resources between the applications and the users. The EPS needs to ensure that
all these different service requirements are supported and that the different services
receive the appropriate QoS treatment in order to enable a positive user experience.

One of the main goals during session management is establishing PDN connectivity
and initial EPS was mainly providing PDN Type IP (IPv4 and IPv6). As different types of
services demand arose for EPS, the need to support other PDN types also became important. Current EPS systems support two additional PDN types known as Non-IP and
Ethernet PDN. The Non-IP and Ethernet PDN types are useful tools, for example,
for low complexity and low throughput UEs specifically designed for Cellular IoT
services.

4.2.4 Control-plane aspects

There are several procedures available in EPS to control the bearers. These procedures are
used to activate, modify, and deactivate bearers, as well as to assign QoS parameters,
packet filters, etc., to the bearer. Note, however, that if the default bearer is deactivated

the whole PDN connection will be closed. EPS has adopted a network centric QoS control paradigm, meaning that it is basically only the PDN GW that can activate, modify, and deactivate an EPS bearer and decide which packet flows are transported over which bearer.

4.2.5 QoS

The EPS only covers QoS requirements for the traffic within the EPS—that is, between UE and PDN GW. If the service extends beyond that, QoS is maintained by other mechanisms that, for example, depend on operator deployments and service level agreements (SLAs) between network operators. The EPS bearer represents the level of granularity for QoS control in E-UTRAN/EPS and provides a logical transmission path with well-defined QoS properties between UE and the network.

The QoS concepts of the EPS bearer are then mapped to the QoS concepts of the underlying transport. For example, over the E-UTRAN radio interfaces, the EPS bearer QoS characteristics are implemented using E-UTRAN-specific traffic handling mechanisms. Each EPS bearer is transported over an E-UTRAN radio bearer with the corresponding QoS characteristics. In the "backbone" network between eNB, Serving GW, and PDN GW, the EPS bearer QoS may be mapped to IP transport layer QoS, for example, using DiffServ. One of the properties of a bearer is the bit rates it is associated with. We distinguish between two types of bearers: GBR bearers and non-GBR bearers, where GBR is short for Guaranteed Bit Rate. A GBR bearer has, in addition to the QoS parameters discussed above, associated bit rate allocations: the GBR and the Maximum Bit Rate (MBR). A non-GBR bearer does not have associated bit rate parameters.

A bearer with an associated GBR means that a certain amount of bandwidth is reserved for this bearer, independently of whether it is utilized or not. The GBR bearer thus always takes up resources over the radio link, even if no traffic is sent. The GBR bearer should not in normal cases experience any packet losses due to congestion in the network or radio link. This is ensured since GBR bearers are subject to admission control when they are set up. A GBR bearer is only allowed by the network if there are enough resources available. The MBR limits the bit rate that can be expected to be provided by a GBR bearer. Any traffic in excess of the MBR may be discarded by a rate shaping function.

4.2.6 The EPS bearer for E-UTRAN access

For E-UTRAN access in EPS, one basic tool to handle QoS is the "EPS bearer". In fact, the PDN connectivity service described above is always provided by one or more EPS bearers (also denoted as "bearer" for simplicity). The EPS bearer provides a logical transport channel between the UE and the PDN for transporting IP traffic. Each EPS bearer is associated with a set of QoS parameters that describe the properties of the transport channel, for example, bit rates, delay and bit error rate, scheduling policy in the radio base station, etc. All conforming traffic sent over the same EPS bearer will receive the same QoS treatment. In order to provide different QoS treatment to two IP packet flows, they need to be sent over different EPS bearers. All EPS bearers belonging to one PDN connection share the same UE IP address.

4.2.7 Default and dedicated bearers

A PDN connection has at least one EPS bearer but it may also have multiple EPS bearers in order to provide QoS differentiation to the transported IP traffic. The first EPS bearer that is activated when a PDN connection is established in LTE is called the "default bearer." This bearer remains established during the lifetime of the PDN connection. Even though it is possible to have an enhanced QoS for this bearer, in most cases the default bearer will be associated with a default type of QoS and will be used for IP traffic that does not require any specific QoS treatment. Additional EPS bearers that may be activated for a PDN connection are called "dedicated bearers." This type of bearer may be activated on demand, for example, when an application is started that requires a specific guaranteed bit rate or prioritized scheduling. Since dedicated bearers are only set up when they are needed, they may also be deactivated when the need for them no longer exists, for example, when an application that needs specific QoS treatment is no longer running.

4.2.8 User-plane aspects

The UE and the PDN GW use packet filters to map IP traffic onto the different bearers. Each EPS bearer is associated with a so-called Traffic Flow Template (TFT) that includes the packet filters for the bearer. These TFTs may contain packet filters for uplink traffic (UL TFT) and/or downlink traffic (DL TFT). The TFTs are typically created when a new EPS bearer is established, and they can then be modified during the lifetime of the EPS bearer. For example, when a user starts a new service, the traffic filters corresponding to that service can be added to the TFT of the EPS bearer that will carry the user plane for the service session. The filter content may come either from the UE or from the PCRF. The TFTs contain packet filter information that allows the UE and PDN GW to identify the packets belonging to a certain IP packet flow aggregate. This packet filter information is typically an IP 5-tuple defining the source and destination IP addresses, source, and destination port, as well as protocol identifier (e.g., UDP or TCP). It is also possible to define other types of packet filters based on other parameters related to an IP flow.

When an EPS bearer is established, a bearer context is created in all EPS nodes that need to handle the user plane and identify each bearer. For E-UTRAN and a GTP-based S5/S8 interface between Serving GW and PDN GW, the UE, eNodeB, MME, Serving

GW, and PDN GW will all have bearer context. The exact details of the bearer context will differ somewhat between the nodes since the same bearer parameters are not relevant in all nodes. Between the core network nodes in EPC, the user-plane traffic belonging to a bearer is transported using an encapsulation header (tunnel header) that identifies the bearer. The encapsulation protocol is GTP-U. When E-UTRAN is used, GTP-U is used on S1-U and can also be used on S5/S8.

4.2.9 Policy control and charging

Policy control is a very generic term and in a network there are many different policies that can be implemented, for example, policies related to security, mobility, use of access technologies, etc. When discussing policies, it is thus important to understand the context of those policies. When it comes to PCC for EPC, policy control refers to the two functions gating control and QoS control:

1. Gating control is the capability to block or to allow IP packets belonging to IP flow(s) for a certain service. The PCRF makes the gating decisions that are then enforced by the PCEF. The PCRF could, for example, make gating decisions based on session events (start/stop of service) reported by an Application Function (AF) via the Rx reference point.

2. QoS control allows the PCRF to provide the PCEF with the authorized QoS for the IP flow(s). The authorized QoS may, for example, include the authorized QoS class and the authorized bit rates. The PCEF enforces the QoS control decisions by setting up the appropriate bearers. The PCEF also performs bit rate enforcement to ensure that a certain service session does not exceed its authorized QoS.

Over the various releases, additional features have been added to the overall PCC, they are very similar to what is described in Chapter 10.

Charging control includes means for both offline and online charging. The PCRF makes the decision on whether online or offline charging will apply for a certain service session, and the PCEF enforces that decision by collecting charging data and interacting with the charging systems. The PCRF also controls what measurement method applies— that is, whether data volume, duration, combined volume/duration, or event-based measurement is used. Again, it is the PCEF that enforces the decision by performing the appropriate measurements on the IP traffic passing through the PCEF.

With online charging, the charging information can affect, in real time, the services being used and therefore a direct interaction of the charging mechanism with the control of network resource usage is required. Online credit management allows an operator to control access to services based on credit status. For example, there has to be enough credit left with the subscription in order for the service session to start or an ongoing service session to continue. The OCS may authorize access to individual services or to a group of services by granting credits for authorized IP flows. Usage of resources is granted in different forms. The OCS may, for example, grant credit in the form of certain amount of time, traffic volume, or chargeable events. If a user is not authorized to access a certain service, for example, if the pre-paid account is empty, then the OCS may deny credit requests and additionally instruct the PCEF to redirect the service request to a specified destination that allows the user to refill the subscription.

The PCRF is the central entity in PCC-making PCC decisions. The decisions can be based on input from a number of different sources, from UE, GWs, RAN, AF.

The PCRF provides its decisions in the form of so-called PCC rules. A PCC rule contains a set of information that is used by the PCEF and the charging systems. First of all, it contains information (in a so-called Service Data Flow (SDF) template) that allows the PCEF to identify the IP packets that belong to the service session. All IP packets matching the packet filters of an SDF template are designated an SDF. The filters in an SDF template contain a description of the IP flow and typically contain the source and destination IP addresses, and the protocol type used in the data portion of the IP packet, as well as the source and destination port numbers. These five parameters are often referred to as the IP 5-tuple. It is also possible to specify other parameters from the IP headers in the SDF template. The PCC rule also contains the gating status (open/closed), as well as QoS and charging-related information for the SDF. The QoS information for an SDF includes the QCI, MBR, GBR, and ARP. However, one important aspect of the QoS parameters in the PCC rule is that they have a different scope than the QoS parameters of the EPS bearer. A single EPS bearer may be used to carry traffic described by multiple PCC rules, as long as the bearer provides the appropriate QoS for the service data flows of those PCC rules.

4.3 (Enhanced) Dedicated Core Networks ((e)DECOR)

(e)DECOR was inspired by the desire and flexibility for the operators to deploy within an operator's network (designated by PLMN ID(s)) multiple core networks and directing users towards specific core networks and thus allowing partitioning off the full core networks. This feature as such enables an operator to deploy such multiple Dedicated Core Networks (DCN) within a PLMN with each DCN consisting of one or multiple CN nodes (e.g., MME only, MME with GWs, MME, GWs, and PCRF). Each DCN may be dedicated to serve specific type(s) of subscriber and the difference between DECOR and (e)DECOR is that the latter requires the UE to provide specific information (i.e., DCN) to facilitate faster and optimal selection of the core network preferred.

EPS already makes it possible to direct the UEs towards specific PLMNs, which in turn takes the UE to the specific network (including Core network). Usage of the concept of APN allows the PLMN to direct the UEs towards specific service networks in a differentiated manner via selecting different user plane entities (i.e., PDN GWs). Fig. 4.1 in Chapter 4.1 shows interconnects of different network elements within a PLMN. Without (e)DECOR enhancement, it is possible for the operator to redirect its users,

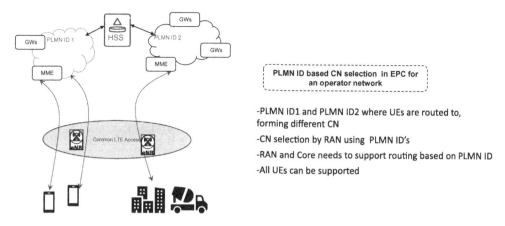

Fig. 4.5 PLMN-based CN selection existing since pre-DECOR.

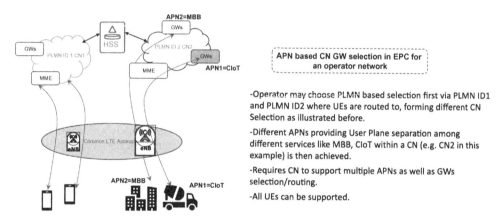

Fig. 4.6 APN-based GW selection existing pre-DECOR.

but it requires that the PDN GWs are capable, within the PLMN, to be able to handle the connectivity towards different service networks (e.g., CIoT, eMBB, VoLTE) for all users.

Figs. 4.5 and 4.6 illustrate how the routing differentiation works in pre-DECOR networks. The EPS system already allowed for directing the UEs towards different CN based on PLMN ID separation within an operator's network (i.e., PLMN). But this is quite static and did not allow separation of a deeper granularity and UEs stayed within the PLMN/CN chosen. Then within a CN, using the concept of APN, which allows for user plane separation towards different GWs (PDN GWs) leading to service differentiation within the CN, such as an APN for MBB allowed UEs to be connected to GWs that support MBB services and that allows operators flexibility to isolate them accordingly. A UE may be connected to multiple APNs within a single CN.

(e)DECOR enables the EPC to "slice" the core network into components that can be tailored to serve specific group of UEs (users), based on their subscription and optionally configured for specific UEs in the UE (only for enhanced DECOR) and in the subscription (HSS). This provides operators additional flexibility to separate users into different core network types (e.g., MBB, IoT) as appropriate for the intended usage. Prior to DECOR being introduced, users (UE) could access different data services using concepts like Access Point Name (APN), which led to the selection of a different edge GW in the core towards a different data network or selecting a PLMN based on supported PLMN-ID which allowed routing the UEs to a specific CN. DECOR allows operators to separate certain types of traffic into specific core network node(s) and if needed, scale them differently than rest of the core network nodes. In this way, the operator is also able to segregate specific users more efficiently and control this from the subscription data. Whereas, if preferred, with enhanced DECOR, which requires update of the UEs as well, users are also able to choose when registering to the network, which type of DCN network it preferred for the connection. Where it may come in handy, as an example, a user coming back into the factory, may choose to use the Dedicated Core Network (DCN) that enables connectivity specific to the factory floor giving access to specific services provided only in that location. In 5GS, support of the network slicing function (a more elaborate version of DECOR) in the UEs is enabled from day one, thus eliminating any issues of different types of devices requiring different network behavior on how to trigger the dedicated core network (aka slicing) separation.

Some of the key principles of DECOR is that millions of already deployed devices must be able to benefit from this feature. That means network entities need to be able to (re)route UEs within the network and using existing system procedures that the UEs already support. At the same time, DECOR must not force an operator to have to handle every UE with DECOR and for that reason the existing core network that is common for all users prior to DECOR must also coexist in a DECOR deployment.

The following principles are applied for DECOR delivering the architectures as illustrated in Fig. 4.7. The main principles to apply for an E-UTRAN DCN, which comprises of MME(s), and associated Serving GWs, PDN GWs, and PCRF(s):

- UEs are not impacted.
- RAN (or Network Node Selection Function (NNSF)) triggers DECOR based on local configuration.
- Operator may indicate in HSS, as part of user subscription, the parameter known as "UE usage type" that provides specific service characteristic that applies for that UE (or set of UEs).

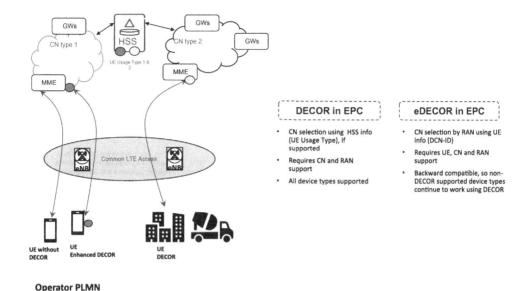

Fig. 4.7 High-level (e)DECOR architectures.

- MME may use the UE usage type, local operator policies configured in the MME, MME Group ID(s) information which indicates to the MME the type of DCN to select the DCN.
- Then MME select appropriate GWs within the DCN.
- The UE usage type may be defined by standardized values or use operator-specific values. In case of roaming, the PLMN operators must have agreements to make use of UE usage type information, otherwise default serving network behavior applies. Such default behavior may imply no DCN selected, or default DCN selected and then further rerouting may take place in the core network to redirect to the appropriate core network.

A step by step illustration is shown in Fig. 4.8 for DECOR and enhanced DECOR for a specific UE.

In case of DECOR:

1. The UE does not provide any DCN-related information, so as a default, the E-UTRAN (NNSF) either selects a default MME or based on configuration may select a dedicated MME.
2. The default MME may, based on UE Usage type, if available, the MMEGI and other policies either continue to support the UE or ask E-UTRAN to reroute (including MMEGI).
3. E-UTRAN then triggers rerouting by selecting a new MME based on the input from the first MME.

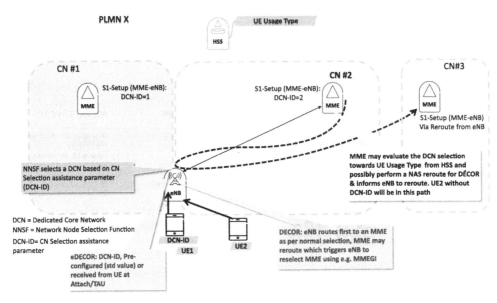

Fig. 4.8 Flow description of DECOR and enhanced DECOR procedure.

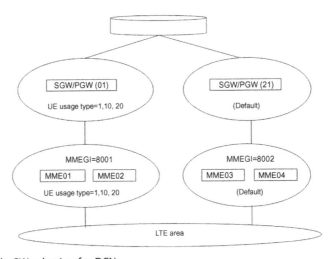

Fig. 4.9 Example GW selection for DCN.

4. The new MME then continues with the DCN selected for that UE and selects appropriate GWs as described in Chapter 4.1 and an example illustration in Fig. 4.9. UE usage type, when configured and available, is a key parameter in differentiating different GWs serving different DCNs.

In case of enhanced DECOR, the UE is configured with the DCN ID which it provides to the E-UTRAN. This allows E-UTRAN (NNSF) to select the appropriate MME associated with the DCN, as configured. Once the MME is selected, MME may verify that the UE can use the selected DCN based on information from HSS.

As a general principle, any UE may be directed towards a common or default core network to get services from an operator's PLMN. Use of DCN may simplify operator's network operation and maintenance and configuration of the GWs as dedicated network nodes can be deployed and management and load balancing can be optimized for that specific DCN.

An example of MME and GWs selection flow for DCN can be as follows, taken from 3GPP TS 29.303 (3GPP TS 29.303). The following figure shows the simple example of DCN deployment.

Fig. 4.9 illustrates the following functionality across the MME and GWs.

- Combined PGW/SGW (01) supports UEs with UE usage type of 1, 10, 20, and Combined PGW/SGW (21) supports all the UEs except UEs associated with UE usage type 1, 10, 20. PGW/SGW (21) is the default GW where all UEs will be routed to, in the absence of the UE usage type in their subscription data.
- Two MME pool areas are defined, and each MME pool has two MMEs. The MME pool with MMEGI of 8001 supports UE usage type of 1, 10, 20 where the other MME pool supports all the UEs except UEs associated with UE usage type 1, 10, 20.

It may not be feasible to deploy DCN throughout the PLMN in a homogeneous manner from the beginning, in such conditions, local configuration can be applied to direct users to specific network segments/nodes without assistance of subscription information. This type of approach may be more suitable within the home PLMN of the UEs, unless other form of roaming agreements is in place.

4.4 Control and User Plane Separation (CUPS)

The feature known as Control and User Plane separation, came from the necessity of independently scaling the Packet Core network's user and control plane function for the Session management and User Plane services. EPC was designed, compared to its predecessor GPRS system, to have separate control plane functions but with the emphasis of separating the mobility management from session management functions with user plane management. But with the Serving and PDN GWs combining the Control and User Plane functions of the session and user plane management functions, it is not possible to deploy GW components with only user data function on the user plane, or independently scale the control and user plane parts in a standardized way. The need for such separation became abundantly clear as operators started considering impacts on their network from in house features such as Narrow Band IoT, MBB, as well as growth of Internet driven OTT services such as video streaming, content sharing, and social media communications. Mobile platforms and devices using, for example, LTE dongles result in that there is a very large number of devices connected via cellular networks using 3GPP defined specifications. These features among themselves can require different type of user data traffic scalability requirements as well as node deployment without requiring the control plane parts to be scaled or deployed in the same manner. For example, an operator may require smaller user data processing foot print for NBIoT compared to MBB or dedicated user plane components for video streaming or gaming requirements. Some of these scenarios may also require the user data processing to be nearer to, for example, where user is connected to. The main EPS network nodes that are possible to apply such separation to are the Serving GW, the PDN GW, and Traffic Detection Function (TDF) when it is deployed as stand-alone function outside of PDN GW.

For the sake of simplicity, the rest of this chapter focuses on the SGW and PGW functions starting from the functions belonging to these two nodes. First of all, in order to separate the Control and User Plane components within a single node, the relationship of these functions needs to be identified and documented. Also, it was evident that not all scenarios require such separation and that most common deployment scenarios will result in coexistence of both split as well as non-split CP and UP nodes in a single network. With that in mind, it becomes clear that this CP and UP separation also must not have any impacts on the surrounding functions such as the MME, the PCRF, the Charging system, and Subscription management system. And there is no question that such network flexibility must not impact any procedures or protocols towards UEs and RAN nodes since the deployment of such nodes cannot be influenced on how UEs interact with the network or the type of RAN node(s) that may exist. The surrounding entities will thus not be aware of whether SGW and PGW have split or non-split CP and UP. Table 4.1, a reduced version taken from the study performed in 3GPP and documented in 3GPP TR 23.714 lists the functionality of SGW and PGW nodes. Another architectural aspect that was important to consider is the combined Serving and PDN GW nodes, i.e., a combo GW with both SGW and PGW functionality. With such deployment option the separation of CP and UP also needs to ensure that a combined control plane entity that includes both the S-GW and P-GW CP functions also must be able to work with a combined S-GW/P-GW UP function as well as separated S-GW UP and P-GW UP.

After the separation of the Control and User Plane functions within each GW node, the next crucial part is to ensure that the selection of the Control Plane GW functions from MME works seamlessly as it does when the CP and UP functions are combined. MME continues to select the Serving GW and PDN GW as before but in a split deployment, the selection leads to the S-GW-CP and P-GW-CP entities. Then it will be up to the Control Plane of the GW function to select the corresponding User Plane GW function. The CP entity will provide the tunnel identifiers (or the user-plane GTP-U tunnel) to the MME as per the pre-CUPS specification, but MME will not be aware whether these tunnel identities belongs to a standalone SGW-U or PGW-U entity or to a non-split SGW or PGW. It may seem as a contradiction that MME has S11-C to SGW-C and S11-U to SGW-U and still not being aware whether SGW has a split or non-split CP and

Table 4.1 Example illustration of EPC S-GW and P-GW functional distribution (without CP and UP separation)

Main functionality in EPS	Sub-functionality	S-GW	P-GW
A. Session management (default & dedicated bearer establishment, bearer modification, bearer deactivation)	1. Resource management for bearer resources	X	X
	2. IP address and TEID assignment for GTP-U	X	X
	3. Packet forwarding	X (DL/UL: GTP-U)	X (DL: GTP-U)
	4. Transport level packet marking	X (DL/UL DSCP marking for QoS in transport)	X (DL DSCP marking for QoS in transport)
B. UE IP address management	1. IP address allocation from local pool		X
	2. DHCPv4/DHCPv6 client		X
	3. DHCPv4/DHCPv6 server		X
	4. Router advertisement, router solicitation, neighbor advertisement, neighbor solicitation as defined in RFC 4861		X
C. Support for UE mobility	1. Forwarding of "end marker" (as long as user plane to source eNB exists)	X	
	2. Sending of "end marker" after switching the path to target node	X (inter-eNodeB and inter-RAT HOV)	X (SGW change)
	3. Forwarding of buffered packet	X	
	4. Change of target GTP-U endpoint (e.g., handover procedures) = mobility anchor	X (intra-3GPP RAT HO with eNB change)	X (intra-3GPP RAT HO with SGW change)
	5. Mobility between 3GPP and non-3GPP access		X

Continued

Table 4.1 Example illustration of EPC S-GW and P-GW functional distribution (without CP and UP separation)—cont'd

Main functionality in EPS	Sub-functionality	S-GW	P-GW
D. S1-Release/ Buffering/Downlink Data Notification	1. ECM-IDLE mode DL packet buffering; Triggering of Downlink Data Notification message generation per bearer (multiple, if DL packet received on higher ARP than previous DDN); Inclusion of DSCP of packet in DDN message for Paging Policy Differentiation	X	
	2. Delay Downlink Data Notification Request (if terminating side replies to uplink data after UE service request before SGW gets updated)	X	
	3. Extended buffering of downlink data when the UE is in a power saving state and not reachable (high latency communication); dropping of downlink data (if MME has requested SGW to throttle downlink low priority traffic and if the downlink data packet is received on such a bearer)	X	
	4. PGW pause of charging procedure based on operator policy/ configuration the SGW (failed paging, abnormal radio link release, number/ fraction of packets/bytes dropped at SGW)	X	X
E. Bearer/APN policing	1. UL/DL APN-AMBR enforcement		X
	2. UL/DL bearer MBR enforcement (for GBR bearer)		X

Table 4.1 Example illustration of EPC S-GW and P-GW functional distribution (without CP and UP separation)—cont'd

Main functionality in EPS	Sub-functionality	S-GW	P-GW
	3. UL/DL bearer MBR enforcement (for nonGBR bearer on Gn/Gp interface)		X
F. PCC-related functions	1. Service detection (DPI, IP-5-tuple)		X
	2. Bearer binding (bearer QoS & TFT)		X
	3. UL bearer binding verification and mapping of DL traffic to bearers		X
	4. UL and DL service level gating		X
	5. UL and DL service level MBR enforcement		X
	6. UL and DL service level charging (online & offline, per charging key)		X
	7. Usage monitoring		X
	8. Event reporting (including application detection)		X
	9. Request for forwarding of event reporting		
	10. Redirection		X
	11. FMSS handling		X
	12. PCC support for NBIFOM		X
	13. DL DSCP marking for application indication		
G. NBIFOM	Non-PCC aspects of NBIFOM	X	X
H. Inter-operator accounting (counting of volume and time)	1. Accounting per UE and bearer	X	X
	2. Interfacing Off-line Charging System	X	X
I. Load/overload control functions	Exchange of load/overload control information and actions during peer node overload	X	X

Continued

Table 4.1 Example illustration of EPC S-GW and P-GW functional distribution (without CP and UP separation)—cont'd

Main functionality in EPS	Sub-functionality	S-GW	P-GW
J. Legal intercept	Interfacing LI functions and performing LI functionality	X	X
K. Packet screening function	(check that UE uses only assigned IP addresses in uplink packets)		X
L. Restoration and recovery		X	X
M. RADIUS/Diameter interfaces on SGi			X
N. OAM interfaces		X	X
O. GTP bearer and path management	Generation of echo request Handling of echo response Handling of echo request timeout Handling of Error Indication message	X	X

UP. This is however possible since the GTP protocol was designed from the start with an inherent CP-UP split, where CP and UP IP addresses and TEIDs are signaled in separate IEs even for non-split SGW and PGW. This made it possible to introduce CUPS in 3GPP Release 14 as an add-on without impacting MME.

The CP function (SGW-C and PGW-C) needs to take into account various possibilities for selecting the UP function (SGW-U and PGW-U, respectively) that may be tailored for the specific user, the session type/APN, location of the UE, requirement such as DC support for the UE, DCN-related information, need for the UP function's proximity to the RAN node (e.g., if the UP function needs to be closer to the UE's location), distribution of the UP in relation to CP and load conditions, etc. The details of how UP function selection is done will thus depend a lot on operator deployment and use case aspects.

Table 4.1 illustrates an example functionality grouping of SGW and PGW in EPC. Taking an example function group from that table, e.g., Group A, the Session management functions are intertwined between S-GW and P-GW nodes. The specific procedures related to session management include EPS bearer establishment/modification/deletion involving both S-GW and P-GW. When the procedure is triggered from MME to S-GW, the split of the CP and UP function requires that the CP procedure

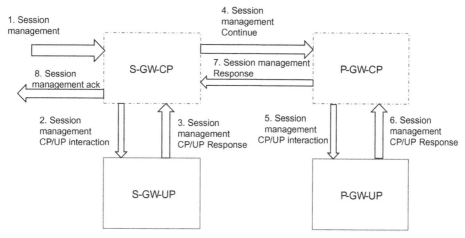

Fig. 4.10 Illustrates procedures for GW functional separation.

needs to trigger the UP procedure, when required, before continuing towards the P-GW-C. Once the SGW-C triggers PGW-C, the PGW-C then needs to ensure that it triggers the appropriate PGW-U function before continuing with the rest of the procedure, as needed. This is illustrated in a call flow in Fig. 4.10. To ensure that a split S-GW can connect to a non-split PGW in a mixed network, the procedures between the CP and UP functions are self-contained without changing the non-split nodes.

Similarly, other functions shown in the table, as they share interdependence between S-GW and P-GW, need to ensure on the high-level principles to follow the same interactions as described in Fig. 4.10.

From this conclusion follows the architectural diagram for the control and user plane separation architecture as described in Fig. 4.11.

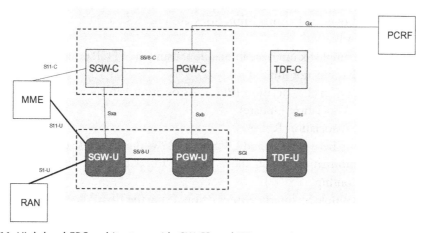

Fig. 4.11 High-level EPC architecture with GW CP and UP separation.

As can be seen, the CP and UP separation for each entity (S-GW, P-GW, and TDF) utilizes the Sx interface to complete the CP and UP functions, as needed. The Sx interface needs to enable establishment/modification/termination procedures in order to provide support for the CP and UP operations between the CP and UP component of each separated node. 3GPP has defined the Packet Forwarding Control Protocol (PFCP) to support the functionality over Sx. It can be noted that the Packet Forwarding Control Protocol was also reused for the N4 interface between SMF and UPF in the 5G System. More details about CP-UP split in 5G is available in Chapter 6 (Session Management). More details about the Packet Forwarding Control Protocol can be found in Chapter 13 (Protocols).

Table 4.2 shows how the functions are distributed among S-GW-C, P-GW-C, S-GW-U, and PGW-U according to the CP-UP separation of the GWs.

Functions such as load/overload control, restoration and recovery and OAM interfaces, echo message/response depend on the Sx interface and the protocol interactions defined, in addition to GTP procedures. 3GPP TS 29.244 described the functions and protocol aspects of Sx interface. Specifically, the load condition information of the UP when made available to the CP may be very useful during UP node selection process for a session by the CP node. Similarly overload conditions may be reported as well. These types of information is exchanged over PFCP Association procedures as described next. Load information from the UP which reflects the UP node operating resource situation, allows the CP nodes to better manage sessions towards any UP node and as a result may avert overload conditions to occur. When supported, sharing load information without any extra signaling is of course the best approach and thus supported by piggybacking via existing/ongoing PFCP messages instead of triggering new messages/signaling. Overload control information from the UP allows the CP(s) to gradually reduce signaling load and reduce or eliminate new sessions towards that specific UP and thus gradually allow the status to stabilize. Load and overload control functions are controlled (i.e., activation/deactivation) independently.

Fig. 4.12 shows simple Sx interface interaction, using the PFCP node association procedures that includes (but not limited to):

1. Establishing Node level association (PFCP Association Setup)
2. Update (PFCP Association Update)
3. Release (PFCP Association Release)

This association setup between CP and UP nodes allows the CP node to learn about the UP node related information which allows the CP to establish the appropriate session related PFCP relationship.

Fig. 4.13 shows simple Sx interface interaction, using the PFCP Session Related Procedures that includes (but not limited to):

4. Establishing PFCP Session

Table 4.2 Functional split of S-GW and P-GW

Main functionality	Sub-functionality	SGW-C	SGW-U	PGW-C	PGW-U
A. Session management (default & dedicated bearer establishment, bearer modification, bearer deactivation)	1. Resource management for bearer resources	X	X	X	X
	2. IP address and TEID assignment for GTP-U	X	X	X	X
	3. Packet forwarding		X		X
	4. Transport level packet marking		X		X
B. UE IP address management	1. IP address allocation from local pool			X	
	2. DHCPv4/DHCPv6 client			X	
	3. DHCPv4/DHCPv6 server			X	
	4. Router advertisement, router solicitation, neighbor advertisement, neighbor solicitation			X	
C. Support for UE mobility	1. Forwarding of "end marker" (as long as user plane to source eNB exists)		X		
	2. Sending of "end marker" after switching the path to target node	X	X	X	X
	3. Forwarding of buffered packet	X	X		
	4. Change of target GTP-U endpoint within 3GPP accesses	X		X	
	5. Change of target GTP-U endpoint between 3GPP and non–3GPP access			X	
D. S1–Release/ Buffering/Downlink Data Notification	1. ECM-IDLE mode DL packet buffering; Triggering of Downlink Data Notification message generation per bearer (multiple, if DL packet received on higher	X	X		

Continued

Table 4.2 Functional split of S-GW and P-GW—cont'd

Main functionality	Sub-functionality	SGW-C	SGW-U	PGW-C	PGW-U
	ARP than previous DDN); Inclusion of DSCP of packet in DDN message for Paging Policy Differentiation				
	2. Delay Downlink Data Notification Request (if terminating side replies to uplink data after UE service request before SGW gets updated)	X			
	3. Extended buffering of downlink data when the UE is in a power saving state and not reachable (high latency communication); dropping of downlink data (if MME has requested SGW to throttle downlink low priority traffic and if the downlink data packet is received on such a bearer)	X	X		
	4. PGW pause of charging procedure based on operator policy/configuration the SGW (failed paging, abnormal radio link release, number/fraction of packets/bytes dropped at SGW)	X		X	
E. Bearer/APN policing	1. UL/DL APN-AMBR enforcement				X
	2. UL/DL bearer MBR enforcement (for GBR bearer)				X

Table 4.2 Functional split of S-GW and P-GW—cont'd

Main functionality	Sub-functionality	SGW-C	SGW-U	PGW-C	PGW-U
	3. UL/DL bearer MBR enforcement (for nonGBR bearer on Gn/Gp interface)				X
F. PCC-related functions	1. Service detection (DPI, IP-5-tuple)				X
	2. Bearer binding (bearer QoS & TFT)			X	
	3. UL bearer binding verification and mapping of DL traffic to bearers				X
	4. UL and DL service level gating				X
	5. UL and DL service level MBR enforcement				X
	6. UL and DL service level charging (online & offline, per charging key)			X	X
	7. Usage monitoring			X	X
	8. Event reporting (including application detection)			X	X
	9. Request for forwarding of event reporting				
	10. Redirection			X	X
	11. FMSS handling				X
	12. PCC support for NBIFOM			X	
	13. DL DSCP marking for application indication				
	14. Predefined PCC/ADC rules activation and deactivation			X	X
	15. PCC support for SDCI			X	X
G. NBIFOM	Non-PCC aspects of NBIFOM	X		X	

Continued

Table 4.2 Functional split of S-GW and P-GW—cont'd

Main functionality	Sub-functionality	SGW-C	SGW-U	PGW-C	PGW-U
H. Inter-operator accounting (counting of volume and time)	1. Accounting per UE and bearer 2. Interfacing Off-line Charging System	X	X	X	X
J. Lawful interception	Interfacing LI functions and performing LI functionality	X		X	X
K. Packet screening function					X
M. RADIUS/ Diameter on SGi				X	X

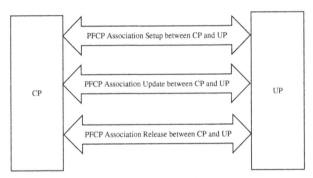

Fig. 4.12 PFCP Association node level between CP and UP.

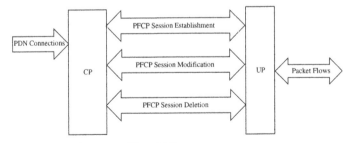

Fig. 4.13 PFCP Session setup between CP and UP.

5. Modification of PFCP Session
6. Deletion of PFCP Session

These are used to set up session (i.e., PDN connection, IP session) related procedures between CP and UP nodes and it installs rule for the UP function to process the packets.

There are general procedures for both PFCP Association and PFCP Session management, for example, error handling, node level management such as heartbeat, load control, overload management, message priority handling, throttling due to node conditions, etc.

An example PDN Connection Establishment flows with Sx as illustrated in 3GPP TS 23.214 shows how CP and UP separation is embedded within the existing session management procedure, for example, PDN Connection in this case.

Fig. 4.14 illustrates the contents in steps 1, 4 shows the Sx Termination interaction for E-UTRAN Initial Attach between the CP and UP functions of SGW and PGW respectively to release old S-GW/PGW entities, without impacting these procedures themselves. Whereas steps 7, 9, 11, 13, 15 & 17 are complete procedures including UE initiated PDN Connectivity with Sx Modification procedures illustrated between the new SGW CP/UP components using Sx Modification to establish the new SGW for that PDN connection.

One definite enhancement that SGW control and user plane separation achieve is that there may be multiple SGW-U entities for a UE in the path, which would not be feasible with SGW without CP and UP separation. In this case, there is still a single SGW-C, which in turn can enable connection to multiple SGW-U entities. Now putting these principles together, we can see that the architecture provides significant flexibility when the CP and UP component of the GWs are separate.

Fig. 4.15 shows an example architecture scenario combining the DECOR feature, in combination with control and user plane separation of the SGW and PGW, connected to a E-UTRAN with dual connectivity via NR as the secondary RAT (also referred to as EN-DC, as described in Chapter 12). This is another architecture principle that has been carried towards the 5GC, as explained in the Session Management Chapter 6.

An evolution of EPC including support for NR NSA has enabled early 5G deployment leading to better preparedness for a later full deployment of 5G system with the new 5GC. EPC has continued to evolve during 3GPP Release 13/14/15 with features like Proximity services, V2X services, Mission Critical services, Enhanced TV services, CIoT enhancements, and Network Exposure support to name a few, in addition to CUPS and (e)DECOR like features. All these features are available in EPC for 5G as well, though mainly operating on LTE and not on NR.

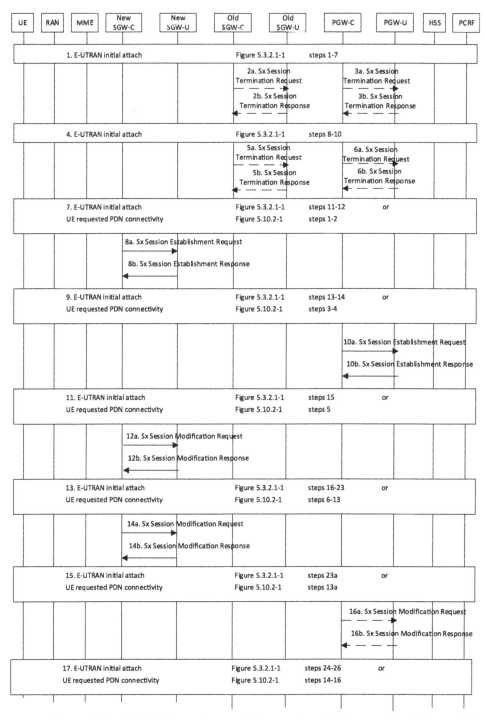

Fig. 4.14 Abstracted flow for PDN Connection establishment when CP UP separation of SGW and PGW applied.

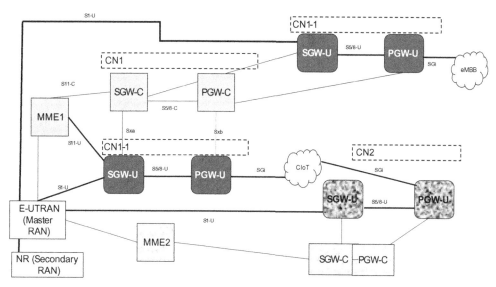

Fig. 4.15 Example network architecture deployment with DECOR, CUPS, and EN-DC components.

CHAPTER 5

Key concepts

This chapter introduces concepts, constructs and identifiers that are useful to understand before reading the later chapters.

5.1 Architecture modeling

As also outlined in Chapter 3, the 5G Core Network has a new network architecture. A major change compared to EPC is that the 5GC Control Plane functions interact in a new paradigm where Service Consumers in Network Functions (NFs) consume Services exposed by Service Producers in other NFs. This design principle gave the new architecture the name Service Based Architecture (SBA).

The related service-based architecture depicts those service based principles by showing the Network Functions, primarily Core Network Control Plane functions, with a single interconnect to the rest of the system. Reference point-based architecture figures are also provided by the specifications see e.g. 3GPP TS 23.501, which represent more specifically the interactions between Network Functions for providing system level functionality and to show inter-PLMN interconnection across various Network Functions.

Compared to the EPC architecture that has more persistent UE specific transport associations between Access Network and Core Network, new functionality simplifying changing the AMF instance that serves a UE has been introduced. This includes functionality for releasing the UE specific Access Network – Core Network transport associations from one AMF and re-binding with another AMF. This together with an "AMF set" concept that allows AMF instances in a set to share UE context data enables new flexibility as every AMF from a set of AMFs deployed for the same network slice can handle procedures of any UE served by the set of AMFs.

5.2 Service Based Architecture

In the Service Based Architecture the services that are provided by the NF Service producer are accessed over a service interface, the SBI (Service Based Interface). Each Network Function instance may expose one or several instances of a given NF Service, as illustrated in Fig. 5.1. The ambition when defining NF Services was to create self-contained, reusable and independently manageable NF Services. This was to some extent achieved, but there are still several NF Services that share data or have dependencies on

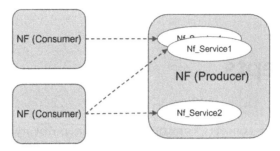

Fig. 5.1 NFs and NF Services.

other services inside a Network Function. Communication between NF Services inside an NF is not specified but left to implementation.

The architecture requires that a Service Consumer must be enabled to select a suitable service producer instance and determine its address. This is supported by the Network Repository Function (NRF) that keeps the repository of all available Network Function instances and their exposed service instances. This is maintained dynamically by NF producers registering their so-called NF profile in the NRF, which then in turn enables the NFs to discover the available Network Function instances, their service instances and status dynamically. The NF profile contains relevant data about the NF including address information.

Communication between the Services on Control plane happens via HTTP2 RESTful APIs. An NF Service consist of operations that are based on either a request-response or a subscribe-notify model. Services are modeled as resources which are either provisioned or can be created, updated or deleted using the RESTful HTTP2 based procedures.

Once an NF consumer has discovered NF producer instances it removes the NF producer instances that do not meet the desired service criteria (Network Slice, DNN…). From that smaller set the consumer selects a producer instance considering capacity, load etc. If resources are created as part of a service request, the created resource is assigned a unique URI pointing to the created resource. The consumer receives the URI in the response and shall use it for all future communication related to the resource (except in case of failure).

In an ideal service-based architecture all instances of service producers of a specific version can be used interchangeably. This is not the case in the 5GC SBA where the suitable service producers for a given request must have certain capabilities e.g. serving a Network Slice, serving a particular DNN or range of SUPI (see Section 5.3).

Therefore, a procedure narrows down the set of instances to the set supporting the required capabilities. Other factors such as load, and capacity of a service may also be considered when an NF producer service is selected. 3GPP call this set of procedures discovery and selection. Discovery and selection are made in the context of the service consumer and the network service use case (e.g. UE procedures).

In many cases the HTTP request of a service request is almost immediately answered by an HTTP response, but sometimes the service producer needs to do additional steps including external NF communication in order to send back a proper response on the 3GPP procedure level. The response is then sent from original service producer instance to the original service consumer instance in a new HTTP request. The address is still derived from the discovery information, but since it should be sent back to the original NF instance only a service instance needs to be selected.

In the subscribe/notify communication pattern the service consumer subscribes to events from a service provider. The consumer subscribes by posting to a subscription resource where it provides also a notification URI where the provider should send the notifications. For the notifications the consumer then acts as a HTTP server and the producer as a HTTP client. To find the subscription service operation in the producer, discovery and selection may be needed.

5.3 Identifiers

Identifiers play an important role in the 5G System, for example, the permanent and temporary subscriber identities are constructed to identify not only a particular subscriber, but also the network function(s) where the permanent and temporary subscriber records are stored. In this chapter we take a brief look on some of the most important identifiers in 5GS.

The main permanent subscription identifier is the Subscription Permanent Identifier (SUPI) that is allocated to each subscriber to the 5G System. The Subscription Concealed Identifier (SUCI) is a privacy preserving identifier containing a concealed SUPI. In addition, temporary identifiers (5G-GUTI, 5G-S-TMSI) are used in the vast majority of signaling flows in order to support user confidentiality protection. The equipment is identified separately from the subscription and each 5G UE has a Permanent Equipment Identifier (PEI).

Subscription Permanent Identifier – SUPI

The SUPI may either contain IMSI or network-specific identifier (used for private networks). The SUPI is privacy protected over the radio using the Subscription Concealed Identifier (SUCI).

Subscription Concealed Identifier – SUCI

The Subscription Concealed Identifier (SUCI) is a privacy preserving identifier containing the concealed SUPI. The SUCI is a one-time use subscription identifier and a different SUCI is generated after the SUCI has been used.

The SUPI and SUCI are represented in the form a Network Access Identifier (NAI). The username part of the NAI representation of a SUCI can take the following forms:

(a) for the null-scheme:

type<supi type>.hni<home network identifier>.rid<routing indicator>. schid<protection scheme id>.userid<MSIN or Network Specific Identifier SUPI username>

(b) for the Scheme Output for Elliptic Curve Integrated Encryption Scheme Profile A and Profile B:

type<supi type>.hni<home network identifier>.rid<routing indicator>. schid<protection scheme id>.hnkey<home network public key id>.ecckey<ECC ephemeral public key value>.cip<ciphertext value>.mac<MAC tag value>

(c) for HPLMN proprietary protection schemes:

type<supi type>.hni<home network identifier>.rid<routing indicator>. schid<protection scheme id>.hnkey<home network public key id>. out<HPLMN defined scheme output>

The SUPI Type identifies the type of the SUPI concealed in the SUCI.

Home Network Identifier identifies the home network of the subscriber.

Routing Indicator is set to 0 unless the Home Network operator partitions AUSF and UDM where the routing indicator helps identify the AUSF and UDM to use.

Protection Scheme Identifier Identifies the protection scheme.

Home Network Public Key Identifier is used to identify the key used for SUPI protection.

Scheme Output, it represents the output of a public key protection scheme or a HPLMN specific protection scheme.

For further details on SUPI and SUCI see Chapter 8, 3GPP TS 33.501 and 3GPP TS 23.003.

Permanent Equipment Identifier – PEI

A Permanent Equipment Identifier (PEI) is allocated to each 5G UE. The PEI parameter consist of a PEI type and either IMEI or IMEISV.

The International Mobile Station Equipment Identity (IMEI) and International Mobile station Equipment Identity and Software Version Number (IMEISV) are the defined the same way as in EPS, For further details see 3GPP TS 23.003.

5G Globally Unique Temporary Identifier – 5G-GUTI

5G-GUTI is assigned to the UE by the 5GC (AMF). The 5G-GUTI can be re-assigned by the AMF at any time.

As detailed in 3GPP TS 23.003 the 5G-GUTI is structured as:

<5G-GUTI> := <GUAMI> <5G-TMSI>

5G-TMSI is a temporary subscriber identifier assigned by an AMF and unique within the GUAMI.

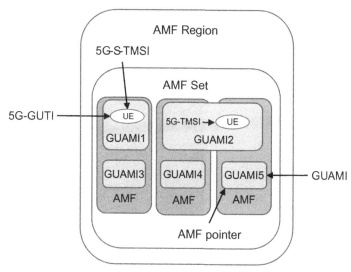

Fig. 5.2 Relation between identifiers.

The Globally Unique AMF ID (GUAMI) identifies one or more AMF(s) and is structured as:

$<$GUAMI$>$:= $<$MCC$>$ $<$MNC$>$ $<$AMF Region ID$>$ $<$AMF Set ID$>$ $<$AMF Pointer$>$

The AMF Region ID identifies the region, AMF Set ID uniquely identifies the AMF Set within the AMF Region and AMF Pointer identifies one or more AMFs within the AMF Set.

5G-S-TMSI is the short form of the 5G-GUTI that is used e.g. during Paging and Service Request for more efficient radio signaling:

$<$5G-S-TMSI$>$:= $<$AMF Set ID$>$ $<$AMF Pointer$>$ $<$5G-TMSI$>$

The relations between AMF Region, AMF Set, GUAMI and temporary identifiers are illustrated in Fig. 5.2.

Generic Public Subscription Identifier – GPSI

The Generic Public Subscription Identifier (GPSI) is a public identifier e.g. used for addressing a 3GPP subscription from an external network. The GPSI can be an MSISDN (a phone number) or an External Identifier in form of username@realm.

CHAPTER 6

Session management

6.1 PDU Session concepts

6.1.1 Introduction

One of the key tasks of the 5G System is to provide the UE with data connectivity toward a Data Network (DN). The Data Network could e.g. be the Internet, an operator specific Data Network for IMS or a Data Network dedicated to e.g. a factory (in a vertical scenario). The Session Management functionality of 5GS has responsibility for setting up connectivity for the UE toward Data Networks, as well as managing the User Plane for that connectivity. Session Management is thus one of the key components of 5GS.

One of the design goals with Session Management in 5G is flexibility to support diverse 5G use cases. As we will see below, this has resulted in Session Management supporting e.g. different PDU Session protocol types, different options for how to handle session and service continuity, as well as a flexible User Plane architecture.

6.1.2 Connectivity service to DN

6.1.2.1 Basic PDU Session connectivity

In order to connect to a DN, the UE requests the establishment of a PDU Session. Each PDU Session provides an association between the UE and a specific DN. When the UE requests establishment of a PDU Session, it may provide a Data Network Name (DNN) that informs the 5G core network what DN the UE wants to connect to. The DNN may e.g. be "Internet" in order to get general Internet connectivity or "IMS" to establish a PDU Session toward the IMS domain. The DN Names used in a network are operator specific except in some special cases like for IMS where the operator community has agreed on well-known DNNs. Fig. 6.1 below illustrates a simplified PDU Session Establishment call flow, highlighting the key Network Functions involved as well as the steps taken in the process.

During PDU Session Establishment, the corresponding User Plane connection between the UE and the DN is activated. The User Plane connection provides transport of PDUs (Protocol Data Units). The "PDU" is the basic end–user protocol type carried by the PDU Session and it depends on the PDU Session type, as explained further below, but it can e.g. be IP packets or Ethernet frames.

5G Core Networks
https://doi.org/10.1016/B978-0-08-103009-7.00006-5

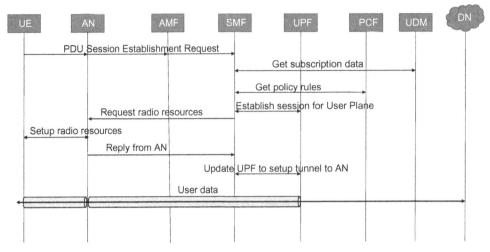

Fig. 6.1 Simplified PDU Session Establishment procedure.

6.1.2.2 Relation between transport network, PDU Session and application traffic

The PDU Session is a logical connection between a UE and a specific DN and it provides the user with a User Plane connection to a DN. The 5GS is concerned with this "PDU Session layer" and associated functions such as IP address management, QoS, mobility, charging, security, policy control, etc. When e.g. NG-RAN is used, the user data belonging to the PDU Session is transported between the UE and the gNB/ng-eNB over the underlying radio connection. Between the NG-RAN and the UPF, and between the UPFs, the PDUs are carried over an underlying transport network, a.k.a. transport layer. The PDUs as such (e.g. IP packet) will in general carry application traffic (e.g. application traffic carried over IP). This application traffic depends on the actual end-to-end service that is running, but can e.g. be HTTP, FTP, SMTP etc. and in case of IP there is usually a transport layer protocol such as UDP or TCP between IP and the application protocol. Here, and in the next few sections, we use the term "application" in a generic manner, including all protocol layers on top of the PDU layer.

The User Plane connection (the PDU Session) is separated from the transport connection between the networks nodes in the 5G system (the transport layer). This is a common feature in mobile networks where the User Plane is tunneled over a transport network in order to provide per-user security, mobility, charging, QoS etc. A reason for tunneling the user plane connection is to decouple the end-user PDU Session "layer" from the underlying transport and allow operators to deploy any transport technology independently of the end-user "PDU layer". Fig. 6.2 below illustrates this concept.

The transport network provides IP transport that can be deployed using different technologies such as MPLS, Ethernet, wireless point-to-point links, etc. The IP transport layer entities in the backbone network, such as IP routers and layer 2 switches, are not

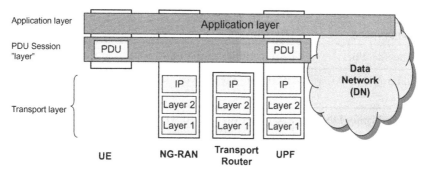

Fig. 6.2 Schematic figure of the user plane protocol stack, with application layer, PDU layer, transport layer.

aware of the PDU Sessions as such. In fact, these entities are typically not aware of per-user aspects at all. Instead, they operate on traffic aggregates and if any traffic differentiation is needed, it is typically based on Differentiated Services (DiffServ) and techniques operating on traffic aggregates.

6.1.2.3 Multiple PDU Sessions

A UE may request establishment of multiple PDU Sessions in parallel. This is useful e.g. in cases where a UE wants both Internet connectivity as well as IMS services at the same time. But as we will see below, it is also possible for a UE to request establishment of multiple PDU Sessions to a single DN at the same time. Fig. 6.3 illustrates the situation where a UE has three simultaneous PDU Sessions to different Data Networks.

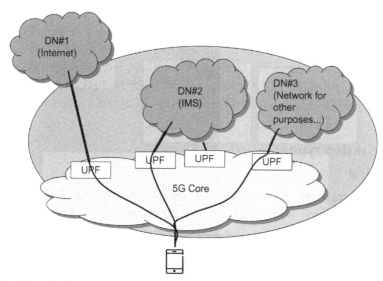

Fig. 6.3 UE with multiple PDU Sessions to different DNs.

Table 6.1 Main properties that characterize a PDU Session

PDU Session property	Description
PDU Session Identifier	An identifier of the PDU Session in the UE and network
Slice identifier (S-NSSAI)	Refers to the network slice in which the PDU Session is established
Data Network Name (DNN)	Name of the DN to which the PDU Session provides connectivity
PDU Session type	The basic end-user protocol type carried by the PDU Session. It can be IPv4, IPv6, dual-stack IPv4/IPv6, Ethernet or Unstructured
Service and Session Continuity (SSC) mode	Refers to the longevity of the User Plane anchor point of the PDU Session, whether it can be re-allocated or not
User Plane Security Enforcement information	Information indicating whether user-plane ciphering and/or use-plane integrity protection is to be activated for the PDU Session

6.1.2.4 PDU Session properties

As mentioned above, a PDU Session is associated with a DNN that describes what DN it connects to. There are however several additional parameters that describe properties of a PDU Session. Table 6.1summarizes some of the most important PDU Session properties. They are explained in more detail later in this chapter.

The PDU Session parameters are determined at the time of PDU Session Establishment and do not change during the PDU Session lifetime. In addition, there are also several additional PDU Session "properties", and they can typically change dynamically during the lifetime. For example, there are PCC rules applied to each PDU Sessions. These are described in more detail in the relevant Chapters 10 and 9 respectively for PCC and QoS. In the scenario that Non-3GPP (N3GPP) access is supported by the operator, the access type of the PDU Session, i.e. whether the PDU Session is active over 3GPP access or N3GPP access, is another property of the PDU Session.

6.2 PDU Session types

6.2.1 General

The 5G System supports different PDU Session types to cater for different use cases:
- IP based PDU Session types: IPv4, IPv6 and dual-stack IPv4v6
- Ethernet PDU Session type
- Unstructured PDU Session type

Readers familiar with EPC will recognize the IP based PDU Session types and they indeed have similar properties also in 5GS, even though a few additional features for

IPv6 are available in 5GS. Unstructured PDU Session type is similar to non-IP PDN type in EPS while Ethernet PDU Session type did not have any counterpart in EPS initially (but has later been added also to EPS). More details on the different PDU Session types are provided below.

6.2.2 IP based PDU Session types
6.2.2.1 General
When it comes to IP, 5GS supports the same set of PDU Session types as EPS, i.e. IPv4, IPv6 and IPv4v6. However, especially for IPv6, more features like e.g. IPv6 multi-homing are supported for 5GS compared to EPS (further described in Section 6.4.3.3 below). As the name implies, these PDU Session types provide respectively IPv4, or IPv6, or both IPv4 and IPv6 services to the UE.

PDU Sessions with PDU Session types IPv4, IPv6 and IPv4v6 can have any of the SSC modes 1, 2 or 3 (SSC modes are further described in Section 6.4.2). They also support the full range of QoS features. Further details on QoS features can be found in Chapter 9.

6.2.2.2 IP addressing for IP-based PDU Session types
For IP based PDU Session types, the 5GC is responsible for allocating the IPv4 address and/or IPv6 prefix for the UE. The IP address allocated to the UE belongs to the DN where the UE is accessing. It should be noted that this UE IP address, and the IP address domain of the DN, is different from the IP network (or backbone) that provides the IP transport between entities within the 5GC, or between the (R)AN and 5GC. The back-bone providing the IP transport can be a purely private IP network used solely for the transport of User Plane traffic, either for a single operator in non-roaming cases or between operators in roaming scenarios. The DN is, however, an IP network where a user gains access and is provided services, for example the Internet. This section is only concerned with the IP addresses allocated to the UE.

Each DN may provide services using IPv4 and/or IPv6. A PDU Session must thus provide connectivity using the appropriate IP version. While most IP networks where end-users gain access, for example using 4G or fixed broadband accesses, are still based on IPv4 the number of services e.g. on Internet that support IPv6 is continuously increasing. Usage of IPv6 instead of IPv4 is primarily motivated by the vast number of IPv6 addresses available for allocation to devices and terminals. The shortage of IPv4 addresses is imminent for most operators today and use of various forms of private IPv4 addresses and Network Address Translators (NATs) are common. The amount of IPv4 addresses available does however differ a lot depending on country and organization. IPv6 does not have this problem since the addresses are 128 bits long, in theory providing 2^{128} addresses (this is more than 3.4×10^{38} or 340 undecillion IPv6 addresses). In comparison IPv4 uses 32 bits and thus in theory provides 2^{32} addresses (in total almost 4.3×10^{9}, or 4.3 billion addresses). IPv6 therefore provides significantly more addresses.

Introduction of IPv6 can be a great challenge in terms of migration and smooth introduction. This is because IPv4 and IPv6 are not interoperable protocols; IPv6 implements a new packet header format, designed to reduce the amount of processing an IP header requires. Due to this fundamental difference in headers, workarounds are needed to enable them to function on the same network. An option is to use IP version translation or transition techniques e.g. to enable devices using an IPv6 connection to communicate with applications based on IPv4. Such translation and transition technologies are however not specified by 3GPP and are beyond the scope of this book.

When the UE requests an IP based PDU Session, the UE sets the requested PDU Session Type during the PDU Session Establishment procedure based on its IP stack capabilities as follows:

- A UE which supports IPv6 and IPv4 shall set the requested PDU Session Type according to UE configuration or policy received from the operator (i.e. IPv4, IPv6, or IPv4v6).
- A UE which supports only IPv4 shall request for PDU Session Type "IPv4".
- A UE which supports only IPv6 shall request for PDU Session Type "IPv6".
- When the IP version capability of the UE is unknown in the UE (e.g. if the IP stack is implemented on a separate device from the 5G modem), the UE shall request for PDU Session Type "IPv4v6".

It is SMF that is responsible for assigning an IP address to the UE. When receiving the PDU Session Establishment request from the UE, the SMF selects the PDU Session Type of the PDU Session based on the IP versions that the DN supports (e.g. in case it is an IPv4-only DN, or IPv6-only DN) as well as based on configuration and operator policies configured in the SMF. This means that if the UE requests "IPv4v6" the PDU Session may be granted as "IPv4" or "IPv6" only.

5GS supports different ways to allocate an IP address. The detailed procedure for allocating an IP address also depends on deployment aspects as well as the IP version (v4 or v6). This is explained in more detail in the following sections.

6.2.2.3 IP address allocation

The methods used to allocate IPv4 addresses and IPv6 prefixes are different. Below we will describe how IPv4 addresses and IPv6 prefixes are allocated in 5GS.

There are two main options for allocating an IPv4 address to the UE:

1. One alternative is to assign the IPv4 address to the UE during the PDU Session Establishment procedure itself. In this case the IPv4 address is sent to the UE as part of the PDU Session Establishment accept message. This is a rather 3GPP-specific method of assigning an IP address and this is the way it works in basically all existing 3G/4G networks. The terminal will also receive other parameters needed for the IP stack to function correctly (e.g. DNS address) during the PDU Session Establishment. These parameters are transferred in the so-called Protocol Configurations Options (PCO) field.

2. The other alternative is to use DHCPv4 (often referred to as just DHCP). In this case the UE does not receive an IPv4 address during PDU Session Establishment. Instead, the UE uses DHCPv4 to request an IP address after session establishment is completed. This method to allocate IP addresses is similar to how it works in e.g. Ethernet and WLAN networks, where terminals use DHCP after the basic layer 2 connectivity has been set up. When DHCP is used, the additional parameters (e.g. DNS address) are also sent to the UE as part of the DHCP procedure.

Whether alternative 1 or 2 is used in a network depends on what is requested by the UE, as well as what is supported and allowed by the network. It should be noted that both these alternatives are supported already in 2G/3G/4G core network standards, even though alternative 1 is used in most existing mobile networks.

We now proceed to the IP address allocation procedure for IPv6. The primary method supported in 5GS in Rel-15 is Stateless IPv6 Address Auto Configuration (SLAAC). When SLAAC is used, a /64 IPv6 prefix (i.e. a 64-bit prefix) is allocated for each PDU Session and UE. The UE can utilize the full prefix and can construct the IPv6 address (i.e. 128-bit address) by adding an Interface Identifier to the IPv6 prefix. Since the full /64 prefix is allocated to the UE and the prefix is not shared with any other devices, the UE does not need to perform Duplicate Address Detection (DAD) to verify that no one else is using the same IPv6 address. With stateless IPv6 address auto configuration, PDU Session Establishment is completed first. The SMF then sends an IPv6 Router Advertisement (RA) to the UE via the User Plane to the UE. The RA contains the IPv6 prefix that is allocated to this PDU Session. The RA is sent over the already established PDU Session User Plane and is therefore sent only to a specific terminal. This is different to some non–3GPP access networks, where many terminals share the same layer 2 link (e.g. Ethernet). In these networks, the RA is sent as a broadcast message to all connected terminals. After completing the IPv6 stateless address auto configuration, the terminal can use stateless DHCPv6 to request other necessary parameters, for example DNS address. Alternatively, the UE gets these parameters in the PCO, as described above for IPv4 address allocation.

For Rel-16 additional IPv6 address allocation methods are being discussed, driven by the work to integrate fixed accesses with 5GC. IPv6 prefix delegation (PD) using DHCPv6 is one such feature being discussed for Release-16. Additionally, the option to allocate individual 128-bit IPv6 addresses using stateful DHCPv6 (DHCPv6 NA) is also discussed as part of Release-16.

6.2.3 Ethernet PDU Session type

6.2.3.1 General

The Ethernet PDU Session type is new in 5GS and does not have a direct counterpart in EPS. At the time of writing this book, however, the Ethernet PDN type is being added by 3GPP to EPS as well. The intent with this PDU Session type is to provide an Ethernet service to the UE, i.e. to connect the UE to a Layer 2 Ethernet Data Network. The use cases for this could e.g. be that the UE is connecting a remote office to a corporate network, or that the UE is an industrial device that is connected to the LAN of a factory. Additional use cases are to support e.g. fixed (wireless) access where a Residential Gateway (RG) is providing bridged Layer 2 services to a fixed (wireless) broadband customer.

For a PDU Session set up with the Ethernet PDU Session type, the PDU Session carries Ethernet frames between UE and the DN.

PDU Sessions with PDU Session type Ethernet may use SSC modes 1 or 2. SSC mode 3 is not supported for this PDU Session type (see Section 6.4.3.3 for more info on SSC modes).

6.2.3.2 MAC addressing

5GC does not allocate any Ethernet addresses (usually called MAC addresses) to the UE. The main reason is that MAC addresses are normally encoded into the devices themselves at manufacturing and thus dynamic address allocation is not used in Ethernet networks. The 5GC does also not allocate any IP addresses to the UE for Ethernet PDU Session. If IP address allocation is needed, it can be supported by deploying e.g. a DHCP server on the DN that is accessible to the UEs over the Ethernet PDU Session.

Since no Ethernet (or IP address) is allocated by 5GC, it raises the question on how the 5GC (and UPF in particular) can route down-link Ethernet frames received from the DN to the right UE. If the UPF does not know what Ethernet address belongs to what PDU Session, it is not possible to map down-link frames to the right PDU Session. There are several solutions supported by 5GC for handling such scenarios:

- One basic feature is MAC address learning in SMF/UPF. With this approach the UPF inspects the source MAC addresses received on a PDU Session in up-link traffic and configures down-link filters with this MAC addresses. Then the UPF will send all down-link traffic that contains such a MAC address in the destination address field to this specific PDU Session. The SMF instructs the UPF to perform such MAC address learning for a PDU Session. Alternatively, the SMF can instruct the UPF to report all detected source MAC addresses in up-link traffic, and then SMF will provide down-link filters in UPF for the MAC addresses that should be forwarded to this PDU Session.

- As an additional option, when an Ethernet PDU Session is authorized by a DN-AAA server, the DN-AAA server may, as part of authorization data, provide the SMF with a list of allowed MAC addresses and/or a list of allowed VLAN IDs for this PDU Session (maximum 16 MAC addresses and 16 VLAN IDs). This option is useful e.g. if a specific set of MAC addresses and/or VLAN IDs have been provided for a PDU Session subscription. This option also allows 5GC to authorize the MAC addresses used on a PDU Session, i.e. only the set of MAC addresses received from DN-AAA can be forwarded to the UE over that PDU Session. In addition, the DN-AAA can also provide a set of allowed VLAN IDs.

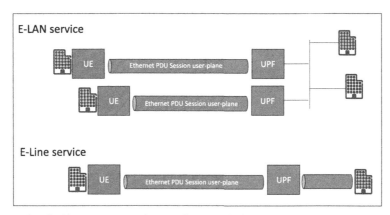

Fig. 6.4 Example of Ethernet services that can be provided using Ethernet PDU Session type.

– A third option is an alternative to MAC address learning in SMF/UPF, and is that there is a PDU Session specific point-to-point tunnel on N6, between the UPF and an entity on the DN. The UPF will forward all down-link traffic received over the tunnel to the UE over the PDU Session. This option leaves it up to the DN to determine what down-link traffic should be sent to which UE.

The first approach where UPF learns what MAC addresses are available on a PDU Session could be a way to provide a so called "E-LAN" service where multi-point to multi-point connectivity is provided. The latter approach where a PDU Session specific point-to-point tunnel is established on N6 on the other hand can be a way to provide an "E-Line" service for point-to-point connectivity e.g. between two enterprise sites. E-LAN and E-Line are two service types for carrier Ethernet networks defined by the Metro Ethernet Forum. Fig. 6.4 illustrates these two cases.

6.2.3.3 Support for Virtual LANs
Ethernet PDU Sessions requires handling Ethernet Virtual LANs (VLANs). VLANs are typically used on Ethernet LANs to provide traffic separation and divide the Layer 2 network into logically separate (virtual) networks. The UPF can e.g. remove or reinsert VLAN tags on N6 interface for downlink and uplink frames, respectively, as instructed by the SMF. The UPF may also transparently forward VLAN tags sent by and received from the UE.

The network may also authorize the set of VLAN IDs used on a PDU Session by providing a set of VLAN IDs to the SMF as part of the PDU Session authorization with a DN-AAA Server, similar to how MAC addresses can be authorized as described above.

6.2.3.4 QoS and charging aspects
5GC can support some similar QoS and flow-based charging features for Ethernet PDU Sessions as for IP based PDU Sessions. For example, the use of GBR and non-GBR QoS

flows can be supported with the difference that the packet filters (SDF filters) used to map traffic onto each QoS flow can also contain parameters from the Ethernet header such as source/destination MAC addresses, VLAN ID etc.

6.2.3.5 Handling of broadcast

One specific aspect that introduces some challenges for mobile systems is the frequent use of broadcast in Ethernet networks. Ethernet broadcast frames are e.g. used by the ARP (Address Resolution Protocol) and IPv6 ND (Neighbor Discovery) protocols to discover what MAC address corresponds to a certain IPv4 or IPv6 address. In general, if a UE or a peer on the DN issues a broadcast, it would be replicated onto all Ethernet PDU Sessions belonging to the same DN. Local policies in the UPF can indicate whether broadcast replication is allowed.

In case a broadcast is due to ARP or ND protocols, only one of the UEs would reply to such broadcast message and the rest would discard it. Not only would this flood the NG-RAN for little benefit, it would also wake up all UEs in CM-IDLE state for no real reason. Therefore, it is possible for the SMF/UPF to reply to an ARP/ND message on behalf of the UE owning the MAC address and thus avoid sending the ARP/ND message to any UE.

It can be noted that a prerequisite for the SMF/UPF to be able to reply to an ARP/ND on behalf of the UE is that SMF/UPF knows the mapping between IP address and MAC address and has stored this mapping. The ARP/ND proxy feature thus requires that IP address allocation to the UE and devices behind the UE is handled by some protocol running over the user plane (e.g. DHCP) and that SMF/UPF can inspect that traffic to deduce IP address to MAC address mapping.

6.2.4 Unstructured PDU Session type

For a PDU Session established with the Unstructured PDU Session type, 5GC does not assume any specific format of the PDU, i.e. user data. 5GC basically treats the PDUs as unstructured bits and therefore has very limited possibility to do packet classification or differentiated treatment of different traffic flows.

The Unstructured PDU Session may be used to carry any protocol, also IP and Ethernet, but the primary use case for this type of PDU Session is to support protocols typically used for IoT deployments such as 6LoWPAN, MQTT, CoAP etc.

Since 5GC is not interpreting the PDUs carried over Unstructured PDU Session it also does not allocate any protocol addresses or other protocol parameters to the UE. In addition, since there is no mechanism to differentiate traffic within the PDU Session based on packet filters there is only a single QoS flow supported. This QoS flow will have the default QoS class.

PDU Sessions with PDU Session type Unstructured may have SSC modes 1 or 2. SSC mode 3 is not supported for this PDU Session type (see Section 6.4.3.3 for more info on SSC modes).

6.3 User plane handling

6.3.1 General

The main task of the Session Management functionality is to manage the User Plane for the PDU Session. The User Plane is where the actual end-user data is carried between the UE and the DN. The User Plane consists of a concatenation of multiple legs. From the UE side, it is first a User Plane connection over the access technology used (e.g. NG-RAN). From the AN there is then a User Plane connection toward a UPF in the core network (over the N3 reference point), and then possibly additional hops between UPFs in the core network (over the N9 reference point) before the User Plane connection continues into the DN (over the N6 reference point).

In case of NG-RAN, the User Plane consists of one or more Data Radio Bearers (DRBs) managed by the NG-RAN. It is beyond the scope of this book to go into details for how NG-RAN manages the User Plane (Dahlman et al., 2018). Over the N3 and N9 reference points, the User Plane data is carried in GTP-U tunnels. This is the same User Plane tunneling protocol used in EPC. So even if GTP-C is not used in 5GC, the User Plane encapsulation is still based on GTP-U. Alternative options for User Plane in 5GC was discussed and analyzed (e.g. GRE and other variants) but in the end it was decided to maintain GTP-U e.g. due to its flexibility. GTP-U has however been enhanced to support new requirements in 5G, such as the new 5G QoS model. The User Plane protocol stack between UE and UPF is shown in Fig. 6.5 (from 3GPP TS 23.501).

Another key aspect of the User Plane architecture in 5GS is that a CP-UP separation has been included from the start, i.e. it is in a sense a mandatory part of the 5G architecture. With EPC the CP-UP split (usually referred to as "CUPS"; Control and User Plane Separation of EPC nodes) was added in Rel-14 and is thus an optional add-on to EPC. There are several reasons why a CP-UP split is an integral part of 5GC. CP-UP split enables flexible network deployment and operation, by distributed or centralized deployments and the independent scaling between control plane and user plane functions. Furthermore, with the ever-increasing amount of a traffic in the mobile operator's networks,

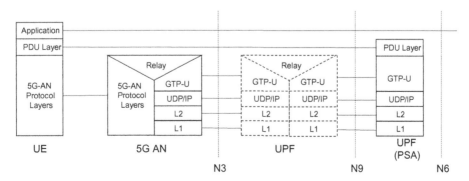

Fig. 6.5 User Plane protocol stack.

there is a strong need for cost efficient User Plane solutions that can meet end-users demand on bit-rates and delays and at the same time be sustainable for the mobile operator.

6.3.2 User Plane path and UPF roles

The User Plane architecture in 5GC has been made very flexible, to allow new use cases e.g. for Edge computing. In EPC, the User Plane architecture is quite fixed; there is always one SGW (or SGW-U in case of CUPS) and one PGW (or PGW-U in case of CUPS) on the User Plane path of a PDN Connection. The SGW (or SGW-U) and PGW (or PGW-U) have well defined roles and functionality. In 5GC there is only one User Plane entity; the UPF. The User Plane path for a PDU Session may however consist of a single UPF or multiple UPFs in a chain. The standard does not restrict the number of UPFs that can be chained for a PDU Session. The standard also allows the user plane path to fork in order to e.g. route certain traffic to a more local DN/N6 connection and other traffic to a different (more central) DN/N6 connection. Such functionality can e.g. be used to support edge computing or CDNs. We will talk more about this forking when we discuss edge computing in Sections 6.4.3 and 6.5.

The general functionality of UPF is described in the list below. The functionality that a specific UPF instance serving a PDU Session provides depends however on where in the User Plane chain this UPF is located, the UPF capabilities and what rules the SMF has provided to a specific UPF. In principle, with a few exceptions, the standard allows the SMF to invoke e.g. packet buffering for IDLE mode UE, or charging, or other functionality, in any one of the UPFs on the path. This is a difference from EPC where the SGW-U and PGW-U have clearly defined functionality and e.g. buffering is always performed by SGW. Even if the 5GC standard is very flexible, real world deployments (at least initially) will most likely be rather simple with one or maybe two UPFs on the path depending on use case in order to ensure User Plane efficiency. Further aspects on User Plane paths are provided in Section 6.3.3.2.

The general functionality of UPF is classified as follows:
- Anchor point for Intra-/Inter-RAT mobility.
- External PDU Session point of interconnect to Data Network (i.e. N6).
- Packet routing and forwarding.
- Packet inspection (e.g. Application detection).
- User Plane part of policy rule enforcement, e.g. Gating, Redirection, Traffic steering.
- Lawful intercept (UP collection).
- Traffic usage reporting.
- QoS handling for user plane, e.g. UL/DL rate enforcement, Reflective QoS marking in DL.

- Uplink Traffic verification (SDF to QoS Flow mapping).
- Transport level packet marking in the uplink and downlink.
- Downlink packet buffering and downlink data notification triggering.
- Sending and forwarding of one or more "end marker" to the source NG-RAN node.
- Functionality to respond to ARP and IPv6 ND requests for the Ethernet PDUs.

Even though the standard only defines a single User Plane function (UPF), it has defined a few functional roles that a UPF can perform on the User Plane path:

- PDU Session Anchor (PSA): This is the UPF that terminates the N6 interface toward the DN.
- Intermediate UPF (I-UPF): This is a UPF that has been inserted on the UP path between the (R)AN and a PSA. It forwards traffic between (R)AN and the PSA.
- UPF with UP-link Classifier (UL-CL) or Branching Point (BP): This is a UPF that is "forking" traffic for a PDU Session in up-link, and "merging" UP paths in down-link.

Note that UL-CL/BP roles are not mutually exclusive with the PSA and the I-UPF – i.e. a UPF acting as PSA or UPF acting as I-UPF can at the same time act as UL-CL/BP for the PDU Session. Also note that these roles are not to be interpreted as different UPF types. A single UPF entity can act in different roles for different PDU Sessions, e.g. be a UL-CL/BP for one PDU Session and a I-UPF for another PDU Session.

Fig. 6.6 illustrates three different User Plane scenarios for a PDU Session:

(a). In the simplest scenarios, only a PSA is needed.

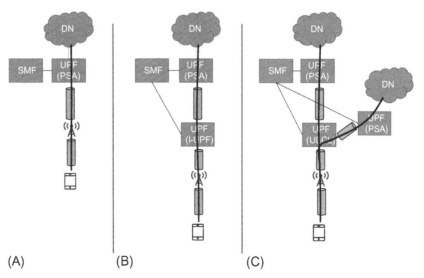

(A)   (B)   (C)

Fig. 6.6 Example of UPF configurations and functionality performed by a UPF. Fig. A: PSA only. Fig. B: PSA + I-UPF. Fig. C: UL-CL/BP + two PSAs.

(b). If, due to mobility, the UE moves to a new RAN node and the new RAN node cannot support N3 tunnel to the old PSA, an I-UPF needs to be inserted.

(c). In case of traffic breakout (e.g. edge computing), a UL-CL/BP can be inserted to fork/merge the UP traffic.

In general, the standard is flexible to what UP functionality is executed where. For example, in (b), data buffering for UE in IDLE state can be done in either the I-UPF or the PSA. More details on the scenario (c) is described in the sections on selective routing of traffic to DN (Section 6.4.3) and Edge Computing (Section 6.5).

6.3.3 Control-plane and user-plane separation and the N4 interface
6.3.3.1 General
The separation between CP and UP in 5GS follows many of the same principles as CUPS for EPC as described in Chapter 4. For example, the functional split between control plane and user plane, i.e. the functionality placed in the CP side and the functionality placed in the UP side, is very similar to the functional split specified for CUPS. Also, the N4 protocol between SMF and UPF is re-used from CUPS, i.e. the Packet Forwarding Control Protocol (PFCP) specified for CUPS has been re-used and evolved also for N4 in 5GC. This is an important aspect as it allows a UP entity to easily support both EPC and 5GC which simplifies e.g. in interworking and migration scenarios between EPC and 5GC. More details on PFCP is available in Chapter 14, Section 14.6.

6.3.3.2 UPF discovery and selection
The SMF is in charge of selecting the UPF. The details for how this is done is not standardized, and is dependent on several aspects, for example deployment aspects related to the network topology of the deployed UPFs as well as the requirements on the service to be delivered (e.g. User Plane delays, reliability etc.).

A key use case for supporting flexible User Plane paths, deployment and selection of UPFs etc. is to ensure an efficient User Plane path between the UE and the place where the application is deployed e.g. for Edge computing. This will be covered in in a dedicated section later. In this section we will describe the general aspects of UPF selection that will then form the basis for more advanced scenarios.

When SMF is to make UPF selection, a prerequisite is that SMF is aware of what UPFs are available and their respective properties such as UPF capabilities, load status etc. There are different ways this can be done.

- Firstly, the SMF can be configured with the available UPFs via O&M. This configuration may include topology related information so that SMF is aware about UPF location and in what way UPFs are connected (e.g. properties of the links between). This allows the SMF to select suitable UPFs e.g. depending on UE location.

- It is also possible to discover available UPFs with the NRF. In this case the SMF can query the NRF and in the reply receive a list of UPFs together with some basic information about each UPF, such as what DNN(s) and network slice(s) (S-NSSAI) that each UPF supports. This reduces the need for preconfigured information in SMF. On the other hand, the information that can be learned from NRF is rather limited and it does e.g. not contain detailed information about UPF topology so for more advanced use cases pre-configuration in SMF may be the preferred choice.
- In addition, the N4 protocol supports exchange of SMF and UPF capabilities when setting up the basic N4 connection between SMF and UPF. The SMF will e.g. learn whether the UPF supports optional features such as traffic steering (service chaining) on N6-LAN, header enrichment, traffic redirection etc., and will also receive information about the load of a UPF.

Once the SMF knows about the available UPF(s), and there is a need for SMF to select one or more UPFs for a PDU Session, e.g. at PDU Session Establishment or at some mobility event, the SMF can take different information into account for selecting a UPF instance. The details here are not standardized but left for implementation and operator configuration. Examples of information that SMF can use for UPF selection are listed below. Some of this information is received from the UPF, others are received from AMF while some can be pre-configured in SMF. The SMF may consider e.g.:

- UPF's dynamic load.
- UPF's relative static capacity among UPFs supporting the same DNN.
- UPF location.
- UE location information.
- Capability of the UPF
- The functionality required for the UE session.
- Data Network Name (DNN).
- PDU Session Type (i.e. IPv4, IPv6, IPv4v6, Ethernet Type or Unstructured Type)
- SSC mode selected for the PDU Session.
- UE subscription profile in UDM.
- DNAI (see Section 6.4.4 for more info).
- Local operator policies.
- S-NSSAI.
- Access technology being used by the UE.
- Information related to user plane topology and user plane terminations.

6.3.3.3 Selective activation and deactivation of UP connections

Similar to EPS, 5GS supports a UE having multiple PDU Sessions active at the same time (e.g. one PDU Session to IMS and one to Internet). In EPS, the UP connections (S1-U tunnel) was established for all active PDN Connections when the UE moves from

IDLE to CONNECTED state. If there was a down-link data for one PDN Connection in EPC when a UE was in IDLE state, and the UE was paged, when the UE enters CONNECTED state also the user-plane of other PDN Connections was activated, even if there was no data to send over these PDN Connections. This was done to simplify the procedures and keep an always-on behavior in the system.

In 5GS on the other hand, this is not necessarily the case. The general behavior in 5GS is that the PDU Session User Plane is only activated for the PDU Session that has pending data. The User Plane connection (N3 tunnel) for the other PDU Sessions will not be activated and will thus remain "idle" even if the UE is in CM-CONNECTED state. A motivation for this was to ensure a better isolation between network slices, i.e. a PDU Session in one slice should not be impacted just because a PDU Session in another slice had to activate the User Plane in order to send data.

5GS thus supports cases where a UE in CM-CONNECTED state has some PDU Sessions with active User Plane (established N3 tunnel) and some PDU Sessions with inactive User Plane (no N3 tunnel). If the UE or the network later needs to send data for a PDU Session with inactive User Plane, the Service Request procedure is used also in CM-CONNECTED state to active User Plane connection of that PDU Session.

A risk with the above principle is that there may be an increased delay in sending data if UE is in CM-CONNECTED state but the User Plane connection for the PDU Session is inactive. In that case the Service Request procedure need to be executed first. In some cases, there may also be race conditions where e.g. a procedure to activate the User Plane for a PDU Session needs to wait in order to complete some other ongoing procedure. This may be especially a concern for PDU Sessions that are sensitive to delays, such as PDU Sessions for IMS or for low-latency services. It has therefore been specified that the UE may decide to request activation of User Plane connection of additional PDU Sessions when the UE moves from CM-IDLE to CM-CONNECTED, even if there is no pending data to be sent. This is done in order to avoid the delay later when data actually needs to be sent.

6.4 Mechanisms to provide efficient user plane connectivity

6.4.1 General

When 5GC Session Management was defined, providing solutions for efficient User Plane connectivity was one of the key goals. As mentioned before, the UP architecture of 5GC was specified in a flexible way, allowing implementations and deployments to utilize the tools and enablers in the standard to achieve specific use cases and requirements. In the same way a set of tools have been defined for User Plane efficiency that can be utilized depending on the use case and scenario.

Maybe the most basic tool for achieving an efficient UP path is the UPF selection taking place at PDU Session Establishment. Here the SMF can e.g. take UE location and other information about User Plane topology into account when selecting UPF. This can e.g. result in a UPF that is located close to the UE. UPF selection at PDU Session Establishment has already been described earlier in the chapter. The tools that are described below are rather relying on UPF-reselection, e.g. during the lifetime of a PDU Session, in order to modify the UP path due to UP mobility. This can be useful if the UE has moved far away from the location where the PDU Session was initially established or due to other triggers (e.g. the user has started an application that requires low latency communication). Below we will look more closely at this set of tools.

6.4.2 Service and Session Continuity (SSC) modes

6.4.2.1 General

When a PDU Session is established, a PDU Session Anchor UPF is selected that remains as the IP anchor point for the PDU Session. At establishment, this PSA UPF may have been selected close to the UE's location. However, in case the UE moves far away, that PSA UPF may no longer be optimally located; there may be other UPFs closer to the UE's new location that could act as PSA UPF. Changing PSA UPF however requires the change of UE IP address, which may or may not cause a problem for applications/services running on the UE. Some applications/services may require IP address continuity in order to run smoothly, while others may handle IP address changes without much impact to the user experience.

5GS supports differentiated session and service continuity to address different IP address continuity requirements the various applications and services in the UE may have. To enable this, three different Service and Session Continuity (SSC) modes have been defined; SSC modes 1, 2 and 3. When a PDU Session is established, it is assigned one of the SSC modes. SSC mode selection is done by the SMF based on the allowed SSC modes in the user subscription, the allowed SSC modes for the specific PDU Session type, and the SSC mode requested by the UE (if any).

Below we describe each SSC mode and its properties.

6.4.2.2 SSC mode 1

With this SSC mode the network preserves the PDU Session connectivity service provided to the UE and the UPF acting as PDU Session Anchor at the establishment of the PDU Session is maintained regardless of UE mobility. In the case of an IP-based PDU Session type (IPv4, IPv6 or IPv4v6), the IP address/prefix is maintained. IP session continuity is thus supported regardless of UE mobility events during the lifetime of the PDU Session. This SSC mode is therefore suitable for applications that require IP address continuity.

6.4.2.3 SSC mode 2

For a PDU Session with SSC mode 2, the network may release the connectivity service delivered to the UE and release the corresponding PDU Session(s), e.g. when the UE has moved away from its original location. In the PDU Session Release message, the network also includes an indication triggering the UE to request establishment of a new PDU Session (for the same DNN and S-NSSAI) to regain PDU Session connectivity to the same DN. With SSC mode 2 there is an interruption in the UE's connectivity after the old PDU Session is release until the new PDU Session has been established, and SSC mode 2 can therefore be described as "break-before-make". At the establishment of the new PDU Session, a new SMF selection and UPF selection takes place, and a PDU Session Anchor UPF closer to the UE's current location may therefore be selected. The SSC mode 2 procedure thus allows the PSA UPF to be "re-located" to a location closer to the UE's current point of attachment. For the case of IPv4 or IPv6 or IPv4v6 type, the release of the PDU Session implies the release of the IP address/prefix that had been allocated to the UE. A new IP address/prefix will then be allocated for the new PDU Session. This SSC mode is thus suitable for applications that can handle short interruptions in User Plane connectivity and IP address changes (in case of IP-based PDU Session types).

6.4.2.4 SSC mode 3

SSC mode 3 is similar to SSC mode 2 in the sense that it allows the PSA UPF to change, but with SSC mode 3 the network ensures that the UE suffers no loss of connectivity during the time that the PSA UPF change takes place. SSC mode 3 can therefore be described as "make-before-break". SSC mode 3 can be supported in two ways:

- Multiple PDU Sessions: In this case the SMF instructs the UE to request establishment of a new PDU Session to the same DN before the old PDU Session is released. This means that the User Plane connectivity via a new PDU Session Anchor is available to the UE for a while before the old PDU Session and its User Plane connection is released.
- IPv6 multi-homing: In this case a single PDU Session (of PDU Session type IPv6) is used and a new UPF PSA (with a new IPv6 prefix) is allocated in that PDU Session, before the old PSA UPF (and old IPv6 prefix) is released. In the same way as when multiple PDU Sessions are used, the new PDU Session Anchor can be used for a while before the old PDU Session Anchor is released.

In both cases above, the IP address/prefix is not preserved. The new PDU Session Anchor will be associated with a different UE IP address/prefix than the old PDU Session Anchor. This SSC mode is thus suitable for applications that need continuous User Plane connectivity but can handle IP address/prefix changes. SSC mode 3 only applies to IP-based PDU Session types.

Fig. 6.7 illustrates the principles of the different SSC modes.

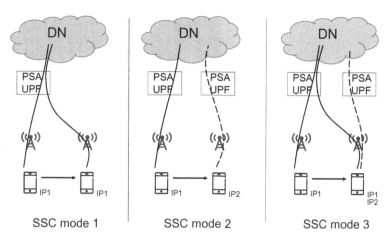

Fig. 6.7 Principles of the SSC modes.

6.4.3 Selective traffic routing to a DN

6.4.3.1 General

As we saw in Section 6.3.2, a PDU Session has in the simplest case a single PSA UPF and thus a single N6 interface to a DN, but a PDU Session may also have more than on PSA UPF and thus multiple N6 interfaces to a DN (see Fig. 6.6C). This latter option can be used to selectively route User Plane traffic to different N6 interfaces, e.g. to one local PSA UPF with N6 interface to a local edge site, and one more central PSA UPF with N6 interface to a central data center or Internet peering point. Such functionality may be used to enable edge computing use cases or to reach distributed content delivery sites.

Two mechanisms have been defined to support selective traffic routing to a DN and we will describe them further below.

6.4.3.2 Up-link Classifier

An Up-link Classifier (UL CL) is a functionality that is supported by a UPF where the UPF diverts some traffic to a different (local) PSA UPF. The UL CL provides forwarding of up-link traffic toward different PDU Session Anchors and merge of down-link traffic to the UE i.e. merging the traffic from the different PDU Session Anchors on the link toward the UE. The UL CL diverts traffic based on traffic detection and traffic forwarding rules, with traffic filters provided by the SMF. The UL CL applies the filtering rules (e.g. to examine the destination IP address/prefix of up-link IP packets sent by the UE) and determines how the packet should be routed. The UPF supporting an UL CL may also be controlled by the SMF to support traffic measurement for charging, bit rate enforcement etc. The use of UL CL applies to PDU Sessions of type IPv4 or IPv6 or IPv4v6 or Ethernet, so that SMF can provide suitable traffic filters.

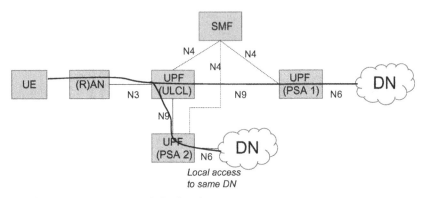

Fig. 6.8 Local access to DN using Uplink Classifier.

When the SMF decides to divert traffic, the SMF inserts a UL CL into the data path, and an additional PSA. This can be done at any time during the lifetime of a PDU Session, e.g. triggered by AF requests as we will see in a subsequent section. The additional PSA may be collocated in the same UPF as the UL CL or it may be a standalone UPF. An example architecture is shown in Fig. 6.8. When SMF determines that the UL CL is no longer needed it can be removed by the SMF from the data path.

It should be noted that the UE is unaware of the traffic diversion by the UL CL, and is not taking part in the insertion and the removal of UL CL. The solution with UL CL does therefore not require any specific functionality in the UE.

6.4.3.3 IPv6 multi-homing

The support of IPv6 multi-homing also enables traffic to be selectively routed to different PDU Session Anchors. IPv6 multi-homing enables a UE to be assigned multiple IPv6 prefixes in a single PDU Session. Each IPv6 prefix will be served by a separate PDU Session Anchor UPF, each with its own N6 interface to the DN. The different user plane paths leading to the different PDU Session Anchors branch out at a "common" UPF referred to as a UPF supporting "Branching Point" (BP) functionality. The Branching Point provides forwarding of UL traffic toward the different PDU Session Anchors and merge of DL traffic to the UE i.e. merging the traffic from the different PDU Session Anchors on the link toward the UE. An example architecture is shown in Fig. 6.9.

Similar to UL CL, the SMF may decide to insert or remove a UPF supporting the Branching Point functionality anytime during the lifetime of a PDU Session. The UPF supporting a BP may also be controlled by the SMF to support traffic measurement for charging, bit rate enforcement etc.

IPv6 multi-homing applies only for IPv6 and only if the UE supports it. When the UE requests a PDU Session for IPv6 the UE also provides an indication to the network whether it supports IPv6 multi-homing.

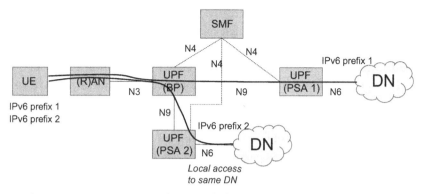

Fig. 6.9 Local access to DN using BP and IPv6 multi-homing.

When IPv6 multi-homing (and BP) is used, it is the UE that selects which IPv6 prefix to use for the source address of the up-link traffic. This will in turn decide which path the packets take, since the BP will forward UL packets based on the source IPv6 address. In order to influence the UE in source address selection and ensure that the UE selects the appreciate IPv6 prefix for a given application traffic, the SMF can configure routing information and preferences into the UE. This is done via Router Advertisement messages as described in IETF RFC 4191 (RFC 4191). The involvement of the UE is one of the key differences compared to the UL CL approach, since in IPv6 multi-homing approach certain UE functionality is required, and it is also the UE that selects traffic path (although based on rules received from SMF) while in the UL CL approach it is a pure network-based feature.

Finally, it should be noted that IPv6 multi-homing is both a tool to provide selective routing of traffic to different PSAs and N6 interfaces (as described in this section) and a tool to implement SSC mode 3 (as described in Chapter 4, Section 4.2.4).

6.4.4 Application Function influence on traffic routing

Application Function influence on traffic routing is a related but somewhat different concept from SSC modes and selective routing to a DN. While e.g. SSC modes and UL CL/BP were mechanisms that help achieve an efficient User Plane path, AF influence on traffic routing is rather a control plane solution for how an AF (e.g. a 3r party AF) can influence the use of traffic routing mechanisms such as SSC modes or UL CL/BF. It allows an AF to provide input to the 5GC for how certain traffic should be routed. It is then up to 5GC (and in particular the SMF) to decide how to do that using the available tools, e.g. UPF selection, SSC modes, UL CL, IPv6 multi-homing etc.

The AF sends the request either directly to the PCF (if the AF can communicate directly with PCF) or via the NEF which in turn sends the request to the PCF. If the request goes via the NEF, the NEF can map external identifiers provided by the AF to internal identifiers known by 5GC. The AF may provide information such as:

- Traffic descriptor (IP filters or Application Identifier). This information describes the application traffic covered by the request from the AF
- Potential locations of applications represented by a list of DN Access Identifiers (DNAI). A DNAI is an identifier representing a user plane access to one or more DN(s) where applications are deployed and can be interpreted as an index that point to one specific access into a Data Network. It can e.g. represent a specific data center. The DNAI values as such are not specified by 3GPP (DNAI data type is a string) but is left to be defined by operator deployment and configuration.
- UE identifier(s), such a GPSI(s) or UE group identifier, for which UEs the request targets.
- N6 traffic routing information, indicating how the traffic should be forwarded on N6. The N6 traffic routing information can contain the target IP address (and port) in the DN to which the application traffic shall be tunneled.
- Spatial and temporal validity conditions. These conditions indicate the time interval(s) and geographic area for when and where the AF request is to be applied.

When PCF receives this information, it creates PCC rules that include relevant information and provides it to SMF. The SMF then acts on the information, e.g. by inserting a UL CL, triggering PSA relocation using SSC mode 2 or 3 procedures or some other action. Fig. 6.10 illustrates an example use case where a UL CL is inserted, and the targeted traffic is redirected to a local data center.

The AF may also request to be notified by the SMF when a UPF related even occurs, e.g. when a UL CL is inserted or a SSC mode 2 or 3 procedure is triggered. The AF can request to be notified just before the event is to take place and/or after the event has taken place. This allows an AF e.g. to take application layer actions such as relocating application state or handle UE IP address changes.

More details on Application Function influence on traffic routing can be found in 3GPP TS 23.501, clause 5.6.7.

6.5 Edge computing

Edge computing is about bringing the services closer to the location where they are to be delivered. Services here includes computing power and memory needed for e.g. running a requested application. Edge computing therefore aims to push applications, data and computing power (services) away from centralized points (central

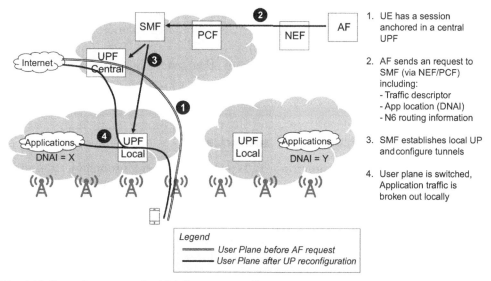

Fig. 6.10 Example use case for AF influence on traffic routing.

data centers) to locations closer to the user (such as distributed data centers). The goal is both to achieve a lower latency and to reduce transmission costs. Applications that use high data volumes and/or require short response times, e.g. VR gaming, real-time facial recognition, video surveillance etc. are some candidates that could benefit from Edge computing.

A lot of work in the industry around Edge computing has been done on the application platform for edge applications and related APIs, e.g. by an ETSI Industry Specification Group called MEC (Multi-access Edge Computing). In 3GPP however, the focus when it comes to edge computing has so far been concerned with the access and connectivity aspects. This may change in the future releases as new work is started, but in Release-15 this was the case.

3GPP does not specify any special solutions or architecture for Edge computing. Instead 3GPP defines several general tools that can be used to provide an efficient User Plane path. These tools, most of which have already been described earlier in this chapter, are not specific to Edge computing but they can be used as enablers in deployments of Edge computing. The main tools for UP path management are listed below, with references to other sections where it is described in further detail:

- UPF selection (see Section 6.3.3.2 for more details).
- Selective traffic routing to DN (see Section 6.4.3 for more details).
- Session and Service Continuity (SSC) modes (see Section 6.4.2 for more details).

- AF influence on traffic routing (see Section 6.4.4 for more details).
- Network capability exposure (see Chapter 3, Section 3.11 for more details).
- LADN (see Section 6.7 for more details).

Edge computing can of course also benefit from other general 5GS features such as differentiated QoS and charging.

6.6 Session authentication and authorization

When establishing a PDU Session toward a Data Network there is sometimes a need to authenticate and/or authorize it against a AAA (Authentication, Authorization and Accounting) Server in the Data Network. This can e.g. be the case if the DN is corresponding to a corporate network or is in some other way provided by a 3rd party. As we will see below it can also be useful in other cases as a way for the operator to manage parameters and properties for the PDU Session. The 5G System supports this via a secondary authentication/authorization with a DN-AAA server during the establishment of a PDU Session. Such secondary authentication and/or authorization is optional. It takes place for PDU Session authorization/authentication toward the DN and is done in addition to the "primary" 5GC access authentication handled by AMF during registration, and in addition to the "primary" PDU Session authorization enforced by SMF using the subscription data retrieved from UDM.

The secondary authentication between the UE and the DN-AAA Server is performed using EAP. The SMF shall perform the role of the EAP Authenticator. This means when the SMF receives a PDU Session Establishment request from a UE and is configured to require secondary authentication/authorization by a DN-AAA Server, the SMF initiates EAP authentication by requesting the UE to provide its DN-specific Identity. This identity may be specific to the DN and unrelated to the SUPI/SUCI. The credentials used are also unrelated to the credentials stored in UDM used for "primary" authentication. After the UE has provided its DN Identity, the UE and DN-AAA exchange EAP authentication messages, forwarded by the SMF. Between the UE and the SMF, EAP messages shall be sent in SM NAS message. Between the SMF and the DN-AAA, the EAP messages are sent via RADIUS or Diameter. Fig. 6.11 shows a simplified call flow for secondary authentication/authorization during PDU Session Establishment.

When the secondary authorization takes place, the DN-AAA is checking that the user is authorized to access the DN (done in addition to the primary authorization based on subscription data in UDM). The DN-AAA may also provide DN authorization data to the SMF that will apply to the established PDU Session. The DN authorization data may include e.g. the list of authorized MAC addresses for an Ethernet PDU Session, or the UE IP address/prefix to be assigned for an IP-based PDU Session.

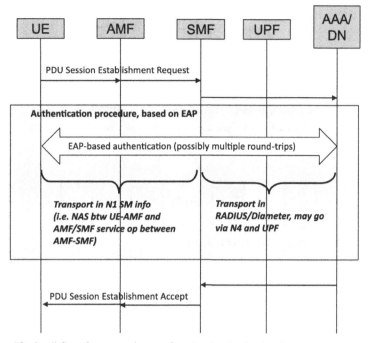

Fig. 6.11 Simplified call flow for secondary authentication/authorization.

6.7 Local Area Data Network

Local Area Data Networks, or LADN for short, is a feature that is new in 5GS. The purpose with LADN is to enable access to a DN (and a DNN) in one or more specific area(s). Outside of that area the UE is not able to access the DNN. This could e.g. be used for special DNNs that are local to a stadium, a shopping center, a campus or similar.

The area where a LADN DNN is available is called a LADN Service Area and is configured in the network as a set of Tracking Areas. The list of Tracking Areas for a LADN DNN is configured in the AMF. DNNs that are not using the LADN feature do not have a LADN Service Area and are not restricted by this feature. The LADN Service Area is provided to the UE when the UE registers. The UE is thus aware of what area a LADN DNN is available and should not try to access that DNN when it is outside that area. An example scenario for LADN is shown in Fig. 6.12.

When the UE sends a PDU Session Establishment request to the network for a LADN DNN, the AMF will provide an indication to the SMF, telling SMF whether the UE is inside or outside the LADN area. This allows the SMF to determine whether to accept or reject the request. If the UE is inside the LADN Service Area, SMF can accept the request, otherwise the SMF will reject the request.

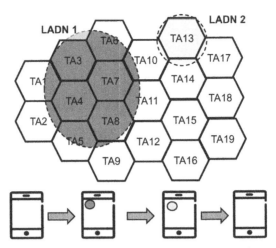

Fig. 6.12 Example scenario with two LADNs in a city.

In many cases, e.g. when the UE is in CM–IDLE state or when RRC INACTIVE is used, the 5GC will not be aware of the exact location of the UE. In those cases, the LADN Service Area is enforced by the network the next time the UE requests service from the network, i.e. transitions to CM-CONNECTED state or RRC ACTIVE.

The LADN feature is only applicable when UE is using 3GPP access.

CHAPTER 7

Mobility Management

7.1 Introduction

The general principles for Mobility Management in 5GS are similar as for previous 3GPP systems but with some key differences. In this section, we therefore first describe the general principles and then focus on the main differences compared to EPS.

As with previous systems, mobility is a core feature of 5GS. Mobility Management is required to ensure the following:

- That the network can "reach" the user, for example to notify the user about incoming messages and calls,
- That a user can initiate communication toward other users or services such as Internet access, and
- That connectivity and ongoing sessions can be maintained as the user moves, within or between access technologies.

The above is ensured by establishing and maintaining connectivity between the UE and the network through mobility management procedures.

In addition, the Mobility Management functionality enables the possibility for identification of the UE, security, and serves as a generic message transport for other communication between the UE and the 5GC.

The aim for 5GC is to act as a converged core network for any access technology, however, the aim is also to provide a flexible support for a wide range of new use cases as discussed in Chapter 2. As a result, there is a need to be able to select the required functionality in relation to mobility because different users have different mobility requirements. For example, a device that is used in a machine in a factory does not normally move, but other devices may. If there is a need to track and ensure that the device is reachable, then mobility procedures are required. In addition, mobility procedures are also used for the basic registration to the network that are required for enabling security procedures and allowing the UE to communicate with other entities as required.

For certain use cases such as fixed wireless access, however, there is less of a need to provide a full set of mobility procedures: in such cases, the procedures that are not essential for all users can be added or removed as "a mobility related service". During the development of the 5GS specifications, this was referred to as "mobility on demand". While in previous systems no mobility signaling was generated for or by the UEs that did not move (except Periodic Registration Updates), 3GPP developed support for further use cases that would not demand support for mobility or only require limited support for mobility.

5G Core Networks
https://doi.org/10.1016/B978-0-08-103009-7.00007-7

As a result, there are several optional 5GS mobility management related functions that differ from previous 3GPP systems:

- Service Area Restriction: mobility with session continuity is controlled at UE level at certain areas (see Section 7.5.3).
- Local Area Data Network (LADN): mobility with session continuity is controlled at PDU Session level making communication available at certain areas (see Chapter 6).
- Mobile Initiated Connection Only (MICO): paging capability (as part of the mobility service) is optional (see Section 7.3.2).

5G Mobility Management (5GMM) related procedures are divided into three categories depending on the purpose of the procedure and on how they can be initiated:

1. **Common procedures**; can always be initiated when the UE is in CM-CONNECTED state.
2. **Specific procedures**; only one UE initiated specific procedure can be running for each of the Access Types.
3. **Connection management procedures**; used to establish a secure signaling connection between the UE and the network, or to request the resource reservation for sending data, or both.

Table 7.1 lists the Mobility Management procedures according to their category and the relevant Sections and Chapters that cover the procedure in detail.

7.2 Establishing connectivity

7.2.1 Network discovery and selection

Network discovery and selection procedures for 5GS do not differ much from EPS and the principles used when a 3GPP access type is selected, have been maintained.

Before the UE can receive and use the services and capabilities from the 5GS, e.g. Session Management services from SMF, the UE needs to establish a connection to the 5GS. To achieve this, the UE first selects a network/PLMN and a 5G-AN. For 3GPP access i.e. NG-RAN, the UE selects a cell, then the UE establishes a RRC connection to the NG-RAN. Based on the content (e.g. selected PLMN, Network Slice information) provided by the UE in establishing the RRC connection the NG-RAN selects an AMF and forwards the UE NAS MM message to the AMF in the 5GC using the N2 reference point. Using the AN connection (i.e. RRC connection) and the N2, the UE and the 5GS complete a Registration procedure. Once the Registration procedure is completed, the UE is registered in the 5GC i.e. the UE is known and the UE has a NAS MM connection to the AMF, the UE's entry point to the 5GC, which is used as the NAS connection to the 5GC. Further communication between the UE and other entities in the 5GC uses the established NAS connection as NAS transport from that point forward. To save resources the NAS connection is released while the UE is still registered and known in the 5GC, i.e. to re-establish the NAS connection the UE or 5GC initiates a

Table 7.1 Summary of the Mobility Management functionality

Type	Procedure	Purpose	Reference
5GMM common procedures	Primary authentication and key agreement procedure	Enables mutual authentication between UE and 5GC and provides key establishment in UE and 5GC in subsequent security procedures.	See Security Chapter 8
	Security mode control procedure	Initiates 5G NAS security contexts i.e. initializes and starts the NAS signaling security between the UE and the AMF with the corresponding 5G NAS keys and 5G NAS security algorithms.	See Security Chapter 8
	Identification procedure	Requests a UE to provide specific identification parameters to the 5GC.	See Selected Call flows Chapter 15
	Generic UE configuration update procedure	Allows the AMF to update the UE configuration for access and mobility management-related parameters.	See Selected Call flows Chapter 15
	NAS transport procedures	Provides a transport of payload between the UE and the AMF.	See Chapters 10 and 14
	5GMM status procedure	Report at any time certain error conditions detected upon receipt of 5GMM protocol data in the AMF or in the UE.	See 3GPP TS 24.501
5GMM specific procedures	Registration procedure	Used for Initial Registration, Mobility Registration Update or Periodic Registration Update from UE to the AMF.	See Selected Call flows Chapter 15
	Deregistration procedure	Used to Deregister the UE for 5GS services.	See Selected Call flows Chapter 15
	eCall inactivity procedure	Applicable in 3GPP access for a UE conFigured for eCall only mode.	See Selected Call flows Chapter 15
5GMM connection management procedures	Service request procedure	To change the CM state from CM-IDLE to CM-CONNECTED state, and/or to request the establishment of User Plane resources for PDU Sessions which are established without User Plane resources.	See Selected Call flows Chapter 15
	Paging procedure	Used by the 5GC to request the establishment of a NAS signaling connection to the UE, and to request the UE to re-establish the User Plane for PDU Sessions. Performed as part of the Network Triggered Service Request procedure.	See Selected Call flows Chapter 15
	Notification procedure	Used by the 5GC to request the UE to re-establish the User Plane resources of PDU Session(s) or to deliver NAS signaling messages associated with non-3GPP access.	See 3GPP TS 24.501

Service Request procedure. See Chapter 14 for NAS messages used and for further description of the NAS transport usage for communication between UE and various 5GC entities. Interested readers may consult Chapter 15 for further details about the Registration and Service Request procedures.

For 5G-ANs of Access Type untrusted non-3GPP the principles are similar but an N3IWF (see Chapter 3) is also involved. In such cases, the UE first establishes a local connection to a non-3GPP access network (e.g. to a Wi-Fi access point), subsequently a secure tunnel between the UE and the N3IWF (NWu) is established as an AN connection. Using the tunnel, the UE initiates a Registration procedure toward the AMF via the N3IWF.

7.2.2 Registration and Mobility

Idle-mode Mobility Management for 5GS using NR and E-UTRA is built on similar concepts to LTE/E-UTRAN (EPS), GSM/WCDMA, and CDMA. Radio networks are built by cells that range in size from tens and hundreds of meters to tens of kilometers as discussed in Chapter 3 and the UE updates the network about its location on a regular basis. It is not practical to keep track of a UE in idle mode every time it moves between different cells due to the amount of signaling it would cause, nor to search for a UE across the entire network for every terminating event (e.g. an incoming call). In order to create efficiencies, therefore, cells are grouped together into Tracking Areas (TA), and one or more Tracking Areas may be assigned to the UE as a Registration Area (RA). RA is used as a base for the network to search for the UE and for the UE to report its location.

The gNB/ng-eNB broadcast TA identity in each cell and the UE compares this information with the one or more TA it has previously stored as part of the assigned RA. If the broadcasted Tracking Area is not part of the assigned RA the UE starts a procedure – called a Registration procedure – toward the network to inform it that it is now in a different location. For example, when a UE that was previously assigned a RA with TAs 1 and 2 moves into a cell that is broadcasting TA 3, the UE will notice that the broadcast information includes a different TA than those it has previously stored as part of the RA. This difference triggers the UE to perform a Registration update procedure toward the network. In this procedure, the UE informs the network about the new TA it has entered. As part of the Registration update procedure, the network will assign the UE a new RA that the UE stores and uses while it continues to move.

As mentioned above, RAs consist of a list of one or more so called TAs. To distribute the Registration update signaling, the concept of TA lists was introduced in EPS and is also adopted by 5GS. The concept allows a UE to belong to a list of different TAs. Different

UEs can be allocated to different lists of TAs. If the UE moves within its list of allocated TAs, it does not have to perform a Registration update for the purpose of mobility (i.e. using a Registration Type set to Mobility Registration Update). By allocating different lists of TAs to different UEs, the operator can give UEs different RA borders and so reduce peaks in Registration update signaling, for example when a train passes a TA border.

In addition to the Registration updates performed when passing a border into a TA where the UE is not registered, there are also periodic Registration updates. When the UE is in idle state, it performs Registration update periodically based on a timer, even it's still inside the RA. These updates are used to clear resources in the network for UEs that are out of coverage or have been turned off without informing the network.

The network thus knows that an idle state UE is located in one of the TA(s) included in the RA. When a UE is in idle state and the network needs to reach the UE (e.g. to send down-link traffic), the network pages the UE in the RA. The size of the TAs/TA lists is a compromise between the number of Registration updates and the paging load in the system. The smaller the TAs, the fewer the cells needed to page the UEs but on the other hand, there will be more frequent Registration updates.

The larger the TAs, the higher the paging load in the cells, but there will be less signaling for Registration updates due to UEs moving around. The concept of TA lists can also be used to reduce the frequency of Registration updates due to mobility. If, for example, the movement of UEs can be predicted, the lists can be adapted for an individual UE to ensure that they pass fewer borders, and UEs that receive lots of paging messages can be allocated smaller TA lists, while UEs that are paged infrequently can be given larger TA lists. A summary of the Registration Area concept and mobility update procedures for the various 3GPP systems are listed in Table 7.2.

A summary of the idle mobility procedure in 5GS is:

- A TA consists of a set of cells,
- The Registration Area in 5GS is a list of one or more Tracking Areas (TA list),
- The UE performs Registration update due to mobility when moving outside its Registration Area i.e. TA list,
- The UE in idle state also performs periodic Registration update when the periodic Registration update timer expires.

Table 7.2 Registration Area representation for 3GPP Radio Accesses PS domain

Generic concept	5GS	EPS	GSM/WCDMA GPRS
Registration Area	List of Tracking Areas (TA list)	List of Tracking Areas (TA list)	Routing Area (RA)
Registration Area update procedure	Registration procedure	TA Update procedure	RA Update procedure

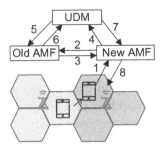

Fig. 7.1 Mobility Registration Update procedure.

A high-level outline of the Registration Update procedure due to mobility (i.e. with Registration type set to Mobility Registration Update – MRU) is shown in Fig. 7.1 and contains the following steps (see Selected Call flows in Chapter 15 for a more detailed call flow):

1. When the UE reselects a new cell and realizes that the broadcast TA ID is not in the list of TAs in the RA, the UE initiates an MRU procedure to the network, the NG-RAN routes the MRU to an AMF serving the new area.
2. Upon receipt of the MRU message from the UE, the AMF checks if a context for that particular UE is available; if not the AMF checks the UE's temporary identity (5G-GUTI) to determine which AMF keeps the UE context. Once this is determined the AMF asks the old AMF for the UE context.
3. The old AMF transfers the UE context to the new AMF.
4. Once the new AMF has received the old UE context, it informs the UDM that the UE context has now moved to a new AMF by registering itself to the UDM, subscribing to being notified when the UDM deregisters the AMF and as well to get the subscriber data for the UE from the UDM.

5–6. The UDM de-registers the UE context (for 3GPP Access Type) in the old AMF.

7. The UDM acknowledges the new AMF and inserts new subscriber data in the new AMF.
8. The new AMF informs the UE that the MRU was successful and the AMF supplies a new 5G-GUTI (where the GUAMI points back to the AMF).

The Registration procedure is also used to communicate information between the UE and the 5GC, which is handled by the AMF. For example, the Registration procedure is used by the UE to provide the UE capabilities, or UE settings such as MICO mode and to retrieve LADN Information. Consequently, if there are changes to such information e.g. the UE capabilities, then the UE initiates a Registration procedure (with the Registration Type set to Mobility Registration Update – MRU).

7.2.3 Cellular connected mode mobility

Great effort has been put into optimized connected mode mobility for cellular systems. The basic concept is somewhat similar across different technologies with some variations in the functional distribution between UE and networks. While in connected mode, the UE has a connected signaling connection and zero, one or more connected user plane resources, and data transmission may be ongoing. To limit interference and provide the UE with good data communication, the UE changes cells through handover when there is a cell that can provide better service than the cell that the UE is currently using. To save on complexity in the UE design and power, the systems are designed to ensure that the UE only needs to listen to a single gNB/ng-eNB at a time. In addition, for inter-RAT handover (e.g. NR to E-UTRAN HO) the UE only needs to have a single radio technology connected at a time. It may need to rapidly switch back and forth between the different technologies, but at any single point in time only one of the radio technologies is connected.

To determine when to perform handover, the UE measures the signal strength on neighboring cells regularly or when instructed to do so by the network. As the UE cannot send or receive data at the same time as it measures neighboring cells, it receives instructions from the network on suitable neighboring cells that are available and which ones the UE should measure. The network (NG-RAN) creates measurement time gaps where no data is sent or received to/from the UE. The measurement gaps are used by the UE to tune the receiver to other cells and measure the signal strength. If the signal strength is significantly stronger on another cell, the handover procedure may be initiated by the NG-RAN.

The NG-RAN can perform direct handover via the direct interface (known as Xn interface) between NG-RAN nodes. In the Xn-based HO procedure, the source NG-RAN node and the target NG-RAN node prepare and execute the HO procedure. At the end of the HO execution, the target NG-RAN node requests the AMF to switch the downlink data path from the source NG-RAN node to the target NG-RAN node. The AMF in turn requests each SMF to switch the data path toward the new NG-RAN node, and the SMF updates the UPF serving the PDU Session.

If downlink packets are sent before the UPF has switched the path toward the new NG-RAN node, the source NG-RAN node will forward the packet over the Xn interface.

If the Xn interface is not available between the NG-RAN nodes, the NG-RAN node can initiate a handover involving signaling via the 5GC network. This is called N2-based handover. The N2-based HO procedure sends the signal via the AMF and may include change of AMF and/or SMF/UPF. See Chapter 15 for a more detailed description of the Xn-based and N2-based handover procedures.

7.3 Reachability

7.3.1 Paging

Paging is used to search for Idle UEs and establish a signaling connection. Paging is, for example, triggered by downlink packets arriving to the UPF. When the UPF receives a downlink packet destined for an Idle UE, it does not have an NG-RAN User Plane tunnel address to which it can send the packet. The UPF instead buffers the packet and informs the SMF that a downlink packet has arrived. The SMF asks the AMF to setup User Plane resources for the PDU Session, and the AMF which knows in which RA the UE is located and sends a paging request to the NG-RAN within the RA. The NG-RAN calculates at which occasion the UE is to be paged using parts of the UE's 5G-S-TMSI (10 bits) as input, and then the NG-RAN pages the UE. Upon receipt of the paging message, the UE responds to the AMF and the User Plane resources are activated so that the downlink packet may be forwarded to the UE. See Chapter 15 for further details of the Network Triggered Service Request procedure which includes paging the UE.

7.3.2 Mobile Initiated Connection Only (MICO) mode

Mobile Initiated Connection Only (MICO) mode was introduced to allow paging resources to be saved for UEs that don't need to be available for Mobile Terminating communication. When the UE is in MICO mode, the AMF considers the UE as unreachable when the UE is in CM-IDLE state. The usage of MICO mode is not suitable for every type of UE and e.g. a UE initiating emergency service shall not indicate MICO preference during Registration procedure.

MICO mode is negotiated (and re-negotiated) during Registration procedures, i.e. the UE may indicate its preference for MICO mode and the AMF decides whether MICO mode can be enabled taking into account the UE's preference as well as other information such as the user's subscription and network policies. When the AMF indicates MICO mode to a UE, the RA is not constrained by paging area size. If the AMF serving area is the whole PLMN, the AMF may provide an "all PLMN" RA to the UE. In that case, re-registration to the same PLMN due to mobility does not apply.

7.3.3 UE's reachability and location

5GS also supports location services in a similar way to EPS (see Chapter 3), but 5GS also provides the possibility for any authorized NF (e.g. SMF, PCF or NEF) in the 5GC to subscribe to UE mobility related event reporting. The NF subscribing to a UE mobility related event can do so by providing the following information to the AMF:

- Whether UE location or the UE mobility in relation to an area of interest is to be reported

- In case an area of interest is requested, then the NF specifies the area as:
  - List of Tracking Areas, list of cells or list of NG-RAN nodes.
  - If the NF wants to get an LADN area, the NF (e.g. SMF) provides the LADN DNN to refer the LADN service area as the area of interest.
  - If a Presence Reporting Area is requested as area of interest, then the NF (e.g. SMF or PCF) may provide an identifier to refer to a predefined area configured in the AMF.
- Event Reporting Information: event reporting mode (e.g., periodic reporting), number of reports, maximum duration of reporting, event reporting condition (e.g. when the target UE moved into a specified area of interest).
- The notification address i.e. address of NF that the AMF is to provide the notifications which can be another NF than the NF subscribing to the event
- The target of event reporting that indicates a specific UE, a group of UE(s) or any UE (i.e. all UEs).

Depending on what information the NF subscribes to, the AMF may need to use NG-RAN to get accurate location information. In such cases the AMF keeps track of the subscribed mobility related events from each NF toward a UE or group of UEs. The AMF then uses NG-RAN location reporting for retrieving location information. NG-RAN location reporting provides identification at a cell level, but the UE is then required to be kept in CM-CONNECTED and RRC-CONNECTED state (e.g. if the UE is in RRC Inactive the NG-RAN may report the location as "unknown"). Typically, cell level accuracy is required for e.g. emergency services and lawful intercept, but can also be used via the AMF if requested by NFs in the 5GC. The AMF can request UE location as described in Table 7.3. When UE presence in an area of interest is requested, the AMF provides one or more areas (up to 64) to NG-RAN in the form of list of TAs, list of cell identities or a list of NG-RAN node identities.

Table 7.3 NG-RAN location reporting options

AMF options for location control	NG-RAN reporting
Direct	NG-RAN directly report the current location of the UE
Change of serving cell	NG-RAN provides UE location at each cell change
UE presence in the area of interest	NG-RAN reports the UE location and the UE's location related to the area of interest as in, out or unknown (NG-RAN does not know whether UE is in the area of interest or outside of it) for each area of interest
Stop or cancel reporting	NG-RAN stops/cancels the reporting

When the UE location is reported by the NG RAN, the UE location is provided as the cell identity, TA and optionally a time stamp of when the UE was last known to be in the reported location (e.g. when the UE is in RRC Inactive state). To ensure the 5GC is aware of the UE's location and state when RRC Inactive is used, the AMF may request the NG-RAN to report the UE's location and RRC state to the AMF when the UE enters or leaves RRC Inactive state.

It is also possible to request location reporting from an N3IWF serving the UE if the UE is in non-3GPP access. In such cases the location report consists of the UE's local IP address used to reach the N3IWF and a UDP or TCP source port number if a NAT is detected.

7.4 Additional MM related concepts

7.4.1 RRC Inactive

Rel-15 includes support for efficient communication with minimal signaling by using a concept called RRC Inactive which affects the UE, NG-RAN and 5GC.

RRC Inactive is a state where a UE remains in CM-CONNECTED state (i.e. at NAS level) and can move within an area configured by NG-RAN (the RAN Notification Area – RNA) without notifying the network. The RNA is a subset within the RA allocated by the AMF. When the UE is in RRC Inactive state the following applies:
- UE reachability is managed by the NG-RAN, with assistance information from 5GC;
- UE paging is managed by the NG-RAN;
- UE monitors for paging with part of the UE's 5GC (5G S-TMSI) and NG-RAN identifier.

In RRC Inactive, the last serving NG-RAN node keeps the UE context and the UE-associated NG (N2 and N3) connections with the serving AMF and UPF. Therefore, there is no need for the UE to signal toward the 5GC before sending User Plane data.

The NG-RAN controls when the UE is put into RRC Inactive state to save RRC resources, and the 5GC provides NG-RAN with RRC Inactive Assistance Information to enable NG-RAN to better judge whether to use the RRC Inactive state. The RRC Inactive Assistance Information is e.g. UE specific DRX values, the RA provided to the UE, the Periodic Registration Update timer, if MICO mode is enabled for the UE, and UE Identity Index Value (i.e. 10 bits of the UE's 5G-S-TMSI) allowing NG-RAN to calculate the UE's NG-RAN paging occasions. The information is provided by the AMF during N2 activation and AMF provides updated information e.g. if the AMF allocates a new RA to the UE.

In Fig. 7.2 the UE has been allocated a RA and within that a RAN Notification Area (dark gray cells). The UE is free to move within the RNA (dark gray cells) without notifying the network, while if the UE moves outside of the RNA while still in the RA (e.g. as shown into a different dark gray cell) the UE performs an RNA update to allow the

CN Registration Area

Fig. 7.2 Relation between Registration Area and RNA.

NG-RAN to update the UE context and the UE-associated connections. If the UE moves outside of the RA (light gray cells) then the UE also needs to notify the 5GC by a Registration procedure with the Registration Type set to Mobility Registration Update.

Despite the fact that the RRC Inactive state is within a CM-CONNECTED state, the UE performs many similar actions as when in RRC Idle state, i.e. the UE does:
- PLMN selection;
- Cell selection and reselection;
- Location registration and RNA update.

The PLMN selection, cell selection and reselection procedures, as well as location registration are done for both RRC Idle state and RRC Inactive state. However, the RNA update is only applicable for the RRC Inactive state, and when the UE selects a new PLMN, the UE transitions from RRC Inactive to RRC Idle state.

When the UE is in RRC Connected state the AMF is informed about the cells that the UE is connected to, but when the UE is in RRC Inactive the AMF is unaware of which cell the UE is connected to and whether the UE is in RRC Connected or RRC Inactive state. However, the AMF can subscribe to being notified about the UE transitions between RRC Connected and RRC Inactive state (both are CM-CONNECTED states), using an N2 Notification procedure (called RRC Inactive Transition Report Request). If the AMF has requested to being continuously notified about the state transitions the NG-RAN continues the reporting until the UE transitions to CM-IDLE or the AMF sends a cancel indication. The AMF can also subscribe to being informed about the UE location, see Section 7.3.3.

If the UE resumes the RRC connection in a different NG-RAN node within the same PLMN or equivalent PLMN, the UE AS context is retrieved from the last serving NG-RAN node and a procedure is triggered toward the 5GC to update the User Plane (N3 connections), see Fig. 7.3 for a description of the procedure and see 3GPP TS 23.502 for further details.

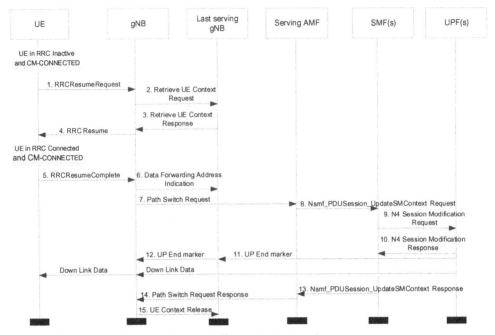

Fig. 7.3 UE resumes RRC connection in a different NG-RAN node.

If the UE resumes the RRC connection in a different NG-RAN node than from where the RRC connection was suspended, the steps to retrieve the UE AS context are:

1. The UE requests to Resume the RRC connection and provides the I-RNTI (Inactive Radio Network Temporary Identifier), allocated by the last serving gNB.
2. The gNB checks if it is possible to resolve the gNB identity as part of the I-RNTI, and in such case the gNB requests the UE AS context from the last serving gNB.
3. The last serving gNB provides the UE context.
4. The gNB resumes the suspended RRC connection.
5. The UE, now in RRC Connected state, confirms the successful completion of the RRC connection resumption. The message may include a NAS message and also a selected PLMN ID.
6. The gNB may provide forwarding addresses, to the last serving gNB, to prevent loss of DL user data buffered in the last serving gNB.
7. The gNB sends a Path Switch Request message to the AMF which is used to inform 5GC that the UE has moved to a new target cell and there is a need to "switch" the CP and UP of the PDU Sessions. The Path Switch Request message includes the NG-RAN DL UP Transport Layer Information and accepted QoS Flows for each PDU Session that is accepted and optionally a list of PDU Sessions that are failed to be setup.

8. For each PDU Session the AMF sends an Nsmf_PDUSession_UpdateSMContext request message to the related SMF, i.e. for accepted PDU Sessions the message includes the NG-RAN DL UP Transport Layer Information and accepted QoS Flows contains, and for failed PDU Sessions the reason for the failure.

9. For accepted PDU Sessions the SMF determines whether the existing UPF can continue to serve the UE and in such case the SMF sends a N4 Session Modification Request to the UPF as to provide the updated DL UP Transport Layer information. If the UPF cannot continue to serve the UE, e.g. the current UPF is not connected to the new gNB, the SMF initiates a procedure to either insert an intermediate UPF between the NG-RAN and the current UPF or to re-allocate a new intermediate UPF in case the current UPF was already an intermediate UPF (steps for such procedure is not included).

10. In case existing UPF can serve the UE, then the UPF replies to the SMF with an N4 Session Modification Response which may include UPF UL UP Transport Layer Information.

11. If indicated by the SMF to assist reordering in the gNB, the UPF sends UP End marker packet(s) on the UP tunnel toward the source gNB (i.e. the last serving gNB) after sending the last PDU on the old UP tunnel, and the UPF then starts sending DL Data packets to the Target gNB.

12. The last serving gNB forwards the UP End marker packet(s) to the gNB.

13. The SMF sends an Nsmf_PDUSession_UpdateSMContext response, including the UPF UL UP Transport Layer Information, to the AMF for PDU Sessions which have been switched successfully, or the message is sent without including any Transport Layer Information for the PDU Sessions for which User Plane resources are deactivated or PDU Session is to be released, and then the SMF releases the PDU Session(s) which is to be released using a separate release procedure.

14. The AMF waits for each SMF to reply and then aggregates the received information from each SMF and sends this aggregated information as part of N2 SM Information along with the Failed PDU Sessions in Path Switch Request Acknowledge to the gNB. If none of the requested PDU Sessions have been switched successfully, the AMF sends a Path Switch Request Failure message to the gNB

15. The gNB confirms the success of the procedure by releasing the UE context in the last serving gNB by using the UE Context Release message.

As described above, the RRC Inactive state is a connected mode state, at NAS level, that enables saving RRC resources while the UE applies similar logic as in idle mode. When the UE is in RRC Inactive state, the UE may resume the RRC Connection due to:
- Uplink data pending;
- Mobile initiated NAS signaling procedure;
- As a response to NG-RAN paging;
- Notifying the network that it has left the RAN Notification Area;
- Upon periodic RAN Notification Area Update timer expiration.

7.5 N2 management

In EPS, when a UE attaches to EPC and is assigned a 4G-GUTI, the 4G-GUTI is associated to a specific MME and if there is a need to move the UE to another MME the UE needs to be updated with a new 4G-GUTI. This may be a drawback e.g. if the UE is using some power saving mechanism or if a large amount of UEs are to be updated at the same time. With 5GS and N2 there is support for moving one or multiple UEs to another AMF without immediately requiring updating the UE with a new 5G-GUTI.

The 5G-AN and the AMF are connected via a Transport Network Layer that is used to transport the signaling of the NGAP messages between them. The transport protocol used is SCTP. The SCTP endpoints in the 5G-AN and the AMF sets up SCTP associations between them that are identified by the used transport addresses. An SCTP association is generically called a Transport Network Layer Association (TNLA).

The N2 (also called NG in RAN3 specifications e.g. 3GPP TS 38.413) reference point between the 5G-AN and the 5GC (AMF) supports different deployments of the AMFs e.g. either

(1) an AMF NF instance which is using virtualization techniques such that it can provide the services toward the 5G-AN in a distributed, redundant, stateless, and scalable manner and that it can provide the services from several locations, or

(2) an AMF Set which uses multiple AMF NF instances within the AMF Set and the multiple AMF Network Functions are used to enable the distributed, redundant, stateless, and scalable characteristics.

Typically, the former deployment option would require operations on N2 like add and remove TNLA, as well as release TNLA and rebinding of NGAP UE association to a new TNLA to the same AMF. The latter meanwhile would require the same but in addition requires operations to add and remove AMFs and to rebind NGAP UE associations to new AMFs within the AMF Set.

The N2 reference point supports a form of self-automated configuration. During this type of configuration the 5G-AN nodes and the AMFs exchange NGAP information of what each side supports e.g. the 5G-AN indicates supported TAs, while the AMF indicates supported PLMN IDs and served GUAMIs. The exchange is performed by the NG SETUP procedure and, if updates are required, the RAN or AMF CONFIGURATION UPDATE procedure. The AMF CONFIGURATION UPDATE procedure can also be used to manage the TNL associations used by the 5G-AN. These messages (see Chapter 14 for a complete list of messages) are examples of non-UE associated N2 messages as they relate to the whole NG interface instance between the 5G-AN node and the AMF utilizing a non-UE-associated signaling connection.

UE-associated services and messages are related to one UE. NGAP functions that provide these services are associated with a UE-associated signaling connection that is maintained for the UE. The Fig. 7.4 shows NG-AP instances in the 5G-AN and in the AMFs,

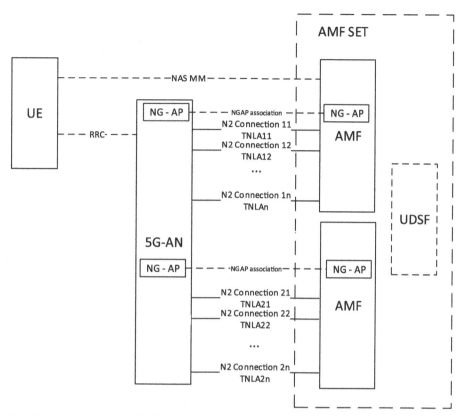

Fig. 7.4 N2 reference point with TNLA as transport.

and these may be either non-UE associated, or UE associated. The NG-AP communication uses the N2 Connection with a TNLA as transport (SCTP is used as transport protocol, see Chapter 14).

The N2 reference point supports multiple TNL associations per AMF (up to 32). TNL associations can be added or removed based on the need and weight factors that can be used to steer to the TNL associations the 5G-AN will use for NGAP communication to the AMF (i.e. achieving load balancing and re-balancing of TNL associations between the 5G-AN and the AMF).

For a specific UE in CM-CONNECTED state the 5G-AN node (e.g. gNB) maintains the same NGAP UE-TNLA-binding (i.e. use the same TNL association and same NGAP association for the UE) unless explicitly changed or released by the AMF.

The AMF may change the TNL association used for a UE in CM-CONNECTED state at any time, e.g. by responding to an N2 message from a 5G-AN node using a different TNL association. This can be beneficial when the UE context is moved to another AMF within the AMF Set.

The AMF may also command the 5G-AN node to release the TNL association used for a UE in CM-CONNECTED state while maintaining user plane connectivity (N3) for the UE at any time. This can be useful if there is a need to change the AMF within the AMF Set (e.g. when the AMF is to be taken out of service) at a later date.

7.5.1 AMF management

The 5GC, including N2, supports the possibility to add and remove AMFs from AMF Sets. Within 5GC, the NRF is updated (and DNS system for interworking with EPS) with new NFs when they are added, and the AMF's NF profile includes which GUAMI(s) the AMF handles. For a GUAMI there may also be one or more backup AMF registered in the NRF (e.g. to be used in case of failure or planned removal of an AMF).

A planned removal of an AMF can be done either through the AMF storing the registered UEs' contexts in a UDSF (Unstructured Data Storage Function), or with the AMF deregistering itself from the NRF, in which case the AMF notifies the 5G-AN that the AMF will be unavailable for processing transactions for the GUAMI(s) configured on this AMF. Additionally, the AMF can initially decrease the load by changing the weight factor for the AMF toward the 5G-AN, e.g. setting it to zero, causing the 5G-AN to select other AMFs within the AMF Set for new UEs entering the area.

If the AMF indicates it is unavailable for processing transactions, the 5G-AN then selects a different AMF within the same AMF Set. If that is not possible, the 5G-AN selects a new AMF from another AMF Set. It is also possible for the AMF to control the selection of new AMF within the AMF Set by the AMF indicates a rebinding of the NGAP UE associations to an available TNLA on a different AMF in the same AMF Set.

Within 5GC, the NRF notifies CP NFs – that have subscribed – that the AMF identified by GUAMI(s) will be unavailable for processing transactions. The CP NF then selects another AMF within the same AMF Set, and the new AMF retrieves the UE context from the UDSF and the new AMF updates the UE with a new 5G-GUTI and peer CP NFs with the new AMF address information.

If there is no UDSF deployed in the AMF Set, then a planned removal of an AMF is done in a similar way as with a UDSF, but with the difference that the AMF can forward registered UE contexts to target AMF(s) within the same AMF set. A backup AMF info can be sent to other NFs during first interaction.

An AMF removal may also be unplanned i.e. due to some failure. To automatically recover to another AMF, then UE contexts can be stored in the UDSF, or per GUAMI granularity in other AMFs (serving as backup AMF for the indicated GUAMI).

7.5.2 5GC assistance for RAN optimizations

As the UE context information is not kept in the NG-RAN when the UE transition to RRC-IDLE, it may be hard for the NG-RAN to optimize the logic related to the UE as

UE specific behavior is unknown unless the UE has been in RRC-CONNECTED state for some time. There are NG-RAN specific means to retrieve such UE information e.g. UE history information can be transferred between NG-RAN nodes. To further facilitate an optimized decision in NG-RAN e.g. for UE RRC state transition, CM state transition decision and optimized NG-RAN strategy for RRC-INACTIVE state, the AMF may provide 5GC assistance information to NG-RAN.

5GC has a better method to store UE related information for a longer time and a means to retrieve information from external entities through external interfaces. When calculated by the 5GC (AMF) the algorithms used and related criteria, and the decision when it is considered suitable and stable to send to the NG-RAN are vendor specific. Therefore, along with the assistance information sent to NG-RAN, it often is accompanied with the information whether it is derived by statistics or retrieved via subscription information (e.g. set by agreements or via an API).

5GC assistance information is divided into 3 parts:
- Core Network assisted RAN parameters tuning;
- Core Network assisted RAN paging information;
- RRC Inactive Assistance Information.

Core Network assisted RAN parameters tuning provides the NG-RAN with a way to understand UE behavior so as to optimize NG-RAN logic e.g. how long to keep the UE in specific states. Besides the content listed in Table 7.4, the 5GC also provides the source of the information e.g. if it is subscription information or derived based on statistics.

Core Network assisted RAN paging information complements the Paging Policy Information (PPI) and QoS information associated to the QoS Flows (see Chapter 9 for more detail on QoS) as well as assisting the NG-RAN to formulate a NG-RAN paging policy. Core Network assisted RAN paging information also contains Paging Priority that provides NG-RAN with a way to understand how important the downlink signaling is. If RRC Inactive is supported and the UE is not required to be in RRC-CONNECTED state e.g. for tracking purposes, the AMF provides RRC Inactive Assistance Information to the NG-RAN, to assist the NG-RAN in managing the usage of RRC Inactive state. Table 7.4 provides an overview of the various CN assistance information, but as part of the N2 messages sent by the 5GC to the NG-RAN further information is available that can also be used by the NG-RAN to optimize its behavior i.e. an interested reader is referred to 3GPP TS 38.413 for a complete set of information.

7.5.3 Service Area and Mobility Restrictions

Mobility Restrictions enables the network, mainly via subscriptions, to control the Mobility Management of the UE as well as how the UE accesses the network. Similar

Table 7.4 Overview of CN assistance information sent to NG-RAN

Assistance information	Information	Values	Description
Core Network assisted RAN parameters tuning	Expected activity period	INTEGER (seconds) (1…30|40|50|60|80|100|120|150|180|181, …)	When set to 181, then expected activity time longer than 181.
	Expected Idle Period	INTEGER (seconds) (1…30|40|50|60|80|100|120|150|180|181, …)	When set to 181, then expected idle time longer than 181.
	Expected HO Interval	ENUMERATED (sec15, sec30, sec60, sec90, sec120, sec180, long-time, …)	The expected time interval between inter NG-RAN node handovers. If "long-time" is included, the interval between inter NG-RAN node handovers is expected to be longer than 180 seconds.
	Expected UE Mobility	ENUMERATED (stationary, mobile, …)	Indicates whether the UE is expected to be stationary or mobile
	Expected UE Moving Trajectory	List of Cell Identities and Time Stayed in Cell (0…4095 seconds)	Includes list of visited and non-visited cells. Time Stayed in Cell included for visited cells (if longer than 4095 seconds, 4095 is set)
RRC Inactive Assistance Information	UE Identity Index Value	10 bits of UE's 5G-S-TMSI	Used by the NG-RAN node to calculate the Paging Frame
	UE Specific DRX	ENUMERATED (32, 64, 128, 256, …)	Input to derive DRX as specified in 3GPP TS 38.304
	Periodic Registration Update Timer	BIT STRING (SIZE(8))	Bits used to derive a timer value as defined in 3GPP TS 38.413
	MICO Mode Indication	ENUMERATED (true, …)	Whether the UE is configured with MICO mode
	TAI List for RRC Inactive	List of TAIs	List of TAIs corresponding to the RA

Assistance Data for Paging	Recommended Cells for Paging	List of Cell Identities and Time Stayed in Cell (0…4095 seconds)	Includes list of visited and non-visited cells. Time Stayed in Cell included for visited cells (if longer than 4095 seconds, 4095 is set)
	Paging Attempt Count	INTEGER (1…16, …)	Increased by one at each new paging attempt
	Intended Number of Paging Attempts	INTEGER (1…16, …)	
	Next Paging Area Scope	ENUMERATED (same, changed, …)	Indicates whether the paging area scope will change or not at next paging attempt.
	Paging Priority	ENUMERATED (8 values)	Can be used for Priority handling of paging in times of congestion at NG-RAN

logic as used in EPS is applied in 5GS, but with some new functionality added as well. The 5GS supports the following:

- RAT restriction:
  - Defines the 3GPP Radio Access Technology(ies) a UE is not allowed to access in a PLMN and may be provided by the 5GC to the NG-RAN as part of the Mobility Restrictions. The RAT restriction is enforced by the NG-RAN at connected mode mobility.
- Forbidden Area:
  - A Forbidden Area is an area in which the UE is not permitted to initiate any communication with the network for the PLMN.
- Core Network type restriction:
  - Defines whether UE is allowed to access to 5GC, EPC or both for the PLMN.
- Service Area Restriction:
  - Defines areas controlling whether the UE is allowed to initiate communication for services as follows:
    - Allowed Area: In an Allowed Area, the UE is permitted to initiate communication with the network as allowed by the subscription.
    - Non-Allowed Area: In a Non-Allowed Area a UE is "service area restricted" meaning that neither the UE nor the network is allowed to initiate signaling to obtain user services (both in CM-IDLE and in CM-CONNECTED states). The UE performs mobility related signaling as usual, e.g. mobility Registration updates when moving out of the RA. The UE in a Non-Allowed Area replies to 5GC initiated messages, which makes it possible to inform the UE that e.g. the area is now Allowed.

RAT, Forbidden Area and Core Network type restrictions works similarly as they do in EPS, but Service Area Restriction is a new concept. As mentioned previously, it was developed to better support use cases that would not require full mobility support. One use case is support of Fixed Wireless Access subscriptions e.g. the user is given a modem or device that supports 3GPP access technologies, but the subscription restricts the usage to a certain area e.g. the users home. The UE is provided with a list of TAs that are indicated as an Allowed Area or a Non-Allowed Area. To dynamically shape the area of the user's home (to become the Allowed Area), the subscription may also contain a number indicating maximum number of allowed TAs. When the UE registers to the network, the TAs to become the Allowed Area is derived while the UE moves around. If the UE de-registers and re-registers, then the Allowed Area is re-calculated. The Service Area Restriction is set in the subscription, but it is also sent to the PCF allowing the PCF to decrease a Non-Allowed Area or increase an Allowed Area e.g. when the user agreed to some sponsoring of access to the network in an area.

As an example of how a subscription for an Allowed Area with a maximum number of allowed TAs can be used is shown in Fig. 7.5, which shows a building which crosses TA1,

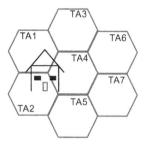

Fig. 7.5 Building crossing Tracking Areas to be used for an Allowed Area.

TA2 and TA4. When the user agrees to a subscription limited to the area of its home, the operator determines a TA to use for the subscription (e.g. TA2 which is determined based on the user's home address) and may in addition add a maximum number of allowed TAs, e.g. set to three, as to ensure that the complete area of the building is covered. When the user registers at home e.g. in TA1, the UE may get a RA of TA1 and an Allowed Area of TA1 and TA2 as a reply to the Registration. The network keeps an internal count of the number of TAs included in the Allowed Area compared to the maximum number of allowed TAs that now include TA1 and TA2 (i.e. two TAs). When the user moves to TA2 and TA4 the UE may get a RA of TA1, TA2, TA4 and Allowed Area of the same TAs. When the user moves outside its home e.g. to TA5, then the UE gets a new RA set to TA5 and the information that TA5 is a Non-Allowed Area. The user cannot use its UE for normal services therefore. When the user moves to their home again, then the UE is likely to get a RA of TA1, TA2 and TA4 and the information that TA1, TA2 and TA4 are parts of the Allowed Area. The user can now use their UE again.

The size of a TA is dependent on many aspects e.g. number of cells used and if higher radio frequencies are used. While the example of a building crossing three TAs might not be that common as a TA is usually larger than one building, the usage of high radio frequencies and the fact that 5GS uses three octets for identifying TAs compared to two octets for EPS enables 5GS a larger possibility to deploy smaller TAs.

7.6 Control of overload

5GS supports the ability to control the amount of load UEs produce toward the 5GS through different mechanisms.

Mechanisms for 5GC to balance load across NFs and also to scale the amount of resources consumed for the NFs are often enough to handle normal fluctuations of load impacting the 5GC. In order to protect itself from overload situations, the 5GC supports a number of mechanisms including instructing UEs to back-off through NAS back-off timers (for Mobility Management as well as Session Management messages) such that the UE does not re-attempt to connect while the back-off timer is running. 5GS also supports the possibility to indicate to the NG-RAN that the load toward the AMF needs to be reduced using different criteria in an NGAP Overload Start message sent to the NG-RAN. This is similar to what is specified for EPC.

The NG-RAN supports possibilities to steer UEs using Radio Resource Management (RRM) techniques so that NG-RAN resources can be efficiently utilized. 5GC can also influence the RRM strategies of the NG-RAN by providing a RAT/Frequency Selection Priority (RFSP) to the NG-RAN per UE, which can be used to tell the NG-RAN how to prioritize the NG-RAN resources for the specific UE in question. In addition, NG-RAN supports different techniques to handle UE traffic at overload situations as illustrated in Fig. 7.6.

Different methods can be used to handle possible bottlenecks in the NG-RAN Control Plane which also protects the 5GC. The mechanism used often depends on the load in the system. These are summarized below.

Congestion in control channel resources: 5QI-based scheduling controls cases when e.g. the number of users awaiting scheduling exceeds the number of users that can be admitted such that the random access procedure fails. The random access procedure is a lower layer procedure used when the UE wants to initiate communication e.g. the UE gets synchronized with the network from a timing perspective, see 3GPP TS 38.321 for a description of the procedure.

Congestion in random access channel (RACH) resources: random access back-off. This pushes some UEs into a longer back-off. This is when there are so many access attempts on the RACH that the UE provided preambles cannot be detected anymore.

Release/reject UE RRC connection: If there are not enough resources to process RRC connection requests, Releasing RRC connection or rejecting RRC connection

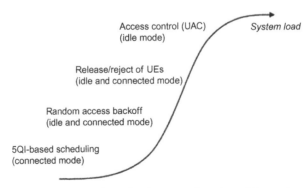

Fig. 7.6 Access and congestion control mechanisms as a function of the system load.

attempts can be used. The RRC connection release is executed to alleviate congestion in the NG-RAN by releasing already established RRC connections. RRC rejection may be applied on a UE attempting initial access, after the UE has successfully completed the random access procedure and has sent an RRC connection request. Note, however, that in 5GS the RRC connection request does not provide enough information about the UE's GUAMI such that the AMF serving the UE can be identified, as in 5GS the identity been increased from 40 bits to 48 bits. This means that if the NG-RAN attempts to control overload of the AMF (e.g. due to an NGAP Overload Start message been received from the AMF) then an RRC connection release will be performed as the identity of the AMF is not known until the RRC connection is regarded as established.

Severe and uncontrollable congestion: in extreme cases, e.g. if random access overload or Control Plane overload in the NG-RAN are at a level that cannot be reduced by the mechanisms outlined above, then it is possible to prevent UEs from even attempting to establish a connection. This is accomplished using a mechanism called Unified Access Control (UAC), which is an evolution of the various barring mechanisms in EPS, into a single and more flexible solution. The UAC is the most drastic measure in the congestion-prevention toolset of the NG-RAN, and therefore UAC should be invoked only if all other means are exhausted as the UAC affects a wider range of UEs e.g. it cannot normally be used to reduce the load toward a specific AMF.

7.6.1 Unified Access Control

EPS supports multiple variants of access barring mechanisms as they were developed in different releases to address different needs for congestion control. The 5GS supports one mechanism called Unified Access Control (UAC) which is extensible, flexible (e.g. each operator can define their own category when to apply access control) and supports a variety of scenarios. The UAC affects UEs in all RRC states i.e. RRC_IDLE, RRC_INACTIVE and RRC_CONNECTED state. In the case of multiple 5GC sharing the same NG-RAN, the NG-RAN provides UAC for each PLMN individually.

When the UE wants to access the 5GS, the UE first performs access control checks to determine if access is allowed. This is done by the UE NAS layer performing a mapping of the request to one or more Access Identities and one Access Category, NAS layer informs the lower layers (AS layer) of the Access Identities and the Access Category. The AS layer will then perform access barring checks for that request based on the determined Access Identities and Access Category. If the AS layer indicates that the access attempt is allowed, the NAS initiates the procedure, but if the AS indicates that the access attempt is barred, the NAS does not initiate the procedure. The AS layer runs barring timers, on a per Access Category basis. At expiry of the barring timers, the AS layer indicates to the NAS layer of the alleviation of access barring on a per Access Category basis.

The following are the defined Access Identities:

0. UE is not configured with any parameters from this list.

1. UE is configured for Multimedia Priority Service (MPS).

2. UE is configured for Mission Critical Service (MCS).

3–10. Reserved for future use

11. Access Class 11 (i.e. For PLMN Use) is configured in the UE.

12. Access Class 12 (i.e. Security Services) is configured in the UE.

13. Access Class 13 (i.e. Public Utilities (e.g. water/gas suppliers)) is configured in the UE.

14. Access Class 14 (i.e. Emergency Services) is configured in the UE.

15. Access Class 15 (i.e. PLMN Staff) is configured in the UE.

The Access identities 11 and 15 are valid in HPLMN or the Equivalent HPLMN. Access Identities 12, 13 and 14 are valid in HPLMN and visited PLMNs of the UE home country only.

The following is a summary of the defined Access Categories:

0. Response to paging or NOTIFICATION over non-3GPP access, or for LPP message

1. Access attempt for delay tolerant service

2. UE is attempting access for an emergency session

3. Access attempt is for MO signaling

4. MO MMTel voice call

5. MO MMTel video call

6. MO SMS over NAS or MO SMSoIP

7. Access attempt is for MO data

32–63. Access attempt for operator-defined access category

Operator-defined Access Categories can be signaled to the UE using NAS signaling and are defined as:

(a) a precedence value that indicates in which order the UE shall evaluate the operator-defined category definition for a match;

(b) an operator-defined access category number i.e. access category number in the 32–63 range that uniquely identifies the access category in the PLMN in which the access categories are being sent to the UE;

(c) one or more access category criteria type and associated access category criteria type values. The access category criteria type can be set to one of the following:

    (1) DNN;

    (2) 5QI (3GPP has not yet decided if 5QI is to be used as criteria);

    (3) OS Identity + OS Application Identity triggering the access attempt; or

    (4) S-NSSAI (see Chapter 11).

(d) optionally, a standardized access category can be used in combination with the Access Identities of the UE to determine the RRC establishment cause.

7.7 Non-3GPP aspects

The supported non-3GPP Access Type in Rel-15 is the Untrusted Non-3GPP access as described in Chapter 3. The UE uses Mobility Management procedures for both the 3GPP access and the non-3GPP access, but there are some differences.

To setup a NAS signaling connection and transition to CM-CONNECTED state, the UE establishes an NWu connection over an Untrusted Non-3GPP access to an N3IWF.

The UE can be connected to the 5GS over both 3GPP and Non-3GPP access at the same time. In such a case, if the UE is registered to the same PLMN for both Access Types then the UE is normally registered to one AMF. The UE may be temporarily connected to two different AMFs for the same PLMN after a mobility from EPS while the UE has PDU Sessions associated with non-3GPP access. If the UE is registered to different PLMNs for the two Access Types, then the UE is registered to two different AMFs, one AMF per serving PLMN.

The UE and the AMF performs separate Registration procedures for each Access Type, and there are separate Registration and CM-IDLE/CONNECTED states for each access, but to ensure that the same AMF is selected for both Access Types, when connected to the same PLMN, the AMF assigns a 5G-GUTI to the UE that is common for both Access Types.

There is one operator-specific (i.e. decided per PLMN) TA Identity used for the Untrusted non-3GPP access that means the UE's RA for the non-3GPP access is one TA and therefore the UE does not perform any Mobility Registration Update when the UE change the non-3GPP access point of attachment e.g. change of WLAN Access Point.

The UE does not use Periodic Registration Updates when using non-3GPP access, i.e. the Periodic Registration Update timer only applies to the UE registered to the 5GS over 3GPP access. When the UE enters the CM-IDLE state for the non-3GPP access (i.e. when the N1 NAS signaling connection over non-3GPP access is released), the UE starts the non-3GPP Deregistration timer (indicating when the UE is considered implicitly Deregistered for the non-3GPP access) and the AMF starts the non-3GPP Implicit Deregistration timer. The AMF provides the value of the non-3GPP Deregistration timer to the UE during a Registration procedure, and the value of the non-3GPP Implicit Deregistration timer is with a value longer than the UE's non-3GPP Deregistration timer. This means that the UE needs to enter CM-CONNECTED state before the timers expires e.g. by re-establishing the User Plane resources for a PDU Session, as otherwise the UE will be de-registered.

There is no support for paging over Untrusted Non-3GPP access, which means there is also no need for MICO mode (see Section 7.3.2). However, a UE that is in CM-IDLE and registered for the non-3GPP access and registered over 3GPP access in the same

PLMN, can be reached via the 3GPP access for procedures related to the non-3GPP access, or for which the last communication was using non-3GPP access. In such case, the AMF provides an indication that the procedure is related to non-3GPP access and the UE replies over 3GPP access. In a similar way, the UE can be reached over non-3GPP access for a PDU Session associated in the SMF (i.e. last routed) to the 3GPP access, while in such case the UE replies over 3GPP access.

In Release-16 support for Trusted non-3GPP access and Wireline access is added. In general they follow the same principles for non-3GPP access described above, but there are some differences, e.g. when it comes to the Access Network specific handling. See Chapter 16 for additional information on Wireline access connected to 5GC.

7.8 Interworking with EPC

7.8.1 General

As described in Chapter 3, interworking with EPC is foreseen to be used for some time and dependent on frequency allocation to NR and time it takes to build out NR coverage. Chapter 3 gives an overview of the reasons for why there is a need for interworking with EPC, a high-level architecture and the high-level principles and options for the interworking. In this section we will go into more details and describe the interworking aspects related to mobility.

It is worthwhile to provide some more details to the architecture diagram already shown in Chapter 3 in order to highlight that the SMF and UPF need to support EPC PGW logic and functionality across the S5-C and S5-U interfaces. Therefore, they are referred to as PGW-C+SMF and UPF+PGW-U respectively, see Fig. 7.7. To ensure successful interworking with the appropriate EPS functionality, only one PGW-C+SMF is allocated per APN for a given UE, and that is enforced e.g. by the HSS+UDM providing one PGW-C+SMF FQDN per APN to the MME.

Interworking with EPC while using non-3GPP access in 5GS is also applicable and in such cases, NR would be replaced with N3IWF and access specific entities underneath e.g. Wi-Fi Access Point. Furthermore, it is also possible to interwork between EPC connected to non-3GPP while using 3GPP access toward the 5GC, and in such case the MME and SGW would be replaced with an ePDG and the HSS with a 3GPP AAA server (while possible, those options are not further described, an interested reader is encouraged to read the 3GPP specifications e.g. 3GPP TS 23.501).

For interworking to be possible it is required that the UE supports both EPC NAS procedures as well as 5GC NAS procedures. If this is not the case, then the UE will be directed toward the Core Network that the UE supports, and no interworking will be applicable.

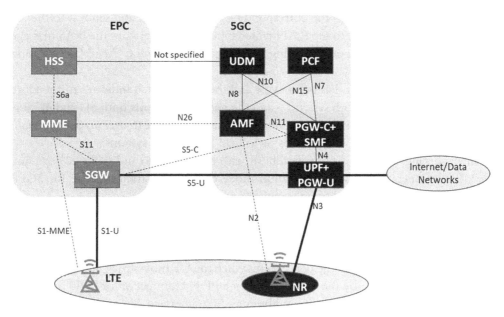

Fig. 7.7 Detailed architecture for interworking between EPC and 5GC.

7.8.2 Interworking with EPC using 3GPP access

7.8.2.1 General

When a UE is selecting networks – or PLMNs – or camping on a cell that is connected to both EPC and 5GC (i.e. the cell broadcast that it is connected to both EPC and 5GC), the UE needs to select which Core Network to register with. That decision can be operator controlled or user controlled. The operator can control the decision e.g. by influencing the network selection using an operator controlled prioritized list in the USIM by which the operator is able to steer the network selection including which Access Technology e.g. NG-RAN or E-UTRAN to prioritize, or the operator can set the subscription to only allow either EPC, 5GC or both, or the operator can control RRM procedures per UE as to prioritize certain radio access to be used. The user can control the decision by making manual selection of network (which creates a user controlled prioritized list of networks including Access Technology), or the user can indirectly influence the selection by requiring to use a certain service that is not (yet) supported by the 5G System which causes the UE to disable the related radio capabilities that allow the UE to access the 5G System such that the UE instead selects e.g. a 4G system. As different NAS protocols are used toward 5GC and EPC, the NAS layer in the UE indicates to the AS layer whether a NAS signaling connection is to be initiated toward the 5GC or the EPC and the NAS layer issues a NAS message to the corresponding Core Network and sends it to the AS layer that indicates in RRC to the RAN which type of Core Network the NAS message is for. The RAN selects a corresponding Core Network entity i.e. AMF for 5GC and MME for EPC.

Once an initial selection has been made and the UE – which indicates to the Core Network that it supports both systems – and the network supports both 5G and 4G systems the system to be used at a certain point may change e.g. due to user invoking certain services or due to radio coverage issues or to load balance the systems.

Interworking with EPC is specified both with usage of N26 and without N26, and the UE may operate in single-registration mode or dual-registration mode for 3GPP access (when N26 is used only single-registration mode applies) i.e.:

In **single-registration mode**; the UE has one active Mobility Management state for 3GPP access toward the Core Network and is either in 5GC NAS mode or in EPC NAS mode dependent on which Core Network the UE is connected to; the UE context information is transferred between the two systems when the UE moves back and forth, which is either done via N26 or by the UE moving each PDN Connection or PDU Session to the other system when interworking without an N26 interface. To enable the RAN in the target system to select the same Core Network entity which the UE was registered to in the source system (if it is available) and to enable the retrieval of UE context over N26, the UE maps the 4G-GUTI to 5G GUTI during mobility between EPC and 5GC and vice versa as described in Fig. 7.8. For handling of security contexts, Chapter 8 describes how to enable an efficient re-use of a previously established 5G security context when returning to 5GC.

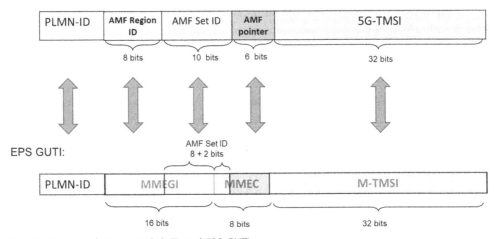

Fig. 7.8 Mapping between 5G-GUTI and EPS GUTI.

In **dual-registration mode**; the UE maintains independent Mobility Management states for 3GPP access toward the 5GC and EPC using separate RRC connections. In this mode, UE maintains 5G-GUTI and 4G-GUTI independently, and the UE may be registered to 5GC only, EPC only, or to both 5GC and EPC.

It should be noted that N26 is used only for 3GPP access. Mobility of PDU Sessions between 3GPP access and non-3GPP access in the EPC and 5GC systems are driven by the UE and is supported without N26. The rest of the description in this section focuses on interworking for 3GPP accesses.

When the UE moves from one system to the other, the UE provides its UE temporary identity in the format of the target system. If the UE has previously been registered/attached to another system or has not registered/attached at all in the target system and does not hold any UE temporary identity of the target system, the UE provides a mapped UE temporary identity as described in Fig. 7.9.

When the UE initially Attaches to EPS the UE uses its IMSI as UE identity toward both E-UTRAN (in RRC) and EPC (in NAS). However, in 5GS, the UE uses a SUCI toward 5GC (in NAS) which conceals the UE's Identity (see Chapter 8 for more information about SUCI). In both cases, there is no stored UE context in the network i.e. the network creates the UE context.

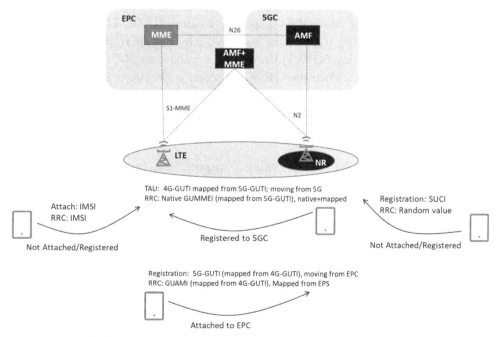

Fig. 7.9 UE provided UE identity at NAS and RRC.

When the UE has been registered in one system and moves to the other, and the UE has no native UE Identity for the target system, the UE maps the UE temporary identity of the source system to the format of target system which enables the RAN to select a Core Network which served the UE last time, if available.

When the UE moves from 5GS to EPS, the UE sets in RRC the GUMMEI (i.e. MCC, MNC, MME Group ID, MME Code) as a native GUMMEI. Otherwise, any non 5G-upgraded eNB would have treated a "mapped GUMMEI" as identifying an SGSN. The UE indicates that the GUMMEI is mapped from 5G-GUTI to enable a 5G-upgraded eNB to differentiate MME addresses from an AMF address. In the TAU message the UE includes the 4G-GUTI mapped from the 5G-GUTI and indicates that the UE is moving from 5G, the MME then retrieves the UE context from 5GC via N26.

When the UE moves from EPS to 5GS, the UE sets in RRC the GUAMI (i.e. MCC, MNC, AMF Region ID, AMF Set ID and AMF Pointer) mapped from the 4G-GUTI, and indicates it as mapped from EPS. This enables the gNB to select the same Core Network Entity e.g. AMF + MME, if available. In the Registration message the UE includes the 5G-GUTI mapped from the 4G-GUTI and indicates that the UE is moving from EPC. Also, if the UE has a native 5G-GUTI, the UE includes it as an "additional GUTI" and in this case the AMF tries to retrieve the UE context from old AMF or from UDSF. Otherwise, the AMF retrieves the UE context from MME using the 5G-GUTI mapped from the 4G-GUTI.

The scenario above for which the UE also has a native 5G-GUTI is that the UE is registered toward 5GC using 3GPP access, and the UE in addition registers toward 5GC over non-3GPP access (using N3IWF) i.e. the UE is using both 3GPP access and non-3GPP access toward 5GC. Then the UE 3GPP access connectivity is moved to EPC while the non-3GPP access connectivity is kept toward 5GC. After that, the 3GPP access connectivity is moved back from EPC to 5GC to which the UE is registered already over non-3GPP access i.e. the UE has a native 5G-GUTI already and consequently indicates it as an "additional GUTI".

As described, when the UE provides a mapped UE temporary identity the E-UTRAN or NG-RAN, can select the same Core Network entity as the UE was registered/attached to before e.g. combined AMF + MME in case such entity is available. The UE temporary identity provided in the NAS message is used by the MME or AMF to retrieve the UE context from the old entity that the UE was registered in before (e.g. over N26 or internal to the entity if a combined AMF + MME was used).

The selection by the UE of the registration mode to use, i.e. single- or dual-registration mode, is decided based on the steps below:

1. When registering to the network i.e. either EPC or 5GC (including Initial Registration and Mobility Registration Update toward 5GC and Attach and TA Update toward EPC) the UE indicates that it supports the mode of the "other" system i.e.

toward 5GC the UE indicates that it supports "S1 mode" i.e. that the UE supports EPC procedures, and toward EPC the UE indicates that it supports "N1 mode" i.e. that the UE supports 5GC procedures.

2. A network that supports interworking indicates to the UE whether the network supports "Interworking without N26"

3. The UE then selects the registration mode as follows:
   a. if the network indicated that it does *not support* interworking without N26 then the UE operate in single-registration mode, and
   b. if the network indicated that it does *support* interworking without N26, the UE decides whether to operate in single- or dual-registration mode based on UE implementation (UE support for single-registration mode is mandatory while dual-registration mode is optional).

There is no support for interworking between 5GS and GERAN/UTRAN, this means that e.g. IP address preservation for IP PDU Sessions cannot be ensured on subsequent mobility from or to GERAN/UTRAN for a UE that has been registered in 5GS or EPS.

The high-level principles specifically for interworking with N26 and without N26 are described in following sections.

7.8.2.2 Interworking using the N26 interface

When N26 interface is used for interworking procedures, the UE operates in single-registration mode, and the UE context information is exchanged over N26 between AMF and MME. The AMF and MME keeps one MM state (for 3GPP access) for the UE, i.e. either in the AMF or MME (and the MME or AMF is registering in the HSS+UDM when it holds the UE context). The interworking procedures provide IP address continuity at inter-system mobility between 5GS and EPS and are required to enable seamless session continuity (e.g. for voice services). The PGW-C+SMF keeps a mapping between PDN Connection and PDU Session related parameters e.g. PDN Type/PDU Session Type, DNN/APN, APN-AMBR/Session AMBR and QoS parameter mapping.

To ensure interworking is possible from 5GS to EPS, the AMF assigns an EPS Bearer Identity (EBI) to QoS Flow(s) of a PDU Session already while the UE is using 5GC (EPS Bearers are used for QoS differentiation, see Chapter 9, and at least one EBI is required for the default EPS Bearer of each PDN Connection in EPS). The AMF keeps track of assigned EBI, ARP pairs to the corresponding PDU Session ID, and the SMF address. The AMF updates the information, when a PDU Session is established, modified (e.g. new QoS Flows are added), released or when PDU Sessions are moved to or from using non-3GPP access. Fig. 7.10 shows at a high level the interactions.

When N26 is supported, the AMF in conjunction with PGW-C+SMF decides, based on operator policies e.g. DNN is equal to IMS, that QoS Flow(s) of a PDU Session requires to be enabled for interworking with EPS and initiates a request (1) toward the AMF for

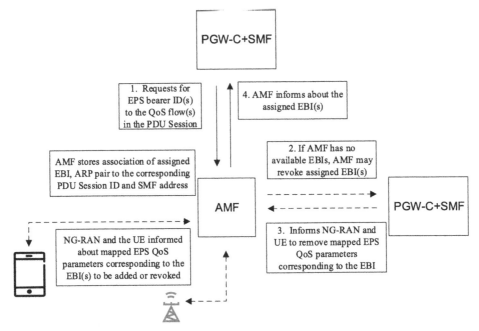

Fig. 7.10 EBI allocation and revocation.

getting EBI(s) assigned to one or more QoS Flows. The AMF keeps track of EBI(s) assigned for the UE and decides whether to accept the request for EBI(s) (4). Due to restrictions in EPS e.g. number of EPS Bearers supported or that not more than one PGW-C + SMF can serve PDN Connections toward the same APN, the AMF may need to revoke (2) previously assigned EBI(s) e.g. in case the new requested QoS Flows have higher ARP priority compared to the QoS Flows that already been assigned EBIs. In such case, the PGW-C + SMF that gets EBI(s) revoked will need to inform the NG-RAN and the UE about the removal of the mapped EPS QoS parameters corresponding to the revoked EBI (3). Once a QoS Flow has been assigned an EBI the SMF informs the NG-RAN and UE about the added mapped EPS QoS parameters corresponding to the EBI.

See Chapter 15 for procedure description of mobility between 5GS and EPS.

7.8.2.3 Interworking without an N26 interface
When interworking without N26 interface it is not possible to retrieve the UE context from the last serving MME/AMF and therefore the HSS + UDM is used for some additional storage and the principle is that the UE makes Attach or Initial Registration and the MME and AMF indicates to the HSS + UDM to not cancel the AMF or MME registered via the other system and thereby the HSS + UDM maintains both an MME and an AMF until the UE successfully transfers all the PDU Sessions/PDN connections. The PGW-C + SMF also makes use of the HSS + UDM for storing its own address/FQDN and

corresponding APN/DNN to support IP address preservation as it enables the MME and AMF to select the same PGW-C + SMF for a PDN Connection/PDU Session that has been moved from the other system.

The AMF indicates to the UE, during Initial Registration, that interworking without N26 is supported and the MME may provide such indication to the UE during the Attach procedure. The UE, operating in dual-registration mode, can use the indication to register as early as possible in the target system to minimize any service interruptions and use the Attach procedure toward EPS as to avoid the MME rejecting the TAU such that the UE needs to retry with an Attach. Toward the 5GS the UE uses the Registration procedure which the AMF treats as an Initial Registration.

As previously explained, UEs in single-registration mode moves any remaining PDU Sessions after the Attach using the UE requested PDN connectivity establishment procedure with Request Type "handover" and moves the PDN Connections after the Registration using the UE initiated PDU Session Establishment procedure with "Existing PDU Sessions" flag. UEs operating in dual-registration mode can selectively decide to move PDN Connections and PDU Session accordingly as the UE is registered in both systems.

CHAPTER 8

Security

8.1 Introduction

Security is a critical aspect of any communications system, and more so for mobile radio networks. One of the more obvious reasons is that the wireless communication can be intercepted by anyone within a certain range of the transmitter and with the technical skills and equipment to decode the signaling. There is thus a risk that the transmission is eavesdropped on, or even manipulated, by third parties. There are also other threats; for example, an attacker may trace a user's movement between radio cells in the network or discover the whereabouts of a specific user. This may constitute a significant threat to users' privacy. Apart from security aspects directly related to end-users, there are also security issues related to network operators and service providers, as well as security between network operators in roaming scenarios. For instance, there should be no doubt regarding which user and roaming partner were involved in generating certain traffic in order to assure correct and fair charging of subscribers.

Security is an important part of the 4G system as well and many aspects are in fact quite similar in 4G and 5G systems. There are however a few new challenges in the 5G era. For example, it is expected that the variety of end-devices used in 5G systems will be significantly more diverse, e.g. with new kinds of simple devices, connected appliances, industrial applications etc. in addition to the well-known mobile broadband for end-user consumers. Privacy aspects are expected to take a more central role in the 5G era, as more and more of our daily lives take place on the Internet while at the same time computing and storage capacity (commonly referred to as "big data") have made it feasible to track and store almost anything that happens. The number and type of devices an end-user has in their home connected to wireless systems is increasing and in combination with the new storage and compute capacities, end-users need assurance and protection from privacy-invasive behavior and security challenges.

Security can be provided on many layers in a system. Application layer security is what most people notice when they use the Internet. This includes web-browsing using HTTPS and secure access to different platforms and servers available over the Internet. However, providing application layer security is not enough to protect against tracking a user's movement between radio cells, or against denial-of-service attacks against devices or the network. Therefore, security in the underlying mobile access and mobile network is a key part to enable a trustworthy 5G system.

5G Core Networks
https://doi.org/10.1016/B978-0-08-103009-7.00008-9
171

There are also regulatory requirements related to security and these may differ between countries and regions. Such regulations can, for example, be related to exceptional situations where law enforcement agencies can request information about the activities of a device and user as well as intercept the telecommunications traffic. The framework in a communications system for supporting this is called "lawful intercept". There may also be regulations to ensure that end-users' privacy is protected when using mobile networks. Requirements like these are generally captured in the national and/or regional laws and regulations by the responsible authorities for that specific nation or region. The 5G standard however needs to provide enough features so that regulatory requirements can be fulfilled.

Below we discuss different aspects of security in mobile networks, starting with a brief discussion on key security concepts and security domains. Then security aspects relating to end-users as well as within and between network entities are discussed. We conclude this chapter with a description of the framework for lawful intercept. The focus is on 5G security as defined by the 5G standards in 3GPP. There are many other aspects of security in a software-based communication system, not covered by 3GPP standards, including product implementation, virtualization and cloud security aspects, etc. Those aspects are equally important, but not specific to 3GPP standards and thus only mentioned very briefly below.

8.2 Security requirements and security services of the 5G system

8.2.1 Security requirements

When designing the 5G system, 3GPP agreed on overall security requirements for the 5G standard. These include overall requirements on the system to support e.g. authentication and authorization of subscribers, usage of ciphering and integrity protection between the UE and the network etc. There are also security requirements on each entity such as the UE, base station (gNB, eNB), AMF, UDM etc., and these include requirements for secure storage and processing of subscription credentials and keys, support for specific ciphering and integrity protection algorithms etc. Some of the security requirements will be described in more detail below, when we discuss about the different security features in the 5G system. Interested readers may also have a look at the detailed security requirements as described in 3GPP TS 33.501.

8.2.2 Security services

Before we go into the actual security mechanisms of 5GS, it may be useful to briefly go through some basic security concepts that are important in cellular networks.

Before a user is granted access to a network, authentication in general must be performed (though exceptions can be made for regulatory services such as emergency calls, depending on the local regulations). During authentication the user proves that he or she is who he/she claims to be. In 5GS, mutual authentication is required, where the network authenticates the user and the user authenticates the network. Authentication is generally done via a procedure where each party proves that it has access to a secret known only to the participating parties, for example a password or a secret key.

The network also verifies that the subscriber is authorized to access the requested service, for example to get access to 5G services using a particular access network. This means that the user must have the right privileges (i.e. a subscription) for the type of services that are requested. Authorization for an access network is often done at the same time as authentication. It should be noted that different kinds of authorization may be required in different parts of the network and at different instances depending on what service is requested by a user. The network may, for example, authorize the use of a certain access technology, a certain Data Network, a certain QoS profile, a certain bit rate, access to certain services, etc.

Once the user has been granted access, there is a desire to protect the signaling traffic and User Plane traffic between the UE and the network, and between different entities within the network. Ciphering and/or integrity protection may be applied for this purpose. Ciphering and integrity protection serve different purposes and the need for ciphering and/or integrity protection differs depending on what traffic it is. With **ciphering** we ensure that the information transmitted is only readable to the intended recipients. To accomplish this, the traffic is modified so that it becomes unreadable to anyone who manages to intercept it, except for the entities that have access to the correct cryptographic keys. **Integrity protection**, on the other hand, is a means of detecting whether traffic that reaches the intended recipient has or has not been modified, for example by an attacker between the sender and the receiver. If the traffic has been modified, integrity protection ensures that the receiver is able to detect it. Furthermore, the data protection may be done on different layers in the protocol stack and, as we will see, 5GS supports data protection features on both protocol layers 2 and 3 depending on interface and type of traffic. This is explained in more detail below.

In order to encrypt/decrypt as well as to perform integrity protection, the sending and receiving entities need cryptographic keys. It may seem tempting to use the same key for all purposes, including authentication, ciphering, integrity protection, etc. However, using the same key for several purposes should generally be avoided. One reason is that if the same key is used for authentication and traffic protection, an attacker that manages to recover the ciphering key by breaking, for example, the encryption algorithm, would at the same time learn the key used also for authentication and integrity protection. Furthermore, the keys used in one access should not be the same as the keys used in another access. If they were to be the same, the keys recovered by an attacker in one access with weak security features could be reused to break accesses with stronger security features. The weakness of one algorithm or access thus spreads to other procedures or accesses.

To avoid this, keys used for different purposes and in different accesses should be distinct, and an attacker who manages to recover one of the keys should not be able to learn anything useful about the other keys. This property is called key separation and, as we will see, this is an important aspect of 5GS security design. In order to achieve key separation, distinct keys are derived that are used for different purposes. The keys may be derived during the authentication process, at mobility events, and when the UE moves to connected state.

Privacy protection is another important security feature. By privacy protection we mean the features that are available to ensure that information about a subscriber does not become available to others. For example, it may include mechanisms to ensure that the permanent user ID is not sent in clear text over the air link. If for example, such information is sent clear over the air, this would mean that an eavesdropper could detect the movements and travel patterns of a user.

Laws and directives of individual nations and regional institutions (e.g. the European Union) typically define a need to intercept telecommunications traffic and related information. This is referred to as lawful intercept and may be used by law enforcement agencies in accordance with the laws and regulations.

8.2.3 Security domains
8.2.3.1 Overview
In order to describe the different security features of 5GS it is useful to divide the complete security architecture into different security domains. Each domain may have its own set of security threats and security solutions. 3GPP TS 33.501 divides the security architecture into different groups or domains:
1. Network access security
2. Network domain security
3. User domain security
4. Application domain security
5. SBA domain security
6. Visibility and configurability of security.

Groups 1–4 and 6 are very similar to corresponding groups for 4G/EPC. Group 5 is however new compared to 4G/EPC.

The first group is specific to each access technology (NG-RAN, Non-3GPP access), whereas the others are common for all accesses. Fig. 8.1 provides a schematic illustration of different security domains.

8.2.3.2 Network access security
Network access security refers to the security features that provide a user with secure access to the network. This includes mutual authentication as well as privacy features. In addition, protection of signaling traffic and User Plane traffic in the access is also included. This protection may provide confidentiality and/or integrity protection of the traffic. Network

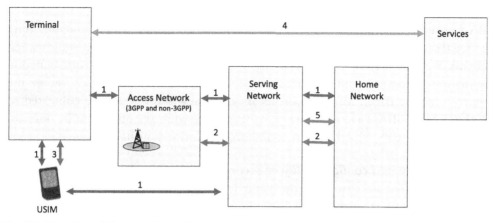

Fig. 8.1 Overview of the security architecture.

access security generally has access specific components – that is, the detailed solutions, algorithms, etc. differ between access technologies. With 5GS, a large degree of harmonization has been done across access technologies, e.g. to use common access authentication. The system now allows authentication over NAS to be used over both 3GPP and Non-3GPP access technologies. Further details are provided later in this chapter.

8.2.3.3 Network domain security

Mobile networks contain many Network Functions and reference points between them. Network domain security refers to the features that allow these Network Functions to securely exchange data and protect against attacks on the network between the Network Functions, both between NFs within a PLMN and in different PLMNs.

8.2.3.4 User domain security

User domain security refers to the set of security features that secure the physical access to terminals. For example, the user may need to enter a PIN code before being able to access the terminal or before being able to use the SIM card in the terminal.

8.2.3.5 Application domain security

Application domain security is the security features used by applications such as HTTP (for web access) or IMS. Application domain security is generally end-to-end between the application in the terminal and the peer entity providing the service. This contrasts with the previous security features listed that provide hop-by-hop security – that is, they apply to a single link in the system only. If each link (and node) in the chain that requires security is protected, the whole end-to-end chain can be considered secure.

Since application-level security traverses on top of the User Plane transport provided by 5GS, and as such is transparent to 5GS, it will not be discussed further in this book.

8.2.3.6 SBA domain security

SBA domain security is the set of security features that enables Network Functions using Service Based interfaces/APIs to securely communicate within a network, and between network domains e.g. in case of roaming. Such features include Network Function registration, discovery, and authorization aspects, as well as the protection of the service-based interfaces. SBA domain security is a new security feature compared to 4G/EPC. Since SBA is a new feature of 3GPP in 5GS, while the other security domains exist also in 4G/EPS, SBA has been considered a security domain on its own.

8.2.3.7 Visibility and configurability of security

This is the set of features that allows the user to learn whether a security feature is in operation or not and whether the use and provision of services will depend on the security feature. In most cases the security features are transparent to the user and the user is unaware that they are in operation. For some security features the user should, however, be informed about the operational status. For example, use of encryption and integrity protection of user data depends on operator configuration and it should be possible for the user to know whether it is used or not, for example using a symbol on the terminal display. Configurability is the property where the user can configure whether the use or provision of a service will depend on whether a security feature is in operation.

8.3 Network access security

8.3.1 General

As mentioned previously, network access security is in many aspects specific to each access technology, but there are also a lot of commonalities. Compared to 4G/EPS, 5GS provides for more commonality between 3GPP and Non-3GPP accesses. For example, the NAS protocol is used over all accesses and therefore the authentication mechanisms can be based on NAS procedures over all accesses. Also, concealment of permanent identifier using SUCI is supported across all accesses in the same manner. In 4G/EPS, even though authentication based on SIM cards are supported in all accesses, the authentication methods differs between accesses and the method of handling protection of the permanent identifier differs. The commonality can however not go all the way since the lower layers differs between access types. Therefore, lower layer security also differs between access types in 5GS (i.e. RAN level security in 3GPP access and IPSec in Non-3GPP).

8.3.2 Flexibility is part of 5GS

Another new aspect of 5GS compared to 4G/EPS is that 5GS supports more flexibility and configurability. For example, 5GS has the capability to support not only IMSI as permanent subscription identifier but also other types and subscription identifier formats. Also, 5GS can support different types of credentials and authentication methods.

Traditional SIM cards are supported as with previous 3GPP generations, but the security framework is now general enough that other types of credentials such as e.g. certificates can be supported as well. It should, however, be pointed out that even though the framework is general, 3GPP Release-15 has focused on the more "traditional" subscription identifiers (i.e. IMSI) and credentials (i.e. SIM card-based credentials). Other types of identifiers, credentials and authentication methods can be supported in Release-15 for private networks, but little work has been done to specify the exact details. It is expected that the 3GPP standard will evolve and more explicitly support new types of identifiers, credentials and authentication methods with later releases as support for e.g. integration with wireline access, enhanced support for industrial use cases etc. are added. The key aspect is however that the Release-15 security framework is already flexible enough to provide a straightforward way to make such additions in the future.

8.3.3 Security entities for network access security

The 5G System architecture introduces a set of security entities in the 5G Core network. These entities are logical entities, and they are contained within the 5GC Network Functions described in Chapters 3 and 13. A reason for defining separate logical entities for security has been to maintain a logical security architecture that can be mapped to the overall 5GC network architecture. The security entities are listed and briefly described below and illustrated in Fig. 8.2, including the relation to 5GC NFs, but more information on how they are used will become clearer in separate sections further below.

- ARPF (Authentication credential Repository and Processing Function). The ARPF contains the subscriber's credentials, i.e. long-term key(s), and the subscription identifier SUPI. The standard associates the ARPF to the UDM NF, i.e. ARPF services are provided via the UDM, and no open interface is defined between UDM and ARPF. It can be noted that, as a deployment option, the subscriber's credentials may alternatively be stored in the UDR.
- AUSF (AUthentication Server Function). The AUSF is defined as a standalone NF in the 5GC architecture, located in the subscriber's home network. It is responsible for handling the authentication in the home network, based on information received from the UE and UDM/ARPF.
- SEAF (SEcurity Anchor Function). The SEAF is functionality provided by the AMF and is responsible for handling the authentication in the serving (visited) network, based on information received form the UE and AUSF.

Fig. 8.2 Logical architecture for network access security.

- SIDF (Subscription Identifier De-concealing Function). The SIDF is a service offered by the UDM NF in the home network. It is responsible for resolving the SUPI from the SUCI.

8.3.4 Access security in 5GS

8.3.4.1 Introduction

Access security in 3GPP mobile networks has evolved continuously with each new generation, from 2G to 3G, then to 4G, and now to 5G. Security features in communication systems may be considered sufficiently secure at one point in time but as computing power increases and attack methods improve, security features need to be upgraded. Therefore, when developing every new 3GPP system, the goal is to provide a level of security that is a step ahead of previous generations in order to meet the new threat landscape.

Taking authentication as an example, when GERAN (2G) was developed, some limitations were purposely accepted. For example, mutual authentication is not performed in GERAN where it is only the network that authenticates the terminal. It was thought that there was no need for the UE to authenticate the network, since it was unlikely that anyone would be able to set up a rogue GERAN network. When UTRAN/UMTS (3G) was developed, enhancements were made to avoid some of these limitations of GERAN. For example, mutual authentication was introduced. These new security procedures are one reason why a new type of SIM card was needed for UMTS: the so-called UMTS SIM (or USIM for short). With the introduction of E-UTRAN (4G), further improvement was done e.g. to allow better separation of keys depending on which serving network was used. That avoids the risk that a key derived in one network can be re-used in another serving network. With 4G/EPS it was however decided that no new SIM card should be needed, i.e. the USIM would be enough to access also E-UTRAN/4G (assuming the subscription allows it). The new features required for 4G were instead supported by software in the terminal. The same applies to 5G, i.e. authentication works with USIM cards together with software enhancements on the terminal.

If we compare 5G access security in general to 4G access security, a few improvements have been made. They are listed on high level below, but we will describe them in more detail later in this chapter:

- Improved privacy protection in that the permanent subscription identifier (SUPI) is never sent in clear text over the air in 5G. In 4G/EPS there is also a large degree of protection of the subscription identifier (IMSI) but, in 4G/EPS, there is a possibility to page the UE using IMSI to recover from certain rare situations.
- Optional integrity protection of the User Plane traffic between UE and the base station. In 4G/EPC ciphering is supported but not integrity protection. One reason for enabling integrity protection of the user plane traffic in 5G was to better serve IoT use cases, where IoT traffic may be more vulnerable to modifications than e.g. voice traffic.

- Improved home-operator control in roaming scenarios. 5G enables both VPLMN and HPLMN to take part in the actual authentication of the UE and allows both operators to verify that the UE is authenticated. In 4G/EPS the actual authentication when using 3GPP access is delegated to the VPLMN based on authentication vectors provided by the HPLMN.
- Improved capabilities to support credentials other than SIM card-based credentials. In 4G/EPC, SIM-based authentication is the only supported authentication method for E-UTRAN, while in 5G it is possible to use non-SIM based credential such as e.g. certificates, also over 3GPP radio access (there are however certain limitations in Release-15, as is further described below).
- Additional configurability of security. In 4G/EPS, User Plane security in 3GPP radio access is always activated in eNB (even though the ciphering algorithm may be NULL). With 5GS, the network decides dynamically at PDU Session establishment, based on subscription data in UDM, what kind of UP security (ciphering and/or integrity protection) should be used.
- Improved protection of initial NAS messages compared to 4G, enabling protection of certain information elements also in the initial NAS messages.

8.3.4.2 Access security overview

Access security in 5GS consists of different components:
- Mutual authentication between UE and network.
- Key derivation to establish separate keys for ciphering and integrity protection, with strong key separation.
- Ciphering, integrity, and replay protection of NAS signaling between UE and AMF.
- Ciphering, integrity, and replay protection of Control Plane signaling between UE and the network. For 3GPP access, the RRC signaling is protected between UE and gNB. For untrusted non-3GPP access, IKEv2 and IPSec is used between UE and N3IWF.
- Ciphering and integrity of the User Plane. For 3GPP access the User Plane can be ciphered and integrity protected between UE and gNB. For untrusted non-3GPP access, the User Plane can be ciphered, and integrity protected between UE and N3IWF.
- Privacy protection to avoid sending the permanent user identity (SUPI) over the radio link.

Fig. 8.3 illustrates some of these components in the network.

We discuss in further detail below how each of these components have been facilitated.

8.3.5 Concealment of permanent subscription identifier

As mentioned above, one security improvement in 5GS compared to EPC/4G is a more complete protection of the permanent subscription identifier. In 4G/EPS, the permanent subscription identifier (IMSI) is sent in clear text over the air in some exceptional cases when MME cannot identify the UE based on GUTI. However, in 5G, the Subscription

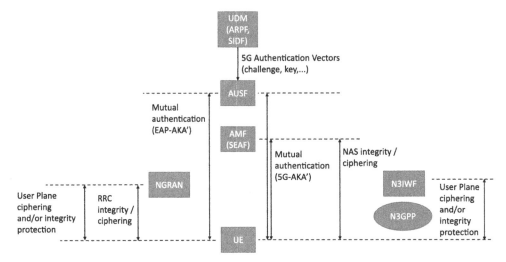

Fig. 8.3 Overview of network access security.

Permanent Identifier (SUPI) is never sent in clear-text over the air. Or, to be more precise, the subscriber-specific part of the SUPI is never sent in clear-text over the air. The Mobile Country Code (MCC) and Mobile Network Code (MNC) must still be sent in clear-text to allow the serving PLMN to find the Home PLMN in roaming cases.

Instead of sending the SUPI over the air, the UE sends either a temporary ID (5G-GUTI) or a concealed version of the SUPI called Subscription Concealed Identifier (SUCI). The 5G-GUTI is sent if the UE has a valid 5G-GUTI from a previous registration, in a similar way as in 4G/EPS. The SUCI is sent if the UE does not have a 5G-GUTI that it can use. The SUCI is created by the UE based on public key cryptography. The UE shall generate a SUCI using a protection scheme with the Home Network Public Key, that was securely provisioned to the UE in control of the home network. The HPLMN (UDM/SIDF) can then derive the SUPI from the SUCI by using the Home Network Private Key. The SUCI format and the use of SUCI between UE and the network is illustrated in the Fig. 8.4.

8.3.6 Primary authentication and key derivation overview

8.3.6.1 Overview

The purpose of the primary authentication and key agreement procedures in 5GS is to enable mutual authentication between the UE and the network and provide keying material that can be used between the UE and the serving network in subsequent security procedures.

As mentioned above, authentication is the process that allows two parties prove to each other that they are who they claim to be. Authentication is usually based on a

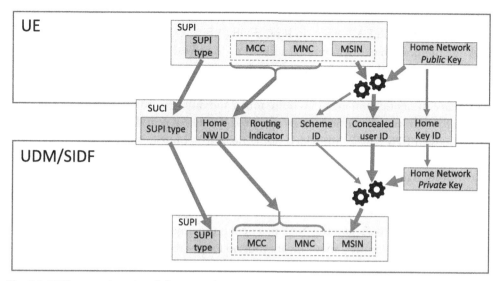

Fig. 8.4 SUPI concealment and de-concealment.

set of credentials known to each party. The credential is a tool for the authentication and is something that each side knows and can use in the authentication process. It may be e.g. that each party has access to the same shared key (as is the case with SIM-based authentication) or it may be that each party has a certificate.

Mutual authentication in 5G public networks in Release-15 is using SIM-based authentication, like in EPS. The same SIM card as in EPS can be used (UMTS SIM cards). However, a key difference to EPS is that the 5GS also supports primary authentication based on credentials other than USIM-based credentials. Credentials could, for example, be based on certificates. However, even though non-USIM based credentials are supported by the 5GS security framework, the only authentication method that has been standardized in Release-15 is based on AKA (i.e. USIM-based). A certificate-based authentication procedure has been described in an informative annex in 3GPP TS 33.501, but since it is informative it is not formally part of the 5GS standard, but rather a description for how it can be done. In later releases further work on non-SIM based authentications may be done.

Before going into how primary authentication for 5GS works, it is worthwhile to recapitulate how primary authentication for 4G/EPS was defined. EPS supports two authentication methods that are used to perform the SIM-based authentication process: EPS AKA and EAP-AKA' and which one is used depends on the access type the UE connects with (this is described in 3GPP TS 33.401 and 3GPP TS 33.402). EPS AKA is the native authentication method used in 3GPP access (E-UTRA, UTRA) while EAP-AKA', which is based on the Extensible Authentication Protocol (EAP) defined by IETF, is used over Non-3GPP access. The EAP framework is defined in IETF RFC 3758 (RFC 3758) and EAP-AKA' in RFC 5448 (RFC 5448). Both EPS AKA and EAP-AKA' are methods to perform mutual authentication based on SIM-card credentials, but they differ in how the actual AKA algorithm is executed between the UE and the network.

Primary authentication in 5GS is both similar and different from how it works in EPS. In 5GS, the primary authentication is supported by 5G AKA and EAP-AKA'. 5G AKA corresponds to EPS AKA with added home network control, while EAP-AKA' is the same EAP method as is used in EPS. On this level it looks very similar to EPS but there are two important differences.

One difference is that in 5GS, both 5G AKA and EAP-AKA' can be used over both 3GPP and non-3GPP accesses. The 5G NAS protocol supports authentication using both 5G AKA and EAP-AKA' and the NAS protocol is used over both 3GPP and non-3GPP accesses. The authentication methods are thus no longer related to a specific access technology, as in EPS/4G. This means that 5G AKA is not only the native authentication method over 3GPP access (NR and E-UTRA) but it can also be used for primary authentication over non-3GPP access. Similarly, EAP-AKA' is not only used over non-3GPP access. 5GS supports EAP authentication via NAS and it is therefore possible to use EAP-based authentication in 3GPP access as well.

Another difference is that 5G AKA is not just EPS AKA with "EPS" replaced with "5G". 5G AKA is an evolution of EPS AKA where home network control is added. With 5G AKA the home-operator receives a cryptographic proof of successful authentication of the UE as part of the authentication procedure, i.e. the HPLMN takes part in the actual 5G AKA authentication. With EPS AKA, it is the MME (in the serving/visited PLMN) that runs the authentication procedure on the network side, and it is only the MME that is verifying the result. The MME then notifies the HPLMN (HSS) about the result, with no possibility for the home PLMN to cryptographically prove that the authentication was successful. With 5G AKA however, as we will see in more detail below, both the AMF/SEAF in the serving/visited PLMN and AUSF in home PLMN are taking active roles in the actual authentication and can verify the authentication result itself. We can also compare this to how EAP-AKA' works. When EAP is used, the authentication signaling is end-to-end between the UE and the AUSF (in the home PLMN). The AMF/SEAF in the serving/visited PLMN is just a pass-through authenticator. In this case it is thus the home PLMN that will verify the outcome of the authentication and notify the AMF/SEAF in the serving PLMN. To summarize, with EAP-AKA' it is the HPLMN that has the cryptographic proof and with EPS AKA it is the VPLMN that has the cryptographic proof that the UE was successfully authenticated, while 5G AKA allows both VPLMN and HPLMN to have cryptographic proof of successful authentication.

Mutual authentication for both 5G AKA and EAP-AKA' is based on the fact that both the USIM and the network have access to the same secret key, K. This is a permanent key that is stored on the USIM and in the UDM/ARPF (or UDR) in the home operator's network. The key K is never sent from UDM to any other NF, and is thus not used directly to protect any traffic and it is also not visible to the end-user or even the terminal. Instead the USIM and UDM/ARPF generate other keys (from the key K) for use during the authentication procedure. Then, during the authentication procedure, additional keys are generated in the terminal and in the network that are e.g. used for ciphering and integrity protection of User Plane and control-plane traffic. For example, one of the derived keys is used to protect the User Plane, while another key is used to protect NAS signaling. One reason why several keys are produced like this is to provide key separation and to protect the underlying shared secret K. For more information on the key derivation and key hierarchy, see Section 8.3.9.

We will soon look more closely how authentication based on 5G AKA works, and then how EAP-AKA' works. But first we will see how the authentication is initiated, including de-concealment of the subscription identifier and selection of authentication method.

8.3.6.2 Initiation of authentication and selection of authentication method

The selection of either 5G AKA or EAP-AKA' depends on operator policy and configuration. According to 3GPP TS 33.501, the UE and serving network shall support both EAP-AKA' and 5G AKA authentication methods. When the UE initiates Registration with the network and authentication is to start, the ARPF/UDM determines which authentication method (5G AKA or EAP-AKA') to use.

During this initial step, the de-concealment of SUCI also takes place, if required. If the UE provided the SUCI instead of 5G-GUTI, the network performs the de-concealment of the SUCI to determine the SUPI. See Section 8.3.5 for more details on SUCI de-concealment.

A simple call flow of authentication method selection and SUCI de-concealment is shown in Fig. 8.5.

Once the SUPI has been determined and an authentication method has been selected, the actual authentication procedure can start. Below we first describe 5G AKA, and then EAP-AKA'.

8.3.7 5G AKA based primary authentication

When the AMF/SEAF initiates the authentication as described above, and the UDM has chosen to use 5G AKA, the UDM/ARPF will generate a 5G Home Environment Authentication Vector (5G HE AV) and provide it to the AUSF. The UDM/ARPF does this by first generating an initial AV in a similar way as HSS generates an AV in 4G/EPS. The UDM/ARP will then derive the 5G-specific 5G HE AV. The "4G" AV consists of five parameters: an expected result (XRES), a network authentication token (AUTN), two keys (CK and IK), and the RAND. UDM/ARPF derives the

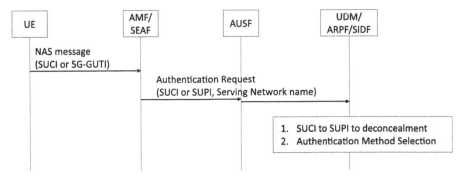

Fig. 8.5 Initiation of authentication.

5G-specific parameters. The K_{AUSF} is derived based on the CK, IK, SQN etc. UDM/ARPF also calculates XRES*. Finally, the UDM/ARPF shall create the 5G HE AV, consisting of RAND, AUTN, XRES*, and $K_{AUSF,}$ and provide this 5G HE AV to the AUSF. Readers familiar with 3G and 4G will recognize the initial Authentication Vector as the parameter that the HSS/AuC would send to the SGSN or MME for access authentication in UTRAN. For 3G/UMTS, the CK and IK were sent to SGSN. For 4G/E-UTRAN, the CK and IK are not sent to the MME and instead, the HSS/AuC generates a new key, K_{ASME}, based on the CK and IK and other parameters such as the serving network identity (SN ID). Now with 5G, UDM/ARPF generates the K_{AUSF} based on the CK and IK and other parameters such as the serving network name. The K_{ASME} and K_{AUSF} are derived in similar ways but using slightly different input values.

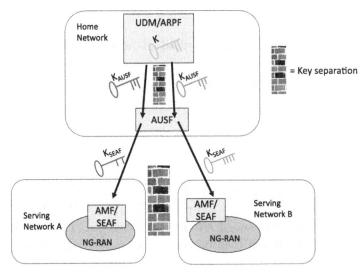

Fig. 8.6 Key separation.

The Serving Network ID includes the Mobile Country Code (MCC) and Mobile Network Code (MNC) of the serving network. A reason for including SN ID is to provide key separation between different serving networks to prevent a key derived for one serving network being (mis)used in a different serving network. Key separation is illustrated in Fig. 8.6.

K_{AUSF}, together with XRES*, AUTN, and RAND, constitutes the 5G HE AV that is sent to the AUSF. The CK and IK never leave the UDM. It can be noted that the "root keys" generated by UDM (for 5G) and HSS (for 3G/4G) are all different, i.e. CK/IK for 3G, K_{ASME} for 4G/E-UTRAN and K_{AUSF} for 5G. Deriving separate root keys ensures a strong key separation between different systems.

Once the AUSF has received the 5G HE AV, it generates a 5G Serving Environment AV (5G SE AV). This is different from EPS AKA where the AV was sent directly to the serving PLMN. However, in 5G, as mentioned above, also the home PLMN can take part in the authentication procedure and this is the reason for AUSF to generate a separate 5G SE AV. The AUSF stores the XRES* and K_{AUSF} and generates a HXRES* value based on the XRES* and a K_{SEAF} key from the K_{AUSF} key. The AUSF then generates a 5G SE AV consisting of the HXRES*, AUTN and RAND and sends it to AMF/SEAF. The key K_{SEAF} is not sent yet to AMF/SEAF but will be sent later if the authentication is successful.

The generation of the AVs are illustrated in Fig. 8.7 below. For more details on the generation of the AV, see 3GPP TS 33.501.

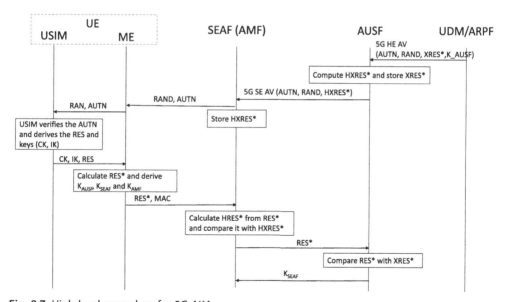

Fig. 8.7 High level procedure for 5G AKA.

The AMF/SEAF stores HXRES* and sends the RAND and AUTN to the UE, so that they can be provided to the USIM. AUTN is a parameter calculated by the UDM/ARPF based on the secret key K and the SQN. The USIM now calculates its own version of the MAC included in the AUTN using its own key K and SQN and compares it with the MAC in the AUTN received from the AMF/SEAF. If they are consistent, the USIM considers the network authenticated. Then the USIM calculates a response RES using cryptographic functions with the key K and the challenge RAND as input parameters. The USIM also computes CK and IK in the same way as when UTRAN is used (it is, after all, a regular UMTS SIM card). When the terminal receives RES, CK, and IK from the USIM, it calculates RES* based on RES and sends the RES* back to the AMF/SEAF. The AMF/SEAF calculates a HRES* based on RES* and authenticates the terminal by verifying that the HRES* is equal to HXRES* that it received from AUSF. The AMF/SEAF then forwards the RES* to the AUSF so that the AUSF (HPLMN) also can perform authentication of the UE. The AUSF compares the RES* with the XRES* that is received from UDM/ARPF and verifies that they are equal. This completes the mutual authentication. The AUSF will now calculate a SEAF key (K_{SEAF}) and send to the SEAF. At the same time, the UE uses the CK and IK and other information to compute K_{AUSF} in the same way as UDM/ARPF did and K_{SEAF} in the same way AUSF did. If everything has worked out, the UE and network have authenticated each other and both UE and the AMF/SEAF now have the same key K_{SEAF} (note that none of the keys K, CK, IK, K_{AUSF} or K_{SEAF} was ever sent between UE and the network). Fig. 8.7 illustrates this flow.

Now all that remains is to calculate the keys to be used for protecting traffic. We will later describe the key hierarchy but first we will look at EAP-AKA' and see how it differs from 5G AKA.

8.3.8 EAP-AKA' based primary authentication

The Extensible Authentication Protocol (EAP), defined by IETF in RFC 3748, is a protocol framework for performing authentication, typically between an end-user device and a network. It was first introduced for the Point-to-Point Protocol (PPP) to allow additional authentication methods to be used over PPP. Since then it has also been introduced in many other scenarios. EAP is not an authentication method per se, but rather a common authentication framework that can be used to implement specific authentication methods. EAP is therefore extensible in the sense that it enables different authentication methods to be supported and allows for new authentication methods to be defined within the EAP framework. These authentication methods are typically referred to as EAP methods. For more details on EAP in general, please see Chapter 14.

EAP-AKA' is an EAP method defined by IETF in RFC 5448 (RFC 5448) for performing authentication based on USIM cards. As mentioned above it is used already in

Fig. 8.8 High level architecture for EAP-AKA'.

EPC/4G for access over non–3GPP access. In 5GS, EAP-AKA' has a more prominent role as it is now possible to use it for primary authentication over any access.

EAP-AKA runs between the UE and the AUSF as shown in Fig. 8.8.

When the AMF/SEAF initiates the authentication as described in section above, and the UDM has chosen to use EAP-AKA', the UDM/ARPF will generate a transformed Authentication Vector (AV') and provide it to the AUSF. This Authentication Vector from UDM/ARPF is the starting point for the authentication procedure. The AV' consists of five parameters: an expected result (XRES), a network authentication token (AUTN), two keys (CK' and IK'), and the RAND. The AV is quite similar to the AV generated in 4G/EPS with the difference that the CK and IK are replaced by CK' and IK' which are 5G variants of the CK and IK, derived from CK and IK and the serving network name. For that reason, the AV is called a "transformed Authentication Vector" and denoted with a prime (AV').

The authentication then proceeds in a similar way as for 5G AKA with the difference that that AMF/SEAF is not actively participating except for forwarding messages. It is only the AUSF that compares the RES received from the UE with the XRES. The AUSF then notifies the AMF/SEAF about the outcome and provides the SEAF key to the SEAF. This procedure is illustrated in Fig. 8.9.

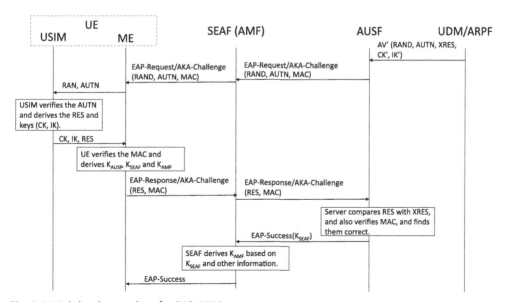

Fig. 8.9 High level procedure for EAP-AKA'.

EAP-AKA′, specified in IETF RFC 5448, is a small revision of EAP-AKA, defined in IETF RFC 4187. The revision made in EAP-AKA′ is the introduction of a new key derivation function that binds the keys derived within EAP-AKA′ to the identity of the access network. In practice, this means that the access network identity is considered in the key derivation schemes. The procedure is thus more aligned with 5G AKA and strengthens key separation.

Now all that remains is to calculate the keys to be used for protecting traffic, which is described in next section.

8.3.9 Key derivation and key hierarchy

Once authentication has been completed and the root keys are established, new keys must be derived for different purposes. As mentioned above, avoiding the use of a single key for multiple purposes is important. The following type of traffic is protected between UE and the network:
- NAS signaling between UE and AMF
- For 3GPP access:
  • RRC signaling between UE and NG-RAN,
  • User Plane traffic between UE and eNB.
- For untrusted Non-3GPP access:
  • IKEv2 signaling between UE and N3IWF,
  • User Plane traffic between UE and N3IWF.

Different ciphering and integrity protection keys are used for each set of procedures above. The key K_{AUSF} is the home network "root key" derived during authentication used by the UE and the network to derive further keys. K_{AUSF} is used to derive a serving network "root key" K_{SEAF}. The K_{SEAF} is then used to derive a K_{AMF}. The UE and AMF uses the K_{AMF} to derive the keys for ciphering and integrity protection of NAS signaling (K_{NASenc} and K_{NASint}). In addition, the AMF also derives a key that is sent to the gNB (the K_{gNB}). This key is used by the gNB to derive keys for ciphering of the User Plane (K_{UPenc}), integrity protection of the User Plane (K_{UPint}) as well as ciphering and integrity protection of the RRC signaling between UE and eNB (K_{RRCenc} and K_{RRCint}). The UE derives the same keys as gNB. For untrusted Non-3GPP access, the AMF also derives a key that is sent to the N3IWF (K_{N3IWF}). This key is used during the IKEv2 procedures to derive keys for IPSec between the UE and N3IWF. The "family tree" of keys is typically referred to as a key hierarchy. The key hierarchy of 5GS is illustrated in Fig. 8.10 (adapted from 3GPP TS 33.501).

Once the keys have been established in the UE and the network, it is possible to start ciphering and integrity protection of the signaling and user data. The standard allows use of different cryptographic algorithms, and the UE and the NW need to agree on which algorithm to use for a particular connection. The encryption

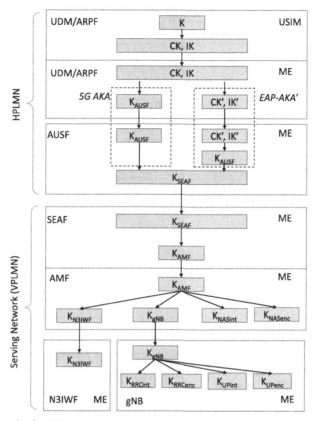

Fig. 8.10 Key hierarchy for 5GS.

algorithms for 5G (NEA) currently supported for NAS, RRC, and UP ciphering in 3GPP access are shown in Table 8.1.

NEA0, 128-NEA1 and 128-NEA2 are mandatory to support in the UE, eNB, and AMF, while 128-NEA3 is optional to support. The 5G integrity protection algorithms (NIA) currently supported for RRC, NAS signaling, and User Plane integrity protection are shown in Table 8.2. The algorithms 128-NIA1 and 128-NIA2 are mandatory to

Table 8.1 Ciphering algorithms for NAS, RRC and UP ciphering in 3GPP access

Name	Algorithm	Comments
NEA0	Null ciphering algorithm	
128-NEA1	128-bit SNOW 3G based algorithm	Also used in EPS
128-NEA2	128-bit AES based algorithm	Also use in EPS
128-NEA3	128-bit ZUC based algorithm	Also used in EPS

Table 8.2 Integrity protection Algorithms for NAS, RRC and UP integrity in 3GPP access

Name	Algorithm	Comments
NIA0	Null integrity protection algorithm	
128-NIA1	128-bit SNOW 3G based algorithm	Also used in EPS
128-NIA2	128-bit AES based algorithm	Also use in EPS
128-NIA3	128-bit ZUC based algorithm	Also used in EPS

support in the UE, eNB, and AMF, while 128-NIA3 is optional to support. Integrity protection for the User Plane is optional to support. The Null integrity protection algorithm NIA0 is only used for unauthenticated emergency calls.

For more details on the ciphering and integrity algorithms supported with 5GS, see 3GPP TS 33.501.

Algorithms for IPSec between UE and N3IWF are described in 3GPP TS 33.501, which refers to relevant IETF RFCs.

8.3.10 NAS security

As mentioned above, the NAS protocol between UE and AMF is ciphered and integrity protected. During the Registration procedure, when authentication and key derivation is done, the keys for protecting NAS messages are derived. NAS security is mostly handled in a similar way as in 4G/EPS, but some enhancements have been made to enable additional protection of initial NAS messages. Initial NAS messages are here the Registration Request and Service Request messages, i.e. messages that are used to initiate communication with the 5GC.

In 4G/EPS, if the UE has an existing security context, the initial NAS messages are integrity protected but not ciphered. This is done to allow the receiving MME to identify the UE (based on e.g. GUTI) even if the MME may have lost the security context for that UE or there is a mismatch between UE and MME. If there is no security context in the UE, the initial NAS messages are neither ciphered nor integrity protected.

In 5GS, support has been added to allow partial ciphering of the initial NAS message to protect information elements that include possibly sensitive information and that are not needed by AMF for the basic processing of the initial NAS message. Such messages will thus contain some information elements in clear text (e.g. 5G-GUTI, 5G-S-TMSI, UE security capabilities) and some information elements that are ciphered (e.g. MM capabilities, Requested S-NSSAI etc.). This can be done if the UE has an existing security context. Without any security context in the UE, the initial NAS messages contain only the clear text information elements and the rest of the information elements are sent ciphered and integrity protected later when NAS security has been established.

8.3.11 Updating of USIM content, including Steering of Roaming

One feature that is also new compared to 4G/EPS is the possibility for UDM to provide, via a secure communication, information to the UE for updating the list of roaming PLMNs in the USIM. This feature is referred to as Steering of Roaming and is described in 3GPP TS 23.122 and 3GPP TS 33.501.

The feature allows the UDM to send the Steering Information List (containing information about preferred and forbidden PLMNs) to the AUSF. The AUSF then calculates a message authentication code (MAC) for the list, based on the K_{AUSF} "root key" and other information and provides the value to UDM. The UDM can then send the Steering Information List to the UE (via AMF) together with the MAC. The UE then verifies the MAC value before accepting and updating the USIM with the new information.

8.3.12 Interworking with EPS/4G

8.3.12.1 General

Interworking with EPS/4G is an important feature and during the 5G work, solutions covering the security aspects of such interworking were developed. In this book we will not describe the security features applicable 4G/EPS in any detail. The interested reader is instead referred to books dedicated to EPS, see for example Olsson et al. (2012). The discussion below focuses on the interworking between EPS/4G and 5GS.

As mentioned in Chapters 3, 7, and 12, the UE can operate in Single Registration or Dual Registration mode when interworking with EPS. The security aspects of interworking depend a lot on whether the UE uses Single or Dual Registration mode. Below we will describe each case separately.

8.3.12.2 Single Registration mode

When operating in Single Registration mode, there are two cases depending on whether the N26 interface between the AMF and the MME is supported in an operator's network.

Single Registration mode with N26

When a UE moves from EPS/4G to 5GS, there are different possibilities to establish the security context to be used in the target access. One possibility is to perform a new authentication and key agreement procedure when the UE enters a new access. In order to reduce the delays during handover between 5GS and EPS/E-UTRAN, however, this may not be desirable. Instead, handovers can be based on native or mapped security contexts. If the UE has previously established a native security context in 5GS access by running 5G AKA or EAP-AKA', then moved to EPS and later returns to 5GS, the UE and network may have cached the native security context for 5GS, including a native K_{AUSF}, from the previous time the UE was in 5GS. In this way a full AKA procedure in the target access is not needed during the inter-RAT handover. If a native context is not available, it is instead possible to map the security context used in the source access to a security context for the target access. This security context mapping is supported when moving between different 3GPP accesses, but not when moving e.g. from 4G/EPS to untrusted non-3GPP access in 5GS. When mapping is performed, the UE and AMF derive keys applicable to the target access (e.g. K_{AMF}) based on the keys used in the source access (e.g. K_{ASME} received by AMF in the 4G security context from the MME). The mapping is based on a cryptographic key derivation function (KDF) having the property that it protects the source context from the mapped target context. This ensure that if attackers compromise a mapped context they get no information about the context from which it was mapped. The AMF may also choose to initiate a primary authentication procedure to create a new native 5G security context.

When the UE moves in the other direction, i.e. from 5GS to EPS, the AMF acts as a source MME to the target MME in EPS (i.e. AMF pretends to be an MME). The AMF will derive a mapped EPS security context (including e.g. K_{ASME} derived from K_{AMF}) and provide to the MME as part of the UE context during the handover.

Single Registration mode without N26
When N26 is not supported there is no interface between AMF and MME for transferring UE security context. In this case there is thus no possibility to use a mapped security context in the target access and instead the UE and network need to re-use an existing native security context in the target access (in case it exists and has been cached from a previous access to the target system) or the UE and network needs to perform a new authentication procedure in the target access (in case a cached native security context does not exist).

8.3.12.3 Dual Registration mode
When using Dual Registration mode, the UE is simultaneously registered to both EPS and 5GS and will consequently use two different security contexts; an EPS security context and a 5G security context. Obviously, the EPS security context is used for accessing EPS and the 5G security context is used for accessing 5GS. When the UE moves between the two systems, e.g. at inter-system mobility, the UE will use the security context that matches the target system, e.g. the UE will start using the EPS security context when the target system is EPS. When accessing EPS, the same security features as defined for EPS/4G applies, as further described in 3GPP TS 33.401 and Olsson et al. (2012).

8.4 Network domain security
8.4.1 Introduction
Most of the text in this chapter has so far concerned network access security, i.e. the security features that support a UE access to the 5GS. However, as mentioned in the introductory

sections of the chapter, it is important to consider security aspects also of network-internal interfaces, both within a PLMN and between PLMNs in roaming cases.

This has however not always been the case. When 2G (GSM/GERAN) was developed, no solution was specified for how to protect traffic in the core network. This was perceived not to be a problem, since the GSM networks typically were controlled by a small number of large institutions and were trusted entities. Furthermore, the original GSM networks were only running circuit-switched traffic. These networks used protocols and interfaces specific for circuit-switched voice traffic and typically only accessible to large telecom operators. With the introduction of GPRS as well as IP transport in general, the signaling and User Plane transport in 3GPP networks started to run over networks and protocols that are more open and accessible to others than the major institutions in the telecom community. This brought a need to provide enhanced protection also to traffic running over core network interfaces. For example, the core network interfaces may traverse third-party IP transport networks, or the interfaces may cross operator boundaries as in roaming cases. 3GPP has therefore developed specifications for how IP-based traffic is to be secured also in the core network and between one core network and another (core) network. On the other hand, it should be noted that even today, if the core network interfaces run over trusted networks, for example a physically protected transport network owned by the operator, there would be little need for this additional protection.

Below we will discuss both the general Network Domain Security (NDS) solution that was specified already for 3G and 4G and is re-used with 5GS, but also look at new 5GS solutions that have been developed specifically for the Service Based interfaces (i.e. the interfaces that use HTTP/2). In this area the interfaces between domains are of special importance, the roaming interface (N32) between PLMNs as well as the interfaces between 5GS and 3rd parties used for Network Exposure.

8.4.2 Security aspects of Service Based interfaces

Service Based interfaces is a new design principle in 3GPP networks, introduced with 5G. Therefore, 3GPP has also defined new security features to accommodate the new type of interactions between core network entities. For example, when a NF Service consumer wants to access a service provided by a NF Service producer, there is support in 5GS to authenticate and authorize the consumer before granting access to the NF Service. These features are optional within a PLMN and an operator may decide to instead rely e.g. on physical security instead of deploying the authentication/authorization framework for NF Services. Below we will describe on high level the general security features for the Service Based interfaces, including the authentication and authorization support.

For protecting the Service Based interfaces, all Network Functions shall support TLS. TLS can then be used for transport protection within a PLMN unless the operator implements network security by some other means. TLS is however optional to use and as alternative an operator could e.g. use Network Domain Security (NDS/IP) within a PLMN, described more in Section 8.4.4. The operator may also decide to not use cryptographic protection at all within the PLMN in case the interfaces are considered trusted, e.g. if they are physically protected operator-internal interfaces.

Authentication between Network Functions within a PLMN is also supported but the method depends on how the links are protected. If the operator uses protection at the transport layer based on TLS as mentioned above, the certificate-based authentication that is provided by TLS is used for authentication between NFs. If the PLMN however does not use TLS-based transport layer protection, authentication between NFs within one PLMN could be considered implicit by using NDS/IP or using physical security of the links.

In addition to authentication between NFs, the Server side of a Service Based Interface also needs to authorize the client for accessing a certain NF Service. The authorization framework uses the OAuth 2.0 framework as specified in RFC 6749 (RFC 6749). The OAuth 2.0 framework is an industry-standard protocol for authorization developed by IETF. It supports a token-based framework in which a service consumer will get a token from an Authorization Server. This token can then be used to access a specific Service at a NF Service producer. In 5GS it is the NRF that acts as the OAuth 2.0 Authorization server and a NF Service Consumer will thus request tokens from the NRF when it wants to access a certain NF Service. The NRF may authorize the request from the NF Service consumer and provide a token to it. The token is specific to a certain NF Service producer. When the NF Service consumer tries to access the NF Service at the NF Service producer, the NF Service consumer provides the token in the request. The NF Service producer checks the validity (integrity) of the token by either using NRF's public key or a shared key, depending on what type of keys have been deployed for the OAuth 2.0 framework. If the verification is successful, the NF Service producer executes the requested service and responds back to the NF Service consumer.

The above framework is the general framework when an NF accesses services produced by any other NF. However, the NRF is a somewhat special NF Service producer in this case since it is the NRF that provides services for NF discovery, NF Service discovery, NF registration, NF Service registration and OAuth 2.0 token request services, i.e. services that support the overall Service Based framework. When an NF wants to consume NRF services (i.e. register, discover or request access token) the above general features for transport security (based on TLS) and authentication (based on TLS or implicit authentication) apply as well. However, the OAuth 2.0 access token for authorization between the NF and the NRF is not needed. The NRF instead authorizes the request based on the profile of the expected NF/NF service and the type of the NF service consumer. The NRF determines whether the NF service consumer can discover the expected NF instance(s) based on the profile of the target NF/NF service and the type of the NF service consumer. When network slicing applies, the NRF authorizes the request according to the configuration of the Network Slice, e.g. so that the expected NF instance(s) are only discoverable by other NFs in the same network slice.

8.4.3 Service Based interfaces between PLMNs in roaming

The internetwork interconnect allows secure communication between service-consuming and service-producing NFs in different PLMNs. Security is enabled by the Security Edge Protection Proxies (SEPP) of both networks, i.e. SEPP(s) in each PLMN. The SEPPs enforce protection policies regarding application layer security thereby ensuring integrity and confidentiality protection for those elements to be protected. The SEPPs also allow topology hiding to avoid that the internal network topology is revealed to external networks.

Between PLMNs with roaming agreements there is, in most cases, an intermediate network that provides mediation services between PLMNs, a so-called roaming IP exchange or IPX. The IPX thus provides interconnect between different operators. Each PLMN has a business relationship with one or more IPX providers. In most cases there will thus be one or more interconnect providers between SEPPs in the two PLMNs. The interconnect provider may have its own entities/proxies in the IPX, that enforce certain restrictions and policies for the IPX provider. Fig. 8.11 shows an example of a serving PLMN where an NF wants to access a service produced by an NF in a home PLMN. The serving PLMN has a consumer's SEPP (cSEPP) and the home PLMN has a producer's SEPP (pSEPP). Each PLMN has a business relation with an IPX operator. The cSEPP's operator has a business relationship with an interconnect provider (consumer's IPX, or cIPX), while the pSEPP's operator has a business relationship with an interconnect provider (producer's IPX, or pIPX). There could be further interconnect providers in between cIPX and pIPX, but that is not shown here.

Interconnect operators (pIPX and cIPX in the figure) may modify the messages exchanged between the PLMNs to provide the mediation services, e.g. to provide value-added services for the roaming partners. If there are IPX entities between SEPPs that want to inspect or modify a message, TLS cannot be used on N32 since it is a transport network protection that does not allow intermediaries to look into or modify a

Fig. 8.11 Overview of security between PLMNs (N32).

message. Instead application layer security needs to be used for protection between the SEPPs. Application layer security means that the message is protected inside the HTTP/2 body which allows some Information Elements in the message to be encrypted while other Information Elements are sent in clear text. The Information Elements that an IPX provider have reasons to inspect would be sent in clear text while other Information Elements, that should not be revealed to intermediate entities, are encrypted. Using Application layer security also allows an intermediate entity to modify the message. The SEPPs use JSON Web Encryption (JWE, specified in RFC 7516) for protecting messages on the N32 interface, and the IPX providers use JSON Web Signatures (JWS, specified in RFC 7515 (RFC 7515)) for signing their modifications needed for their mediation services. It should be noted that even if TLS is not used to protect NF-to-NF messages carried between two SEPPs in this case, the two SEPPs still establish a TLS connection in order negotiate the security configuration parameters for the Application Layer Security.

If there are no IPX entities between the SEPPs, TLS is used to protect the NF-to-NF messages carried over the two SEPPs. In this case there is no need to look inside the messages or to modify any part of the message carried between the SEPPs.

8.4.4 Network Domain Security for IP based communication

The specifications for how to protect general IP-based control-plane traffic is called Network Domain Security for IP-based Control Planes (NDS/IP) and is available in 3GPP TS 33.210. This specification was originally developed for 3G and evolved for 4G to cover primarily IP-based Control Plane traffic (e.g. Diameter and GTP-C). It is, however, also applicable to 5G networks to provide network layer protection. NDS/IP is based in IKEv2/IPSec and is thus applicable to any kind of IP traffic, including HTTP/2 used with 5GS.

NDS/IP uses the concept of security domains. The security domains are networks that are managed by a single administrative authority. Hence, the level of security and the available security services are expected to be the same within a security domain. An example of a security domain could be the network of a single telecom operator, but it is also possible that a single operator divides its network into multiple security domains. On the border of the security domains, the network operator places Security Gateways (SEGs) to protect the control-plane traffic that passes in and out of the domain. All NDS/IP traffic from network entities of one security domain is routed via an SEG before exiting that domain toward another security domain. The traffic between the SEGs is protected using IPsec, or to be more precise, using IPsec Encapsulated Security Payload (ESP) in tunnel mode. The Internet Key Exchange (IKE) protocol version 2, IKEv2, is used between the SEGs to set up the IPsec security associations. An example scenario is illustrated in Fig. 8.12 (adapted from 3GPP TS 33.210).

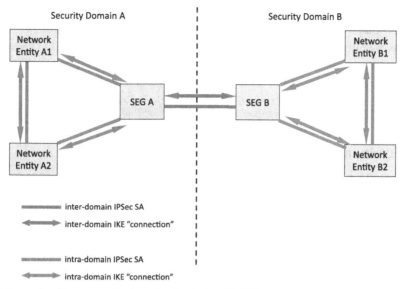

Fig. 8.12 Example of two security domains deploying NDS/IP.

Although NDS/IP was initially intended mainly for the protection of control-plane signaling only, it is possible to use similar mechanisms to protect the User Plane traffic.

Also, within a security domain – that is, between different network entities or between a network entity and an SEG – the operator may choose to protect the traffic using IPsec. The end-to-end path between two network entities in two security domains is thus protected in a hop-by-hop manner.

8.4.5 Security aspects of N2 and N3 interfaces

As described in Chapter 3, N2 is the reference point between the AMF and the 5G-AN. It is used, among other things, to carry NAS signaling traffic between the UE and the AMF over 3GPP and non-3GPP accesses. N3 is the reference point between the 5G-AN and UPF. It is used to carry GTP-tunneled User Plane data from the UE to the UPF.

Protection of N2 and N3 using cryptographic solutions between gNB and the 5GC is important in certain deployments, e.g. if the link to the gNB cannot be assumed to be physically secured. This is, however, an operator's decision. In case the gNB has been placed in a physically secured environment then the 'secure environment' includes other nodes and links beside the gNB.

In order to protect the N2 and N3 reference points using cryptographic solution, the standard requires that IPsec ESP and IKEv2 certificates-based authentication is used between the gNB and the 5GC. On the core network side, a SEG (as described for

NDS/IP) may be used to terminate the IPsec tunnel. This provides integrity, confidentiality and replay-protection for the transport of Control Plane data over N2.

For the N2 interface, as alternative to IPSec, the standard also allows DTLS to be used to provide integrity protection, replay protection and confidentiality protection. The use of transport layer security via DTLS does however not rule out the use of network layer protection according to NDS/IP. In fact, IPsec has the advantage of providing topology hiding.

8.4.6 Security aspects of Network Exposure/NEF

As described in Chapter 3, the NFs can expose capabilities and events to 3rd party Application Functions via the NEF. This exposure includes monitoring of events by an external AF as well as provisioning of session information for Policy and Charging purposes. The NEF also supports provisioning of information to the 5GS allowing an external party to e.g. provision foreseen UE behavioral information to 5G (e.g. mobility patterns) or to provide influence on traffic routing for edge computing use cases. To provide a secure exposure of 5GS capabilities and provisioning of information, these features should only be provided to AFs that have been properly authenticated and authorized, either by explicit procedures or implicitly in case the AF is trusted as part of the network deployment.

For authentication between NEF and an Application Function that resides outside the 3GPP operator domain, mutual authentication based on client and server certificates shall be performed between the NEF and AF using TLS. TLS is also used to provide protection for the interface between the NEF and the Application Function. After the authentication, NEF determines whether the Application Function is authorized to send requests.

8.5 User domain security

User domain security includes the set of security features that secure the user access to the mobile device. The most common security feature in this user domain context is the secure access to the USIM. Access to the USIM will be blocked until the USIM has authenticated the user. Authentication is in this case based on a shared secret (the PIN code) that is stored inside the USIM. When the user enters the PIN code on the terminal, it is passed on to the USIM. If the user provided the right PIN code, the USIM allows access from the terminal/user, for example to perform the AKA-based access authentication.

8.6 Lawful intercept

Lawful Interception (LI) is one of the regulatory requirements operators must satisfy as a legal obligation towards the Law Enforcement Agencies (LEA) and Government

Authorities in most countries where they are operating their businesses. Within 3GPP standards, this is currently defined as: "Laws of individual nations and regional institutions, and sometimes licensing and operating conditions, define a need to intercept targeted telecommunications traffic and related information in communication systems. Lawful Interception applies in accordance with applicable national or regional laws and technical regulations." (as per 3GPP TS 33.126 "Lawful Interception Requirements"). LI allows appropriate authorities to perform interception of communication traffic for specific user(s) and this includes activation (requiring a legal document such as a warrant), deactivation, interrogation, and invocation procedures. A single user (i.e. interception subject) may be involved where interception is being performed by different LEAs. In such scenarios, it must be possible to maintain strict separation of these interception measures. The Intercept Function is only accessible by authorized personnel. As LI has regional jurisdiction, national regulations may define specific requirements on how to handle the user's location and interception across boundaries.

This subsection deals with this aspect on a brief and high level to complete the overall 5GS functionalities; it is intended as a description of the 3GPP LI standards and not of any function implemented in any of the vendors' nodes. The LI function does not place requirements on how a system should be built but, rather, requires that provisions be made for legal authorities to be able to get the necessary information from the networks via legal means, according to specific security requirements, without disruption of the normal mode of operations and without jeopardizing the privacy of communications not to be intercepted. Note that LI functions must operate without being detected by the person(s) whose information is being intercepted and other unauthorized person(s). As this is the standard practice for any communications networks already operating today around the world, 5GS is no exception. The process of collection of information is done by means of adding specific functions into the network entities where certain trigger conditions will then cause these network elements to send data in a secure manner to other specific network entity/entities responsible for such a role. Moreover, specific entities provide administration and delivery of intercepted data to Law Enforcement in the required format. It can be noted that 3GPP dedicates a lot of effort to ensure that when compliance to LI regulation is required, the system is designed to provide the minimum amount of information that is sufficient to achieve compliance, and not more.

As an example, Fig. 8.13 (adapted from 3GPP TS 33.127) shows a simplified view of the LI architecture for the 5G system. The LI-related functions shown in the figure are:

- Law Enforcement Agency (LEA), which in general is the one that submits the warrant to the Service Provider. In some countries the warrant may be provided by a different legal entity (e.g. judiciary).
- The Administration Function (ADMF), responsible for the overall management of the LI system. ADMF uses the LI_X1 interface toward the 5GC NFs for managing the LI functionality.

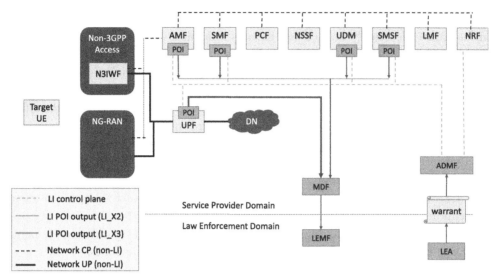

Fig. 8.13 High level LI architecture.

- Point of Interception (POI) is functionality that detects the target communication, derives the intercept related information or communications content from the target communications and delivers the output to the MDF. The POI is located in the relevant 5G NFs. The POI uses the LI_X2 and LI_X3 interfaces for delivering the interception product.
- The Mediation and Delivery Function (MDF) delivers the interception reports to the Law Enforcement Monitoring Facility (LEMF)
- The Law Enforcement Monitoring Facility (LEMF) is the entity receiving the Interception Product. The LEMF is not specified by 3GPP.

Intercept-related information (also referred to as Events) are triggered by activities detected at the network element. Some events applicable to the AMF are:

- Registration.
- Deregistration.
- Location update.
- Start of interception with already registered UE.
- Unsuccessful communication attempt.

Events applicable to SMF include:

- PDU Session establishment.
- PDU Session modification.
- PDU Session release.
- Start of interception with an established PDU Session.

Depending on national regulations, intercept-related information collected may also be reported by the UDM.

This brief overview represents high-level functions supported in 5GS to fulfill the LI requirements. Lawful intercept as such is not directly related to the overall architecture aspects of the new system, and this overview is mainly included for completeness. It does not in any way show the complete possibilities or aspects of this function, 3GPP does not cover the important ethical aspects when providing such sensitive functions. The Telecom Industry is addressing such non-technical issues via other forums, for example the. Telecommunications Industry Dialogue (http://www.telecomindustrydialogue.org/).

CHAPTER 9

Quality-of-Service

9.1 Introduction

Quality of Service is the ability to provide a differentiated packet forwarding treatment of data which may e.g. belong to different users, different applications or even different services or media within the same applications. The differentiated treatment may be to prioritize between the data or to guarantee a certain level of performance to a data flow.

As for 4G, 5G provides support for multiple services e.g. Internet, Voice and Video, but further the 5GS intends to address a wider range of use cases as discussed in Chapter 2. For example, 5G will address new vertical industries which are requiring higher demands when it comes to reliability, latency etc.

In an Evolved Packet System (EPS) Quality of Service is implemented by the Evolved Packet Core through the classification of data and its association to EPS bearers, enforcement of QoS parameters, and the enforcement of a packet forwarding treatment by the Radio Access Network (RAN) scheduler (Downlink and Uplink). QoS Class Identifiers (QCIs) are identifying certain QoS characteristics (i.e. whether it is GBR or non-GBR, Priority Level, Packet Delay Budget and Packet Error Loss Rate), either according to a standardized table in 3GPP TS 23.203 or based on operator configuration in the PLMN.

While maintaining a principle such as network control, some of the design goals when developing the 5G QoS framework were:

Flexibility and support for any Access Type, i.e. 5GC is intended to support any type of access and a wide range of usage.

Separation of concerns between 5GC and the 5G-AN, i.e. if QoS requirements are fulfilled it is up to the 5G-AN how to fulfill them.

Reduce signaling required for QoS establishment and modifications, e.g. previous systems in general require NAS signaling when enabling QoS differentiation.

Fig. 9.1 provides a comparison between the 4G and the 5G QoS frameworks. The 4G System is a connection-oriented transmission network which requires the establishment of a logical connection between two endpoints in the system e.g. between the UE and the PGW. The logical connection is called an EPS Bearer. To each EPS Bearer there is an associated (Data) Radio Bearer, an S1 Bearer and an S5/S8 Bearer. The EPS QoS parameters are associated to an EPS Bearer, and to enable the QoS characteristics described by the EPS QoS parameters there is a need to establish the EPS Bearer i.e. the logical connection and the associated Bearers.

5G Core Networks
https://doi.org/10.1016/B978-0-08-103009-7.00009-0

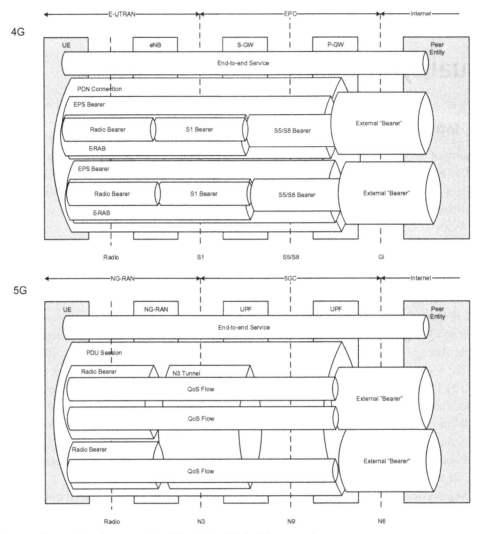

Fig. 9.1 Comparison between the 4G and the 5G QoS frameworks.

In 5G, the QoS framework is based on QoS Flows which is the finest granularity of QoS differentiation i.e. the QoS parameters and characteristics are tied to a QoS Flow. Each QoS Flow is identified by a QoS Flow ID (QFI) which is unique within a PDU Session. The NG-RAN can establish a (Data) Radio Bearer per QoS Flow or NG-RAN may combine more than one QoS Flow into the same (Data) Radio Bearer based on logic in the NG-RAN i.e. there is no strict one-to-one relationship between Data Radio Bearers and QoS Flows (as long as the QoS requested for the QoS Flow is fulfilled the NG-RAN is allowed to handle the NG-RAN resources as deemed most suitable from NG-RAN perspective).

9.2 Flow based QoS framework

The QFI is carried in an (GTP-U) encapsulation header on N3 (and N9) i.e. without any changes to the end-to-end packet header. Data packets marked with the same QFI receives the same traffic forwarding treatment (e.g. scheduling, admission threshold). The QoS Flows can be GBR QoS Flows i.e. that require guaranteed flow bit rate, or QoS Flows that do not require guaranteed flow bit rate (Non-GBR QoS Flows).

Fig. 9.2 illustrates the classification process and the differentiated packet forwarding provided by the NG-RAN of data packets in DL (i.e. packets arriving at UPF which pass through toward the UE) and data packets in UL (i.e. packets generated by the UE e.g. in application layer which are sent to the network). The data packets are shown to be IP packets, but same principles can be applied for Ethernet frames.

In DL, the data packets are compared in UPF towards Packet Detection Rules (PDR), see Chapters 6 and 10, installed by the SMF, as to classify the data packets (e.g. against IP 5-tuple filters in the PDR). Each PDR is then associated with one or more QoS Enforcement Rule(s) (QER) that contains information for how to enforce e.g. bitrates. The QER also contains the QFI value to be added to the GTP-U header (N3 encapsulation header). See Chapter 14 on GTP-U protocol for more information.

In this example, the data packets of five IP flows are classified into three QoS Flows and then sent toward the 5G-AN (in this case NG-RAN) via the NG-U Tunnel (i.e. N3 tunnel). The NG-RAN, based on the QFI marking and the corresponding per QFI QoS Profile received e.g. during the establishment of the PDU Session, decides how to map the QoS Flows to DRBs. The Service Data Adaptation Protocol (SDAP), specified in 3GPP TS 37.324, is used to enable multiplexing if more than one QoS Flow is sent on a DRB, i.e. if the NG-RAN decides to setup a DRB per QFI then the SDAP layer is not needed. Unless Reflective QoS is used. If so the SDAP is used, see 3GPP TS 38.300. For QFI 5, the NG-RAN decides to use a dedicated DRB, but QFI2 and QFI3 are multiplexed on the same DRB. When there is SDAP configured then an SDAP header is added on top of PDCP, i.e. there is some overhead added to the data packets, and the SDAP is used for the QoS Flow to DRB mapping. The QoS Flow to DRB mapping can also be defined using RRC reconfiguration in which case a list of QFI values can be mapped toward a DRB. The NG-RAN then sends the data packets using the DRBs toward the UE. The UE SDAP layer keeps any QFI to DRB mapping rules, and the data packets are forwarded internally toward the application layer's socket interfaces in the UE without any 3GPP specific extensions e.g. as IP packets.

In UL, the UE's application layer generates data packets which first are compared with the set of installed packet filters from the Packet Filter Sets in the UE. The Packet

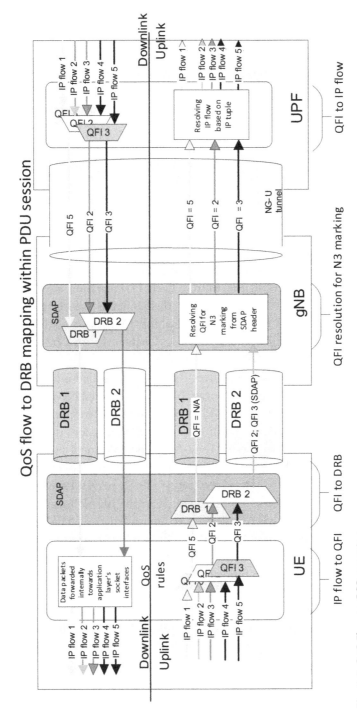

Fig. 9.2 QoS Flow to DRB mapping.

Filter Sets are checked in precedence order and when a match is found the data packet is assigned a QFI. The assigned QFI and the data packet is sent toward the UE's Access Stratum (AS) SDAP layer which performs a QFI to DRB mapping using the available mapping rules. When a match is found the data packet is sent on the corresponding DRB, and if there is no match then the data packet is sent on the default DRB and the SDAP header indicates the QFI such that the NG-RAN can decide whether to move the QFI to another DRB. It is optional to configure a default DRB, but the 5GC may provide additional QoS Flow information indicating that a non-GBR QoS Flow is likely to appear more often than traffic for other QoS Flows established for the PDU Session and such QoS Flows may be more efficient to be sent without any SDAP header e.g. on the default DRB. In Fig. 9.2 the QFI 5 is sent on DRB1 but as it is the only QoS Flow there is no need to include any SDAP header, while QoS Flows 2 and 3 are sent on DRB2 with SDAP header indicating the QFI of the data packet. The NG-RAN uses the available information as to decide how to mark the N3 header of each data packet and forwards the data packet to the UPF. The UPF resolves the data packets into IP flows, and the UPF also performs any bitrate policing and other logic as directed by the various N4 rules provided by the SMF e.g. counting.

9.3 Signaling of QoS

As to enable QoS, the request for QoS and the related QoS parameters needs to be available in the relevant entities. Fig. 9.3 provides a high-level description of the involved entities and the related QoS communication in the CP.

(1) The UE may request for QoS directly toward the 5GC, using NAS SM signaling, or via application layer signaling toward an AF. Typically, a specific QoS is enabled by the subscription in UDM including some special default QoS or that an AF request for QoS based on application layer signaling, toward the PCF, to ensure a better service experience.

(2) When a PDU Session is established the SMF retrieves Session Management subscription data from UDM including default QoS values that the SMF may change based on local configuration or interaction with PCF, the resulting default values are used for the QoS Flow that the default QoS rule is associated with. The default QoS rule is the rule which can contain a Packet Filter Set that allows all UL packets to pass through and is used in case there is no other QoS Rule with a Packet Filter Set matching the UL data packet to be sent by the UE. As a result of a PDU Session establishment the UE gets a default QoS Rule, optionally additional QoS Rules and QoS Flow descriptions. The UE also gets a Session-AMBR that the UE uses for applying UL rate limitation for the Non-GBR QoS Flows for the PDU Session.

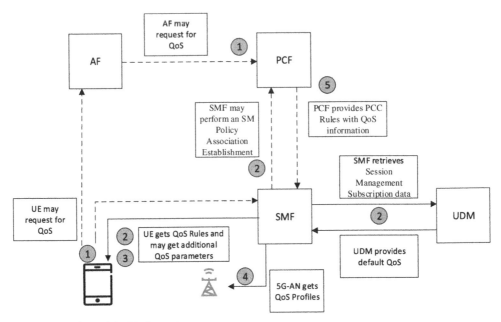

Fig. 9.3 Signaling of QoS information.

(3) A QoS Rule contains e.g. a QFI, a Packet Filter Set and a precedence value, and the UE uses the QoS Rules per PDU Session to decide whether and how to mark and send the UL data packets as previously described in Section 9.2.

Each QoS Flow description sent to the UE contains a QFI, information whether it is creating a new QoS Flow, deleting or modifying an existing one, and optionally an EPS bearer identity (EBI) if the QoS Flow can be mapped to an EPS bearer as described in Chapter 7. If the QoS Flow is a GBR QoS Flow, then Guaranteed and Maximum flow bit rates for UL and DL are sent and optionally an averaging window (see Tables 9.1 and 9.2). If the Averaging Window is not signaled, then the Default Averaging Window defined per standardized 5QI in Table 9.3 is applied.

(4) To enable QoS differentiation in the 5G-AN, the SMF provides QoS Profiles to the 5G-AN. A QoS Profile contains the per QoS Flow QoS parameters described in Table 9.1 and optionally an indication whether the traffic for the QoS Flow is likely to appear more often than traffic for other flows established for the PDU Session (as described in Section 9.2). If the 5QI value is not from the standardized values (see Table 9.3) then the 5G QoS characteristics from Table 9.2 is also included. If the 5QI is from the standardized value range, then SMF can provide values to override the default values of the Priority Level, Averaging Window and Maximum Data Burst Volume (see Table 9.3).

Table 9.1 5G QoS parameters

5G QoS parameter		Description
Per QoS Flow	5G QoS Identifier (5QI)	a scalar that is used as a reference to the 5G QoS characteristics
	Allocation and Retention Priority (ARP)	Includes three parts i.e. • priority level: 1–15 values • pre-emption capability: whether a service data flow may get resources that were already assigned to another service data flow with a lower ARP priority level • pre-emption vulnerability: whether a service data flow may lose the resources assigned to it in order to admit a service data flow with higher ARP priority level
	Reflective QoS Attribute (RQA)	indicates to 5G-AN that traffic carried on this QoS Flow is subject to Reflective QoS i.e. RQA enables Reflective QoS for the QoS Flow
	Notification control	indicates whether notifications are requested from NG-RAN when GFBR can no longer (or can again) be guaranteed for a QoS Flow
	Flow Bit Rates	For GBR QoS Flows following bit rates are indicated Guaranteed Flow Bit Rate (GFBR) – separately for UL and DL Maximum Flow Bit Rate (MFBR) – separately for UL and DL.
	Maximum Packet Loss Rate	indicates maximum rate for lost packets of the QoS Flow that can be tolerated in the uplink and downlink direction
Additional QoS parameters	Aggregate Bit Rates	Each PDU Session is associated with: • per Session Aggregate Maximum Bit Rate (Session-AMBR) which limits aggregate bit rate across Non-GBR QoS Flows for a PDU Session Each UE is associated with: • per UE Aggregate Maximum Bit Rate (UE-AMBR) which limits aggregate bit rate across Non-GBR QoS Flows for a UE

Table 9.2 5G QoS characteristics

5G QoS characteristics	Description
Resource Type	GBR, Delay critical GBR or Non-GBR
Priority Level	indicates a priority in scheduling resources among QoS Flows
Packet Delay Budget (PDB)	defines an upper bound for the time that a packet may be delayed between the UE and the UPF that terminates the N6 interface
Packet Error Rate (PER)	defines an upper bound for the rate of PDUs (e.g. IP packets) that have been processed by the sender of a link layer protocol (e.g. RLC in RAN of a 3GPP access) but that are not successfully delivered by the corresponding receiver to the upper layer (e.g. PDCP in RAN of a 3GPP access). Thus, the PER defines an upper bound for a rate of non-congestion related packet losses
Averaging Window	represents the duration over which the bitrate, i.e. GFBR and MFBR, is calculated
Maximum Data Burst Volume (MDBV)	the largest amount of data that the 5G-AN is required to serve within the period of the 5G-AN part of the PDB. GBR QoS Flows with Delay-critical Resource Type shall be associated with a MDBV. The MDBV aids the 5G-AN to enable low latency requirements as whether a low latency can be achieved with a certain reliability depends on packet size and inter-arrival rate of the packets.

(5) When the PCF gets a request for QoS from an AF, the PCF generates PCC rules sent toward the SMF based on subscription and policies. Based on the PCC rules the SMF generates rules toward the UPF as to enable the UPF to perform classification, bandwidth enforcement and marking of User Plane traffic. See Chapter 10 for more details related to the PCF functionality including SMF and UPF logic.

9.4 Reflective QoS

The concept of Reflective QoS was developed to minimize the need for NAS signaling between the UE and the Core Network when enabling QoS differentiation, and as implied by the name the decision on what QoS to provide is done by a *reflection* to what is previously received i.e. the mirrored data packet is getting the same QoS treatment as the received data packet. In other words, when Reflective QoS (RQ) is enabled for a QFI e.g. QFI 3 as per Fig. 9.4, the UE creates a derived QoS Rule for data classification based on the received DL data packet. When the UE is about to send an UL data packet the UE checks the QoS Rules including the derived QoS Rule and when there is a match

Fig. 9.4 Reflective QoS.

the UE applies the matched QoS Rule's QFI to the UL data packet (i.e. QFI 3 in the Fig. 9.4 example).

Reflective QoS can be enabled for PDU Sessions with IPv4, IPv6, IPv4v6 or Ethernet PDU Session Types, and is especially useful for applications which frequently generate data packets with different header values, e.g. HTTP traffic generating new port numbers as to avoid NAS signaling for updating the UE with new Set of packet filers for each port change.

The Reflective QoS is controlled by the 5GC on a per-packet basis by using the Reflective QoS Indication (RQI) in the encapsulation header on N3 (and N9) reference point together with the QFI, and a Reflective QoS Timer (RQ Timer) as described in Fig. 9.5.

Fig. 9.5 describes the sequence how Reflective QoS is enabled and controlled.

(A) The UE indicates that it supports Reflective QoS during PDU Session establishment, or during PDU Session modification when the UE moved from EPS to 5GS when N26 interface is used (see Chapter 7).

(B) If SMF determines that Reflective QoS is to be used for an SDF corresponding to a specific QoS Flow (e.g. as instructed by the PCF, see Chapter 10), the SMF provides the RQA (Reflective QoS Attribute) within the QoS Flow's QoS profile to the 5G-AN. The SMF includes an indication to use Reflective QoS for this SDF in the corresponding SDF information provided to the UPF.

(C) When the UPF receives an indication to use Reflective QoS for an SDF, the UPF shall set the RQI in the encapsulation header on the N3 reference point for every DL packet corresponding to this SDF. When an RQI is received by 5G-AN in a DL packet on N3 reference point, the 5G-AN indicates to the UE the QFI and the RQI of that DL packet. NG-RAN uses SDAP for the RQI and QFI information.

(D) When the UE receives a DL data packet with RQI, the UE either creates a new UE derived QoS rule with a Packet Filter corresponding to the DL packet and starts a RQ Timer value for the rule, or if the DL packet matches an existing the UE restart the timer associated to the stored UE derived QoS rule.

(E) The UE sends UL data packets corresponding to the UE derived QoS rule with the associated QFI.

(F) When the 5GC determines to no longer use Reflective QoS for a specific SDF, the SMF removes the RQA from the corresponding QoS profile toward the NG-RAN

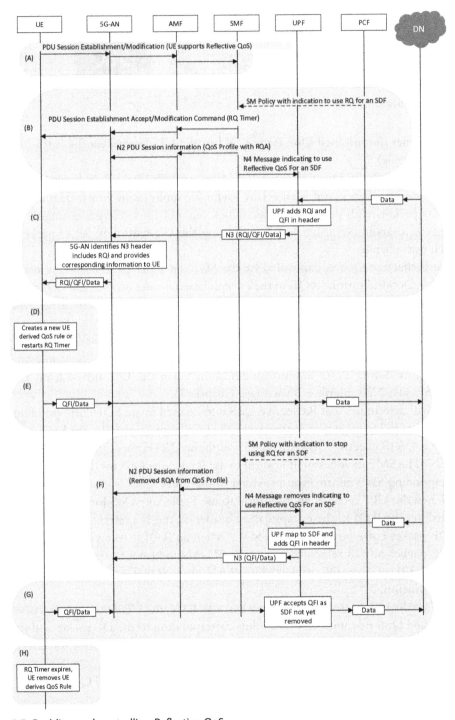

Fig. 9.5 Enabling and controlling Reflective QoS.

and the SMF removes the indication to use Reflective QoS in the corresponding SDF information provided to the UPF. When the UPF receives this instruction for this SDF, the UPF shall no longer set the RQI in the encapsulation header on the N3 reference point.

(G) The UPF shall continue to accept the UL traffic of the SDF for the originally authorized QoS Flow for an operator configurable time.

(H) When the RQ Timer value for the rule expires the UE removes the UE derived QoS rule.

When the UE derives the QoS Rule the UE sets the precedence value to a standardized value defined in 3GPP TS 24.501, which allows precedence values for signaled QoS Rules to be set either with lower or higher precedence value. The UE also starts a timer associated with the derived QoS Rule with a RQ timer value received from the SMF during PDU Session establishment, or modification or a default value in case no RQ timer value been received. When the UE receives a DL data packet matching the derived QoS Rule the UE updates the derived QoS Rule (e.g. if the DL data packet was marked with a different QFI then the UE replaces the derived QoS Rule with the new QFI value) and restarts the RQ timer, and if the RQ timer expires the UE removes the derived QoS Rule. The derived QoS Rules may also be requested to be revoked by the UE e.g. if the UE enters memory issues such that it cannot generate more QoS Rules, and if the SMF accepts the UE's request, the UE removes the derived QoS Rules for the PDU Session.

9.5 QoS parameters and characteristics

5G QoS parameters and 5G QoS Characteristics are signaled to involved entities for describing the QoS requirements and characteristics of the QoS to be enabled.

9.5.1 5G QoS parameters

The defined 5G QoS parameters are described in Table 9.1.

9.5.2 5G QoS characteristics

5G QoS characteristics are associated with a 5QI. The 5G QoS characteristics describe the packet forwarding treatment that a QoS Flow receives end-to-end between the UE and the UPF that terminates N6. The 5G QoS characteristics are used as input to configuration of the entities and connections in the network as to handle each QoS Flow e.g. it is used as input to 3GPP radio access link layer protocol configurations.

5GS supports both standardized and pre-configured 5G QoS characteristics that are indicated through the 5QI value, such that there is no need to signal the actual 5G QoS characteristics values on any interface, unless certain 5G QoS characteristics are modified.

It is also possible to signal the complete 5G QoS characteristics as part of the QoS profile e.g. in case there is no standardized or pre-configured 5QI value suitable for the QoS Flow. The possibility to signal the complete QoS characteristics is not supported in 4G, in which case the values will have to be pre-configured in the network (using operator specific QCI value range).

The standardized 5QI values, see Table 9.3, are specified for services that are assumed to be frequently used and thus benefit from optimized signaling by using standardized QoS characteristics. Also, the standardized QoS characteristics potentially can be supported by a more efficient implementation in the network as the characteristics is known in advance.

9.5.3 Standardized 5QI to QoS characteristics mapping

The mapping of standardized 5QI values to 5G QoS characteristics is specified in Table 5.7.4-1 in 3GPP TS 23.501; Table 9.3 is a simplified version of that table (e. g. not all 5QIs are shown).

Table 9.3 Mapping of standardized 5QI values to 5G QoS characteristics

5QI value	Resource Type	Default Priority Level	Packet Delay Budget (ms)	Packet Error Rate	Default Maximum Data Burst Volume (bytes)	Default Averaging Window (ms)	Example services
1	GBR	20	100	10^{-2}	N/A	2000	Conversational Voice
2		40	150	10^{-3}	N/A	2000	Conversational Video (Live Streaming)
3		30	50	10^{-3}	N/A	2000	Real Time Gaming, V2X messages Electricity distribution – medium voltage, Process automation – monitoring
4		50	300	10^{-6}	N/A	2000	Non-Conversational Video (Buffered Streaming)
65		7	75	10^{-2}	N/A	2000	Mission Critical user plane Push To Talk voice (e.g., MCPTT)

Table 9.3 Mapping of standardized 5QI values to 5G QoS characteristics—cont'd

5QI value	Resource Type	Default Priority Level	Packet Delay Budget (ms)	Packet Error Rate	Default Maximum Data Burst Volume (bytes)	Default Averaging Window (ms)	Example services
66		20	100	10^{-2}	N/A	2000	Non-Mission-Critical user plane Push To Talk voice
67		15	100	10^{-3}	N/A	2000	Mission Critical Video user plane
71		56	150	10^{-6}	N/A	2000	"Live" Uplink Streaming
72		56	300	10^{-4}	N/A	2000	"Live" Uplink Streaming
73		56	300	10^{-8}	N/A	2000	"Live" Uplink Streaming
74		56	500	10^{-8}	N/A	2000	"Live" Uplink Streaming
76		56	500	10^{-4}	N/A	2000	"Live" Uplink Streaming
5	Non-GBR	10	100	10^{-6}	N/A	N/A	IMS Signaling
6		60	300	10^{-6}	N/A	N/A	Video (Buffered Streaming) TCP-based (e.g., www, e-mail, chat, ftp, p2p file sharing, progressive video, etc.)
7		70	100	10^{-3}	N/A	N/A	Voice, Video (Live Streaming) Interactive Gaming
8		80	300	10^{-6}	N/A	N/A	Video (Buffered Streaming) TCP-based (e.g., www, e-mail, chat, ftp, p2p file sharing, progressive video, etc.)
9		90	300	10^{-6}	N/A	N/A	Video (Buffered Streaming)TCP-based (e.g., www, e-mail, chat, ftp, p2p file sharing, progressive video, etc.)

Continued

Table 9.3 Mapping of standardized 5QI values to 5G QoS characteristics—cont'd

5QI value	Resource Type	Default Priority Level	Packet Delay Budget (ms)	Packet Error Rate	Default Maximum Data Burst Volume (bytes)	Default Averaging Window (ms)	Example services
69		5	60	10^{-6}	N/A	N/A	Mission Critical delay sensitive signaling (e.g., MC-PTT signaling)
70		55	200	10^{-6}	N/A	N/A	Mission Critical Data (e.g. example services are the same as 5QI 6/8/9)
79		65	50	10^{-2}	N/A	N/A	V2X messages
80		68	10	10^{-6}	N/A	N/A	Low Latency eMBB applications Augmented Reality
82	Delay Critical GBR	19	10	10^{-4}	255	2000	Discrete Automation
83		22	10	10^{-4}	1354	2000	Discrete Automation
84		24	30	10^{-5}	1354	2000	Intelligent transport systems
85		21	5	10^{-5}	255	2000	Electricity Distribution – high voltage

The 5QI values are as far as possible aligned with the EPS Standardized QCI characteristics, defined in Table 6.1.7-A in 3GPP TS 23.203, which makes mapping of QoS easier e.g. during mobility between 5GS and EPS.

As a comparison between the 5G QoS characteristics with the 4G QoS characteristics, the shortest Packet Delay Budget for 5G is 5 ms while it is 50 ms for 4G, and the Packet Error Rate for 5G is 10^{-8} while it is 10^{-6} for 4G.

CHAPTER 10

Policy control and charging

10.1 Introduction

As the Packet Core architecture has grown more complex and feature rich, so too has the need for differentiated policies to control the service behavior and end user experiences. Applications both within and externally to (e.g. by 3rd party service providers) an operator's network increasingly want access and ability to influence the network, its resources and routing rules in order to provide innovative services based on agreements with the operator. Policy and Charging Control (PCC) enables such innovation within the 3GPP architecture for EPS. It is no different for 5GS, which enables the same level of PCC support as provided in EPS as well as some more features.

With regards to session-based functionality, policy control, PCC rules and the associated charging control has not changed fundamentally from EPS to 5GS. The most fundamental enhancement that has been introduced in 5GS is the non-session related and UE related policy management via PCC. This is more related to the policy management part than the charging control aspects of PCC. As briefly introduced in Chapters 4, 9 and 6, the Policy and Charging Control function provides rules and control over a session between the UE and the 3GPP network and toward external network(s) that the UE is seeking to connect to, such as the Internet or specific applications or services as well as other management functions related to network analytics and packet flow management. These functions are discussed in more detail in subsequent sections.

Another major enhancement in 5G PCC architecture is the support of service based interfaces which is described in more detail in Chapter 13. Even though the 3GPP system allows for some locally configured policies configured in certain network elements (also known as static policy management), we will not discuss them further here. We focus on the dynamic policy control through Policy Control Function (PCF) as already introduced in Chapters 3, 4, 6 and 9.

10.2 Overview of policy and charging control

In 5GS, two distinct policy control and related management functionality has been developed, they are known as:
- Non-Session Management related policy control
- Session Management related policy and charging control

5G Core Networks
https://doi.org/10.1016/B978-0-08-103009-7.00010-7
217

They are introduced below and then further elaborated in the coming sections in this chapter.

1. **Non-Session Management related policy control covers the following aspects**
   (a) Access and mobility related policy control
   (b) UE access selection and PDU Session selection related policy (UE policy) control
   (c) Management of Packet Flow Descriptions (PFDs)
   (d) Network status analytics information requirements (NWDAF)

Fig. 10.1 illustrates the network elements involved in the four non-Session Management related policy control features in 5GS listed above. Policy control related to (a) and (b) involve PCF, UDR, AMF and the UE receiving the policy rules. Policy control related to (c) involve PCF, SMF, optionally NEF and UPF for installation of PFDs. Policies related to (d) involve all relevant NFs in 5GS as described in Chapter 3 including Operations and Maintenance (OAM), NWDAF and PCF.

2. **Session Management related policy control includes both QoS policy control and charging control**
   For Policy handling:
   a. Utilizes subscription information, Access Type and for 3GPP access, specific RAT Type to determine policy rules. In 5G, Network Slice information and DNN are additional input to policy provisioning. DNN-related policy information may be activated into service, and out of service, based on validity conditions of the DNN-related policy information fulfillment, independently of PCC interaction at that point in time.
   b. Performs Gating Control using for example, operator defined policy rules to direct traffic data toward appropriate actions including discarding data that do not match service data flow or redirect to other policy rules.

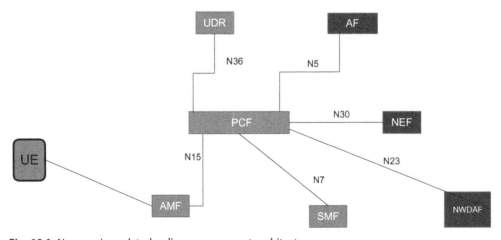

Fig. 10.1 Non-session related policy management architecture.

c. Binding flows that allows the unique association between service data flows and specific QoS Flow.

d. A PCC rule may be predefined (static) or dynamically provisioned at establishment of the PDU session (dynamic) and maintained during the lifetime of a PDU Session. The PCC rules may be activated or deactivated based on for example, specific time of day, independently of PCC interaction at that point in time.

e. Manage a user's guaranteed bandwidth QoS and resolve conflicts, if so arises related to QoS framework.

f. Enables application awareness and session related information access to applications, when allowed by operator policy.

g. Enables 3GPP defined applications like IMS to provide resource and bandwidth management tailored at providing Voice services over IMS.

For Charging policy:

h. Allow the charging control to be applied on a per service data flow and on a per application basis, independent of the policy control.

i. Perform charging control, policy control or both for a DNN access.

j. Using PCC framework, enable usage monitoring control, which allows real-time monitoring of the overall amount of resources that are consumed by a user and to control usage independently from charging mechanisms.

k. Enable policy decisions based on subscriber spending limits.

The overall PCC architecture for Session Management related policy control is shown in Fig. 10.2.

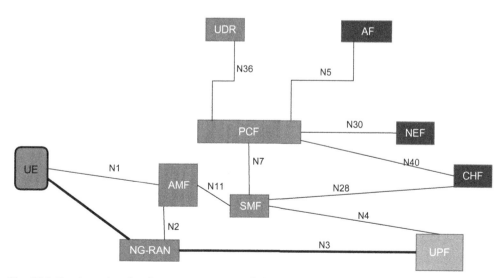

Fig. 10.2 Session related policy management and charging control architecture.

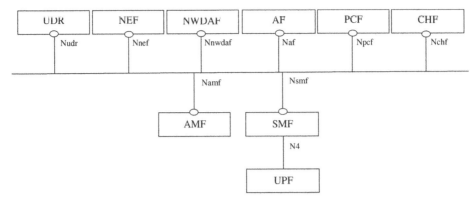

Fig. 10.3 Overall policy management and charging control architecture (SBI).

A description of overall PCC architecture using service based interfaces is shown in Fig. 10.3, as illustrated in Policy Control and Charging architecture specification 3GPP TS 23.503.

In order to support roaming across PLMNs, and to exchange certain non-session and session based policy information, there exists an interface between a PCF in the Visited PLMN and the PCF in the Home PLMN. Figs. 10.4 and 10.5 illustrate point to point reference architecture involving these PCFs in local breakout and home routed connectivity respectively.

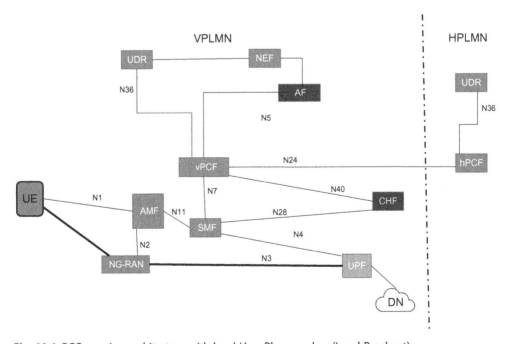

Fig. 10.4 PCC roaming architecture with local User Plane anchor (Local Breakout).

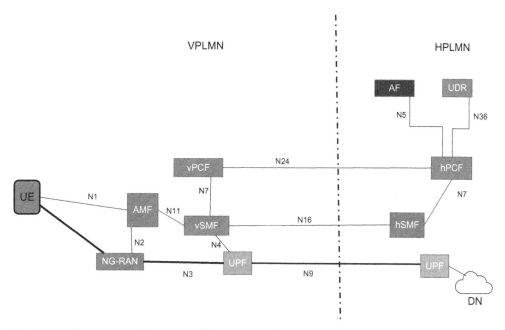

Fig. 10.5 PCC roaming architecture with home routed User Plane anchor.

In case of local breakout, the enforcement entities are in the VPLMN and the data network (DN) terminates locally in the VPLMN. The vPCF in the VPLMN communicates with hPCF in the home PLMN for policies that may be influenced from hPCF. The following functions may be supported in this scenario:

- The PCF in the VPLMN can interact with the AF to generate PCC Rules for services delivered via the VPLMN. The PCF in the VPLMN uses locally configured policies according to the roaming agreement with the HPLMN operator as input for PCC Rule generation. The PCF in VPLMN has no access to subscriber policy information from the HPLMN for PCC Rule generation.
- The PCF in the VPLMN can provide UE access selection and PDU Session selection policy from the PCF in the HPLMN received via N24 or can provide access and motility policy information without contacting the PCF in the HPLMN.
- AF provided routing information for roamers targeting a DNN and S-NSSAI or an External-Group-Identifier (identifying a group of roamers) via the NEF in the VPLMN are stored as Application Data in the UDR located in the VPLMN.

In case of home routed, the User Plane termination, that is the DN, is located in the HPLMN and works in the same manner as in case of non-roaming for session based PCC.

Chapter 6 on Session Management describes different UPF roles depending on type of traffic and routing policies applied and various UPF function deployment such as Uplink Classifier, Branching Point, PDU Session Anchor point. These may influence the role of vPCF and hPCF respectively for the Local Breakout case. In case of non-session based policy, hPCF may provide information related to UE policy about access and PDU session selection, establish association between vPCF and hPCF for UE related AMF policy information sharing and event notification related to such association. These are detailed in Access and mobility related policy control section.

A PCF deployment within a PLMN (session based and non-session based supporting N17) may have different PCF entities serving the non-session based policy than the PCF handling session based policy for a specific user. There is no standards specified mechanism for any coordination between PCFs in such deployments.

10.3 Access and mobility related policy control

10.3.1 Access and mobility management related policies

The access and mobility policy control include two features in Release 15:

- Management of service area restrictions
- Management of the 'Index to RAT/Frequency Selection Priority' (RFSP) functionalities.

As part of the non-session related policy control in the network, PCF may provide, when enabled by the AMF, specific policies derived from the information provided by the AMF based on the subscription data for a specific user as well as policies based on serving PLMN operator configuration. Operator configuration in the PCF may take into account input data such as UE location, time of day, information provided by other NFs, load information accumulated per Network Slice level and accumulated usage for RFSP, etc. These access and mobility restriction policies are applied or enforced in the network, either in the AMF or the NG-RAN serving the UE. The policies associated with access and mobility management are for RFSP and Service Area Restrictions. AMF receives this subscription information from the UDM during Registration procedure and anytime any updates occur at the UDM.

The 'Index to RAT/Frequency Selection Priority' (RFSP Index) is a UE specific index applied to all radio bearers for that UE and is provided by the AMF to the NG-RAN across N2 interface. This RFSP Index is mapped by the NG-RAN to locally defined configuration and applied toward specific Radio Resource Management strategies. The service area restrictions comprise of a list of allowed TAI(s) or a list of non-allowed TAI(s) and optionally the maximum number of allowed TAIs for a UE. Service area restriction is provided to the NG-RAN and to the UE by the AMF. AMF enforces the restriction when the UE changes RA, NG-RAN enforces the restriction during Xn and N2 handover. In case of roaming, the VPLMN vPCF may influence the HPLMN provided subscription data which is received via serving AMF from UDM.

Fig. 10.6 illustrates the high-level procedures for handling the policies.

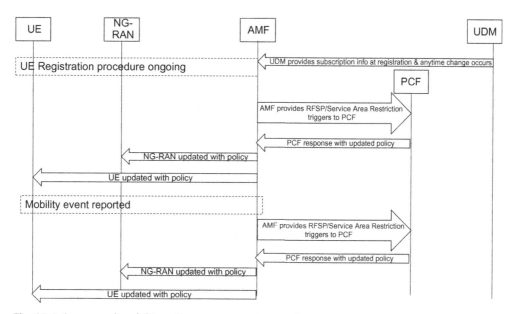

Fig. 10.6 Access and mobility policy management procedure.

In case of Inter-PLMN mobility and AMF change, the AMF notifies the currently serving PCF of the event, which may cause update of the policies related to the Service Area restrictions. In case of change to this data, AMF may have to page the UE to deliver the updated information and AMF also updates NG-RAN with the changed data.

There are 3 procedures defined between AMF and PCF for the policy management, these are:

AM Policy Association Establishment procedure triggered by the AMF, when AMF is first selected for a UE performing initial registration or when an AMF relocation has occurred with PCF change or a UE has moved from EPS to 5GS and thus no AMF – PCF association exists for that UE. This procedure establishes policy association between the AMF and PCF for the UE they are serving.

AM Policy Association Modification can be triggered either by the AMF or the PCF, when local events like UDM update changes AMF current policy related subscription information, when PCF local policy information changes due to local trigger or trigger from UDR, or when AMF relocation causes new AMF to reestablish association with the PCF. This procedure causes updates to all relevant entities associated with the policy.

AM Policy Association Termination triggered due to UE deregistration, AMF change, or 5GS to EPS mobility with N26 connectivity where AMF is no longer available for that UE. This procedure removes the policy association for that UE.

Table 10.1 Policy Control Request Triggers for 3GPP access for AMF

Policy Control Request Trigger	Description	Condition for reporting
Location change (tracking area)	The tracking area of the UE has changed.	PCF (AM Policy, UE Policy)
Change of UE presence in Presence Reporting Area	The UE is entering/leaving a Presence Reporting Area	PCF (AM Policy, UE Policy)
Change of the Allowed NSSAI	The Allowed NSSAI has changed	PCF (AM Policy)

10.3.2 Additional mobility related policy features

The PCF provides Policy Control Request Triggers to the AMF indicating a specific UE (i.e. SUPI or PEI) in the Policy Association establishment and modification procedures. The Policy Control Request Triggers are transferred from the old AMF to the new AMF when the AMF changes. Table 10.1 includes the additional triggers outlined in 3GPP TS 23.503 than ones already described in Section 10.3.1.

When these triggers are activated in AMF, AMF in turn ensures that the notification of change of these triggers are activated appropriately and reported to PCF. Location and Presence Reporting Area information may be used by other services.

10.4 UE policy control

10.4.1 Overview

UE access selection and PDU Session selection related policy (UE policy) control allows for the network and specifically the PCF to configure the UE with two types of operator policies:

- Access Network Discovery and Selection Policy (ANDSP) for non-3GPP access, and
- UE Route Selection Policy (URSP) related to applications and PDU sessions.

The ANDSP includes information for what non-3GPP accesses that UE should prioritize. It is used by the UE for selecting non-3GPP accesses (e.g. Wi-Fi networks) and for selection of the N3IWF in the PLMN.

The URSP includes information mapping certain user data traffic (i.e. applications) to 5G PDU Session connectivity parameters. The user data traffic is defined in the URSP rule by a "traffic descriptor" parameter that can include e.g. IP filter parameters or Application Identity. The URSP is used by the UE to determine if an application started in the UE can be using an already established PDU Session or there is a need to trigger the establishment of a new PDU Session. The URSP also indicates to the UE whether the application traffic can be offloaded to non-3GPP access outside a PDU Session.

A URSP rule includes one Traffic descriptor that specifies the matching criteria and one or more of the components for the policy:

- SSC Mode Selection Policy (SSCMSP) for the UE to associate the matching application with SSC modes.
- Network Slice Selection Policy (NSSP) for the UE to associate the matching application with S-NSSAI.
- DNN Selection Policy for the UE to associate the matching application with DNN.
- PDU Session Type Policy for the UE to associate the matching application with a PDU Session Type.
- Non-Seamless Offload Policy for the UE to determine that the matching application should be non-seamlessly offloaded to non-3GPP access outside of a PDU Session.
- Access Type preference indicating the preferred access (3GPP or non-3GPP) when the UE needs to establish a PDU Session for the matching application.

The UE follows ANDSP and URSP rules provided by the serving PLMN policies, and priority is given to VPLMN policies when roaming for ANDSP, vPCF may receive the hPCF policies via the roaming interface. UEs may be pre-configured with these policies as well but if PCF provides the same policies to the UE then the PCF policies takes precedence over locally pre-configured policies. PCF provides these policies taking into account operator local policies and configuration. The AMF provides the means to transport the PCF provided policies transparently to the UE. PCF may subscribe to the UE connectivity state to ensure PCF is able to deliver the updated policies to the UE as soon as possible. UE shall use the ANDSP and URSP rules with matching traffic description whenever applicable, if no such rule exists then UE may use its local pre-configuration. In the absence of local pre-configuration, UE may use "match all" traffic descriptor.

10.4.2 Delivery of URSP and ANDSP to UE

The PCF may be triggered to provide the UE access selection and PDU Session related policy information during UE Policy Association Establishment and UE Policy Association Modification procedures, these triggers occur during the same events as for access and mobility related policy (i.e. initial UE registration, mobility events, change of policy in the PCF due to local changes).

Fig. 10.7 illustrates how the non-session based ANDSP and URSP UE policies are delivered using NAS Transport, transparently via AMF to the UE. AMF establishes policy association with PCF and then PCF provides the UE policies to be delivered to the UE via the AMF. The AMF delivers the policy information to the UE without any modification.

The following procedure in Fig. 10.8 explains how the policies are transferred/configured from the PCF to the UE.

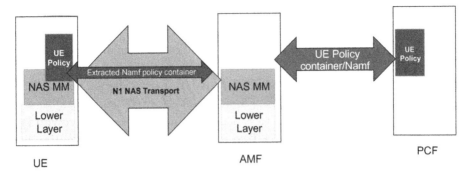

Fig. 10.7 NAS Transport of UE policies from PCF to UE.

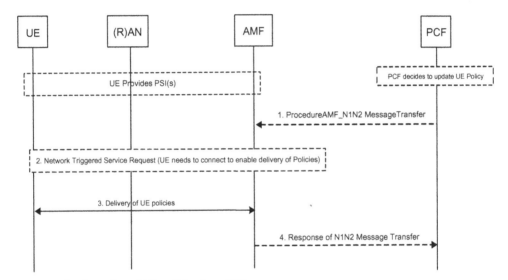

Fig. 10.8 Configuration of UE policies from PCF.

UE provides the Policy Section Identifier(s) to the AMF during registration and mobility events. PSI identifies a Policy Section which consists of one or multiple URSP rule(s) or one or multiple WLANSP rule(s) or non-3GPP access network selection information or a combination of WLANSP rule(s) and non-3GPP access network selection information. PSI(s) allows PCF to determine which policy rules the UE has as well as which ones may need to be updated.

The decision to update UE policy may be triggered at the PCF by comparing the list of PSIs included in the UE access selection and PDU session selection related policy information and determines whether and update is required. When the policies require UE updates, the PDU Session selection related policy information is provided to the UE via the AMF using DL NAS Transport message:

- For the case of initial registration and registration with 5GS when the UE moves from EPS to 5GS; and
- For the network triggered UE policy update case (e.g. the change of UE location, the change of Subscribed S-NSSAIs).

When UE is reachable, the PCF transfers the UE policies to the AMF (step 1). AMF transfers transparently the UE Policy container to the UE via the registered and reachable access (step 3).

In case the UE is determined to be CM-IDLE in AMF, the AMF starts the paging procedure by sending a Paging message in the step 2 (see also Network Triggered Service Request procedure in Chapter 15). Upon reception of paging request, the UE shall initiate the UE Triggered Service Request procedure. AMF then performs step 3 and at the reception of the message, the UE updates the UE policy provided by the PCF and sends the result to the AMF.

The PCF maintains the latest list of PSIs delivered to the UE and updates the latest list of PSIs stored in the UDR to ensure the system is in sync.

The PCF needs to ensure that the size of the UE access selection and PDU Session selection related policy information does not exceed a predefined limit. If the size exceeds the predefined limit, the PCF splits the UE access selection and PDU Session selection related policy information in smaller, logically independent UE access selection and PDU Session selection related policy information enforcing that limit and then sends to the UE in separate N1N2MessageTransfer service operations without exceeding the Radio protocol limits.

10.5 Management of Packet Flow Descriptions

PFD (Packet Flow Description) is a set of information for detection of application traffic. 3GPP specification 3GPP TS 23.503 provides detailed description of the management of PFDs, here we provide a brief overview.

Management of PFDs is a way for a 3rd party (an Application Service Provider, or ASP) to provide packet filters for services related to that 3rd party entity. It may e.g. be a service provider that has an agreement with a mobile operator for treating that service provider's traffic in a specific way, e.g. related to QoS or charging. In order to enable that ASP to update the packet filters related to the ASP's service, the NEF exposes an API where the PFDs for specific Application Ids can be provided. The PFDs are then provided to the SMF and UPFs for use in the actual traffic detection process. PFD management is thus a mix of non-Session Management and Session Management related policy control. The management of PFDs is done outside of any PDU Session, but then the enforcement making use of the PFDs is taking place as part of the regular PDU Session related policy control.

Each PFD may be identified by a PFD id. A PFD id is unique in the scope of a particular application identifier.

A PFD with a PFD id includes one or more of the data:

- a 3-tuple including protocol, server side IP address and port number;
- the significant parts of the URL to be matched, e.g. host name;
- a Domain name matching criteria and information about applicable protocol(s).

The UPF is responsible to perform accurate application detection when PFD(s) are provided by an ASP and then to apply enforcement actions according to the PCC Rule. PFD management is optionally supported in the NEF and the SMF. Depending on the service level agreements between the operator and the Application Server Provider, the ASP may provide individual PFDs or the full set of PFDs for each application identifier maintained by the ASP to the SMF via the PFD Management service in the NEF. The PFDs are part of the application detection filters in the SMF/UPF to detect traffic generated by an application. The ASP may remove or modify some or all of the previously provided PFDs for application identifier(s).

The ASP manages (provision, update, delete) the PFDs through the NEF and NEF is responsible for the transfer of PFDs to the SMF and for storing them in the UDR. When the UPF receives the updated PFD(s) for a specific application identifier, the latest received PFD(s) shall overwrite any existing PFD(s) stored in the UPF. PFDs may also be managed using local O&M procedures, but the PFDs retrieved from NEF overrides any PFDs pre-configured in the SMF.

10.6 Network status analytics

During the initial design of the 5G network, basic functions related to collection and exposure of network analytics information was designed. The 3GPP entity responsible for such analytics is NWDAF which is responsible for operator managed network analytics. NWDAF provides Network Slice specific network data analytics to any NF without any user specific association with that data. The exposure of the information is done when an NF subscribes to such event. PCF and NSSF are two consumers of the NWDAF analytics information, these may be used for shaping policy decisions by the PCF or Network Slice selection by NSSF aided by the load level information it may receive.

The Release 15 analytics framework has rather limited capabilities. In Release 16, 3GPP is expanding the network analytics further to incorporate more extensive data collection and exposure (see Chapter 16 for more details on Release 16 enhancements).

10.7 Negotiation for future background data transfer

Negotiation for future background data transfer is a feature where an AF may contact a PCF to learn about the best time window and related conditions for future background data transfer. This allows an operator to provide information to application providers for when background data transfer is most suitable (e.g. during off-peak hour).

Future background data transfer may thus be valuable information for the AF in order to deliver certain type of data at certain time window when conditions are met. The AF interested in this information negotiates such reporting from the PCF, via NEF, if 3rd party AF is involved and do not have access to the operator's network information. The request includes an Application Service Provider (ASP) identifier, the volume of data to be transferred per UE, the expected amount of UEs, the desired time window and optionally, network area information. The AF provides as Network Area Information either a geographical area, or a list of TAs or list of NG-RAN nodes and/or a list of cell identifiers. When the AF provides a geographical area, then the NEF maps it based on local configuration into a list of TAs and/or NG-RAN nodes and/or cells identifiers that is provided to the PCF. PCF uses information available from UDR such as the ASP identifier, DNN, S-NSSAI and local configuration/information.

A transfer policy for this purpose includes a recommended time window for the background data transfer, a reference to a charging rate for this time window and optionally a maximum aggregated bitrate (indicating that the charging according to the referenced charging rate is only applicable for the aggregated traffic of all involved UEs that stays below this value). The maximum aggregated bitrate (optionally provided in a transfer policy) is not enforced in the network. The operator may apply offline CDRs processing and take action toward the ASP in question and charge the excess traffic differently toward the provider.

PCF provides the candidate list or selected transfer policy(ies) to the AF via NEF together with the Background Data Transfer reference ID. In case the AF is provided with more than one transfer policy, the AF selects one and informs the PCF about the selected transfer policy. The actual selected transfer policy is stored in the UDR for future use.

When the background data transfer is about to start, the AF provides for each UE the Background Data Transfer reference ID together with the AF session information to the PCF. The PCF retrieves the corresponding transfer policy from the UDR and derives the PCC rules for the background data transfer according to this transfer policy.

10.8 Session Management related policy and charging control

10.8.1 Session Management related policy concepts

Session Management related policy control provides operators with advanced tools for service-aware QoS and charging control. In wireless networks, where the bandwidth is typically limited by the radio network, it is important to ensure efficient utilization of the radio and transport network resources. Furthermore, different services have very different requirements on the QoS, which are needed for packet transport. Since a network generally carries many different services for different users simultaneously, it is important to ensure that the services can coexist and that each service is provided with an appropriate transport path. Session Management related policy control also provides the means to control charging on a per-session or per-service basis.

PCF enables centralized control to ensure that the services are provided with the appropriate transport and charging, for example in terms of bandwidth, QoS treatment, and charging method. The PCC architecture for Session Management related policy control enables control of the media plane for both the IP Multimedia Subsystem (IMS) and non-IMS services.

As described in previous chapters, the same 5GC QoS Flow management procedures are used independent of access. In this section we focus on how the operator can control those QoS procedures and the charging mechanisms used for each service session.

The term "service session" is also important here. The QoS Flow concept handles traffic aggregates – that is, all conformant traffic that is transported over the same QoS Flow receives the same QoS treatment. This means that multiple service sessions transported over the same QoS Flow will be treated as one aggregate. This QoS Flow handling still applies when Session Management related PCC is used. As we will see, however, PCC adds a "service aware" QoS and charging control mechanism that in certain aspects is more fine-grained – that is, it operates on a per-service session level rather than on a per-QoS Flow level.

The Session Management related PCC architecture for 5GS is in many ways similar to the PCC architecture for EPS and has inherited most capabilities of EPS PCC. Similar to EPS PCC, 5G PCC is an access-agnostic policy control framework and thus applicable to different kinds of 3GPP and non-3GPP accesses.

Even though PCC is in general considered an optional part of the architecture, some key features now require mandatory use of PCC. Services like IMS voice and Multimedia Priority Services require PCC e.g. as determined by GSMA.

When it comes to Session Management related policy control, policy control refers to the two functions gating control and QoS control:

- Gating control is the capability to block or to allow IP packets belonging to IP flow(s) for a certain service. The PCF makes the gating decisions that are then enforced by the SMF/UPF. The PCF could, for example, make gating decisions based on session events (start/stop of service) reported by the AF via the Rx/N5 reference point.
- QoS control allows the PCF to provide the SMF with the authorized QoS for the IP flow(s). The authorized QoS may, for example, include the authorized QoS class and the authorized bit rates. The SMF enforces the QoS control decisions by setting up the appropriate QoS Flows. The UPF also performs bit rate enforcement to ensure that a certain service session does not exceed its authorized QoS.

Charging control includes means for both offline and online charging. The PCF makes the decision on whether online or offline charging will apply for a certain service session, and the SMF enforces that decision by interacting with the charging systems and requesting UPF to report about data usage. The PCF also controls what measurement method applies – that is, whether data volume, duration, combined volume/duration, or event-based measurement is used. Again, it is the SMF that enforces the decision by requesting UPF to perform the appropriate measurements.

With online charging, the charging information can affect, in real time, the services being used and therefore a direct interaction of the charging mechanism with the control of network resource usage is required. The CHF may authorize access to individual services or to a group of services by granting credits for authorized traffic. If a user is not authorized to access a certain service, then the CHF may deny credit requests and additionally instruct the SMF to redirect the service request to a specified destination.

PCC also incorporates service-based offline charging. Policy control is used to restrict access and then service-specific usage may be reported using offline charging.

For more detailed relationship between policy control and operator's billing and charging, readers may consult Section 10.10.

10.8.2 Policy decisions and the PCC rule

The PCF is the central entity in PCC decisions also for Session Management related policy control. The decisions can be based on input from a number of different sources, including:

- Operator configuration in the PCF that defines the policies applied to given services;
- Subscription information/policies for a given user, received from the UDR;
- Information about the service received from the AF;
- Information from the SMF about applications detected by UPF;
- Information from the charging system about subscriber spending limit status;
- Information from the access network about what access technology is used, etc.

The PCF provides its decisions in the form of so-called "PCC rules". A PCC rule contains a set of information that is used by the SMF, UPF and the charging systems. First of all, it contains information (in a so-called "Service Data Flow (SDF) template") that allows the UPF to identify the User Plane traffic (the PDUs) that belong to the service session. All PDUs matching the packet filters of an SDF template are designated an SDF. The filters in an SDF template depend on the PDU Session type. For IP based PDU Session type, they contain a description of the IP flow and typically contain the source and destination IP addresses, and the protocol type used in the data portion of the IP packet, as well as the source and destination port numbers. These five parameters are often referred to as the IP 5-tuple. It is also possible to specify other parameters from the IP headers in the SDF template. For Ethernet PDU Session type, the SDF filters may also contain parameters from the Ethernet header (MAC source and destination addresses, VLAN tags etc.). The PCC rule also contains the gating status (open/closed), as well as QoS and charging-related information for the SDF. The QoS information for an SDF includes the 5QI, QoS Notification Control (QNC), Reflective QoS Control indication, UL/DL MBR, UL/DL GBR and ARP. The definition of the 5QI is the same as that described in Chapter 5. However, one important aspect of the QoS parameters in the PCC rule is that they have a different scope than the QoS parameters of the QoS Flow. The QoS and charging parameters in the PCC rule apply to the SDF. More precisely, the 5QI, QNC, MBR, GBR, and ARP in the PCC rule apply to the packet flow described by the SDF template, while the corresponding 5G QoS parameters discussed in Chapter 5 apply for the QoS Flow. A single QoS Flow may be used to carry traffic described by multiple PCC rules, as long as the QoS Flow provides the appropriate QoS for the service data flows of those PCC rules. Below we will discuss further how PCC rules and SDFs are mapped to QoS Flows. Table 10.2 lists a subset of the parameters that can be used in a PCC rule sent from PCF to SMF. For a full list of parameters, see 3GPP TS 23.503. As can be seen in Table 10.2, the PCC rule contains not only QoS control related information but also information for controlling charging, usage monitoring and traffic steering. Charging will be described further in Section 10.10 and the other session related PCC functionality in Section 10.9.

The same standardized 5QI values and corresponding 5G QoS characteristics outlined in Chapter 9 apply when 5QI is used in the PCC rule. The standardized 5QI and corresponding characteristics are independent of the UE's current access.

The discussion so far has assumed a case where the PCF provides the PCC rules to the SMF over the N7 interface. These rules, which are dynamically provided by the PCF, are denoted "dynamic PCC rules". There is, however, also a possibility for the operator to configure PCC rules directly into the SMF/UPF. Such rules are referred to as "predefined PCC rules". In this case the PCF can instruct the SMF to activate such predefined rules by referring to a PCC rule identifier. While the packet filters in a dynamic PCC rule are limited to the IP and Ethernet header parameters (e.g. the IP 5-tuple and other IP header parameters, or Ethernet header parameters), filters of a PCC rule that is predefined in the SMF/UPF may use parameters that extend the packet inspection beyond the simple IP and Ethernet header values. Such filters are sometimes referred to as Deep Packet Inspection (DPI) filters and they are typically used for charging control where more fine-grained flow detection is desired. The definition of filters for predefined rules is not standardized by 3GPP.

10.8.3 Use case with application authorization

As a result of the interactions with the PCF, the SMF/UPF perform several different functions. In this subsection we present a use case in order to get an overview of the dynamics of PCC and how PCC interacts with the application level, as well as the access network level. Some of the aspects brought up in the use cases will be discussed in more

Table 10.2 A subset of the elements that may be included in a dynamic PCC rule

Type of element	PCC rule element	Description
Rule identification	Rule identifier	It is used between PCF and UPF via SMF for referencing PCC rules
Information related to Service Data Flow detection	Service Data Flow Template Precedence	List of traffic patterns (packet filters) for the detection of the service data flow Determines the order, in which the service data flow templates are applied at UPF
Information related to policy control (i.e. gating and QoS control)	Gate status 5G QoS Identifier (5QI) QoS Notification Control (QNC) Reflective QoS Control UL and DL Maximum bit rates UL and DL Guaranteed bit rates ARP	Indicates whether a SDF may pass (gate open) or shall be discarded (gate closed) Identifier for the authorized QoS parameters for the service data flow. Indicates whether notifications are requested from 3GPP RAN when the GFBR can no longer (or can again) be guaranteed for a QoS Flow during the lifetime of the QoS Flow. Indicates to apply reflective QoS for the SDF. The maximum uplink (UL) and downlink (DL) bitrates authorized for the service data flow The guaranteed uplink (UL) and downlink (DL) bitrates authorized for the service data flow The Allocation and Retention Priority for the service data flow consisting of the priority level, the pre-emption capability and the pre-emption vulnerability
Information related to charging control	Charging key Charging method Sponsor Identifier Measurement method	The charging system (CHF) uses the charging key to determine the tariff to apply to the service data flow. Indicates the required charging method for the PCC rule. Values: online, offline or neither. An identifier, provided from the AF which identifies the Sponsor, used for sponsored flows to correlate measurements from different users for accounting purposes. Indicates whether the service data flow data volume, duration, combined volume/duration or event shall be measured.
Usage Monitoring Control Traffic Steering Enforcement Control	Monitoring key Traffic steering policy identifier(s) Data Network Access Identifier(s) N6 traffic routing information	The PCF uses the monitoring key to group services that share a common allowed usage. Reference to a pre-configured traffic steering policy at the SMF Identifier(s) of the target Data Network Access. Used for selective routing of traffic to DN. Describes the information necessary for traffic steering to the DNAI.

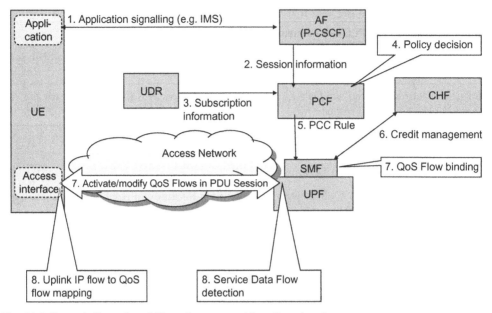

Fig. 10.9 Example flow of an IMS session setup with online charging.

detail later. The intention of placing the use cases first is that a basic overview of the procedures described in the use cases should simplify the understanding of the PCC aspects being discussed in the later subsections.

The following use case is intended to illustrate a service session set up with QoS control and online charging as illustrated in Fig. 10.9.

1. The subscriber initiates a service, for example an IMS voice call, and performs end-to-end application session signaling that is intercepted by the AF (P-CSCF in the IMS case). In the IMS case, the application signaling uses the Session Initiation Protocol (SIP). A description of the service is provided as part of the application signaling. In IMS, the Session Description Protocol (SDP) is used to describe the sessions.

2. Based on service description information contained in the application signaling, the AF provides the PCF with the service-related information over the Rx (or N5) interface. The session information is mapped at the AF from an SDP (e.g. SIP/SDP for IMS) into information elements in the Rx/N5 messages to the PCF. This information typically includes information about the application (type of service, bit rate requirements) as well as traffic parameters (e.g. the IP 5-tuple in case of IP) that allow identification of the packet flow(s) corresponding to this service session.

3. The PCF may request subscription-related information from the UDR. (The PCF may have requested subscription information earlier, but it is shown at this step for illustrative purposes.)

4. The PCF takes the session information, operator-defined service policies, subscription information, and other data into account when building policy decisions. The policy decisions are formulated as PCC rules.

5. The PCC rules are sent by the PCF to the SMF. The SMF will enforce the policy decision according to the received PCC rule, e.g. by providing relevant N4 rules to the UPF.

6. If the PCC rule specified that online charging should be used for this PCC rule, the SMF contacts the CHF to request credit according to the measurement method specified in the PCC rule.

7. The SMF performs QoS Flow binding to ensure that the traffic for this service receives appropriate QoS. This may result in the establishment of a new QoS Flow or modification of an existing QoS Flow. More details on QoS Flow binding are provided below.

8. The media for the service session is now being transported across the network and the UPF performs SDF detection to detect the packet flow for this service. This packet flow is transported over the appropriate QoS Flow. Further details on SDF detection can be found below.

Note that the above example is not exhaustive in any sense. There are many other scenarios and configurations. For example, for services that do not provide an AF or Rx/N5 interface, it is still possible to use PCC. In that case step 2 of the second use case would be omitted and the PCF could e.g. authorize PCC rules based on preconfigured policies without access to dynamic session data. As described further in Section 10.9, the PCF may also authorize PCC/QoS rules based on application detection information provided from SMF/UPF.

10.8.4 QoS Flow binding

The PCC rule needs to be mapped to a corresponding QoS Flow to ensure that the packets receive the appropriate QoS treatment. This mapping is one of the central components of PCC. The association between a PCC rule and a QoS Flow is referred to as QoS Flow binding. (Compare this to EPS where the same procedure also exists but is referred to as bearer binding). The QoS Flow binding is done by the SMF and when SMF receives new or modified PCC rule, the SMF evaluates whether or not it is possible to use the existing QoS Flow. The binding is performed based on the 5QI, ARP and a few other optional parameters in the PCC rule. The SMF ensures that each PCC rule is bound to a QoS Flow that has the same 5QI and ARP as was provided in the PCC rule (also the additional optional parameters are taking into account, if included in the PCC rule).

If one of the existing QoS Flows can be used, for example if a QoS Flow with the corresponding QCI and ARP already exists, the SMF may initiate QoS Flow modification procedures to adjust the bit rates of that QoS Flow. If it is not possible to use any existing QoS Flow, the SMF initiates the establishment of a suitable new QoS Flow. In particular, if the PCC rule contains GBR parameters, the SMF must also ensure the availability of a GBR QoS Flow that can accommodate the traffic for that PCC rule. Further details on the QoS Flow concept can be found in Chapter 5.

10.8.5 Service Data Flow detection

The Service Data Flow detection uses the service data flow template included in a PCC Rule provide by the PCF. The service data flow template defines the data for the service data flow detection as a set of service data flow filters or an application identifier referring to an application detection filter. The SMF maps the service data flow template in the PCC Rule into the detection information in a Packet Detection Rules to the UPF as further described in Chapters 6 and 14.

10.8.6 SMF related policy authorization request triggers

When the PCF makes a policy decision, information received from the access network may be used as input. For example, the PCF may be informed about the current access technology used by the UE, or whether the user is in their home network or is roaming. During the lifetime of a PDU Session, the conditions in the access network may however change. For example, the user may move between different access technologies or different geographical areas. There may also be situations where a certain authorized GBR can no longer be maintained over the radio link. In these cases, the PCF may want to re-evaluate its policy decisions and provide new or updated rules to the SMF. The PCF should thus be able to keep itself up to date about events taking place in the access network. To achieve this, the Npcf service includes functionality for the PCF to inform the SMF about which events the PCF is interested in. (In addition, the PCF may use the general Event Exposure services that have been defined for 5GC NFs). In 5G PCC terminology we say that the PCF provides Policy Control Request Triggers (PCRT) to SMF. The SMF will then Request for policy and charging control decision from the SMF to the PCF when a Policy Control Request Trigger has been met. (PCRT are also defined for non-Session Management related policy control, but then it is for access and mobility related events reported by AMF).

Also, the AF may be interested in notifications about conditions in the access network, such as what access technology is used or the status of the connection with the UE. Therefore, the AF may subscribe to notifications via the Rx/N5 reference point. The notifications over Rx/N5 are not directly related to renewed policy decisions in the PCF, but event triggers also play a role here. The reason is that if the AF subscribes to a notification over Rx/N5, the PCF will need to subscribe to a corresponding PCRT from SMF.

10.9 Additional session related policy control features

There are some additional session related policy control features which may be used in combination with other PCC functions described above. These functions may influence dynamic PCC rules, how a PDU Session may continue, end user's charging and AF interactions. In this section we describe briefly these functions, for more details, readers may consult 3GPP TS 23.503, 3GPP TS 23.502 and 3GPP TS 23.203.

10.9.1 Application detection

PCC supports application awareness when there is no explicit service session signaling like in case of IMS. Application Detection and Control enables detection of specified application traffic in UPF and to report on the start or stop of application traffic to the PCF (via SMF). It is also possible to apply enforcement actions for the application traffic. The enforcement actions may include blocking of the application traffic (gating), bandwidth limitations and redirection of traffic to another address.

In 5GS, the PCF activates PCC rules for a PDU session, in the SMF, for applications that require detection and reporting of the start or the stop event to the PCF. The SMF in turn then instructs the UPF to detect the events in question. When UPF detects the event it reports to SMF, and SMF then reports to the PCF on the event occurrence as per the PCC rule (e.g. start, stop, service data flow descriptions for the detected application traffic, if possible etc.). PCF may take appropriate policy decisions based on the information.

Fig. 10.10 illustrates an example simplified procedure for application detection flow for a PDU session.

10.9.2 Traffic steering control

Traffic steering control provides PCC rules to enable steering of certain traffic that matches certain detection conditions. Traffic steering control function enables policy control for two types of features:

1. N6-LAN traffic steering, where SMF/UPF applies a specific N6 traffic steering policy for the purpose of steering the subscriber's traffic to specific N6 service functions deployed either by the operator or a 3rd party service provider. This feature is also available with EPC (there it is called SGi-LAN traffic steering).
2. Selective traffic routing to a DN, where SMF/UPF diverts traffic matching traffic filters which may include a traffic steering policy ID and/or N6 traffic routing information dynamically provided by the AF for the DN, for example, resulting in Local Breakout. See also Chapter 6 for more information on how Selective traffic routing to a DN and AF influence on traffic routing is supported in the system. This feature has no direct counter-part in EPC.

Traffic steering control is triggered by the PCF via the SMF to the UPF and the steering of the traffic is performed at the UPF. The traffic steering may lead to, for example, for

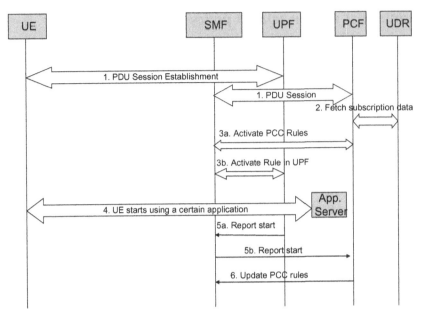

Fig. 10.10 Example flow for application detection activation.

the UPF to perform offload of certain traffic identified by the traffic descriptor to a local tunnel.

The PCF initiates the request of steering the detected service data flows matching application detection filters or service data flow filter(s) in PCC Rules. The PCF may use additional information such as network operator's policies, user subscription, user's current RAT, network load status, application identifier, time of day, UE location, DNN, related to the subscriber session and the application traffic as input for selecting a traffic steering policy. Traffic steering information is, when applicable, included in the PCC rule as indicated in Table 10.2.

10.9.3 Usage Monitoring Control

The Usage Monitoring Control information enables user plane monitoring of resources of both volume and time usage, and report the accumulated usage of network resources to PCF, (1) for individual applications/services, (2) groups of applications/services. Usage Monitoring Control applies to PDU sessions of type IP and Ethernet.

To enable enforcement of dynamic policy decisions based on the total network usage in real-time, usage monitoring needs to be possible for the accumulated usage of network resources on a per Session and user basis.

The PCF provides the SMF with the applicable thresholds based on either time or volume, for the usage monitoring for dynamic policy decisions. The SMF in turn then

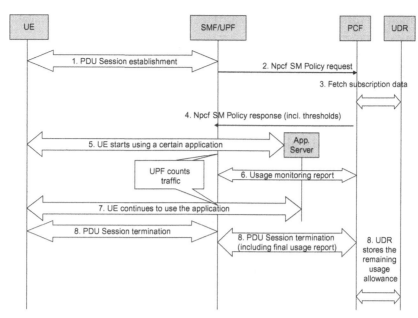

Fig. 10.11 Example flow of Usage Monitoring for a PDU session.

instructs the UPF to provide usage reports to the SMF. The SMF notifies the PCF when a threshold is reached. If the AF specifies a usage threshold, the PCF uses Sponsor Identity for monitoring the volume, time, or both volume and time of user plane traffic, and invoke usage monitoring on the SMF (Fig. 10.11).

The usage monitoring capability can be enabled for a single or a group of service data flow(s), or for all traffic of a PDU Session in the SMF.

10.9.4 Policy decisions based on spending limits

Spending limits for a subscriber enables the 5GS to enforce certain policies, when certain charging related criteria have been met. The Charging Function (CHF) maintains policy counter(s) to track spending for a subscription. Policy decisions based on spending limits can only be made when CHF has such subscription information available. A policy counter may be to determine when a subscriber needs to be restricted from data usage due to a certain limit such as amount of money left unspent, has been reached. A dynamic policy rule then may trigger restricting the user's access.

Policy decisions based on spending limits is a function where the PCF can take actions related to the status of policy counters that are maintained in the CHF. The PCF fetches the subscriber's spending information from the CHF and uses them for dynamic policy decisions for the subscriber, based on spending limit reports the PCF receives. The CHF is responsible for making information regarding the subscriber's spending available to the PCF using spending limit reports.

The CHF selection for a PDU session takes into account that the CHF can provide policy counters for spending limits, when applicable and the CHF address(es) is made available to the SMF for that PDU session by the PCF. The PCF updates the SMF with appropriate PCC rule(s) related to the spending limits. The CHF and the SMF interacts for online and offline charging purposes which may be affected by the spending limits.

The PCF is configured with the actions associated with the policy counter status that is received from CHF.

The PCF:

- may retrieve the status of policy counters in the CHF,
- may subscribe to spending limit reporting for policy counters from the CHF,
- may cancel spending limit reporting for specific policy counter(s).

The PCF uses the status of each relevant policy counter, and any pending policy counter statuses if known, for its policy decision to apply operator defined actions, e.g. change the QoS (e.g. downgrade Session-AMBR), modify the PCC Rules to apply gating or change charging conditions. An example flow Fig. 10.12 illustrates a use of spending limit for Session AMBR update.

When a subscriber is removed from the CHF, the CHF informs PCF to remove the related policy counters for that subscriber.

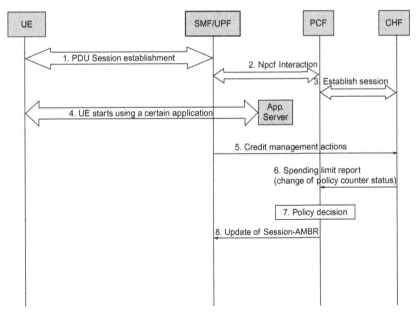

Fig. 10.12 Example spending limit changing Session AMBR for an ongoing PDU session.

10.9.5 Sponsored connectivity

Sponsored connectivity with PCC enables 5GS to allow the operator to generate revenue even if the mobile subscription is flat rate, by enabling additional revenue opportunities for both the Application Service Providers (ASP) and the operators. A user with limited data plans allowing only a nominal data volume per month and the service provider, for example, may dynamically sponsor additional volume allowance for the user to allow access to the services offered by the Application Service Providers. For example, the user may use the limited data plan to browse an online store for interested books; but once a book is purchased, the data usage for downloading the book is not deduced from the user's data plan allowance. The Sponsor may be a different business entity than the ASP. For example, a restaurant chain (Sponsor) could sponsor the mobile data traffic by handing out vouchers to their guests that gives access to content provided by an ASP. When an end-user later is accessing this content using the voucher, the restaurant chain would act as Sponsor. It could also be worth noting that the sponsored traffic may be granted a certain level of QoS (e.g. for video streaming).

In the 5GS architecture, the UDR holds the subscription profile for sponsored connectivity for a subscriber. The PCF enables an ASP to request specific PCC decisions (e.g. authorization to request sponsored IP flows, authorization to request QoS resources) based on this sponsored data connectivity profile and may receive a usage threshold from the AF. If the AF specifies a usage threshold, the PCF monitors the volume, time, or both volume and time of user plane traffic, and invoke usage monitoring on the SMF. The PCF also notifies the AF, when requested by the AF, when the SMF reports that a usage threshold for that specific usage is reached. If the usage threshold is reached, the AF may terminate the AF session or provide a new usage threshold to the PCF. Alternatively, the AF may allow the session to continue without specifying a usage threshold.

If the H-PCF detects that the UE is accessing the sponsored data connectivity in the roaming scenario with home routed access, it may either:

- allow the sponsored data connectivity in the service authorization request,
- reject the service authorization request,
- or initiate the AF session termination based on home operator policy.

If the PDU session terminates and the AF has specified a usage threshold then the PCF notifies the AF of the accumulated usage (i.e. either volume, or time, or both volume and time) of user plane traffic since the last usage report.

10.9.6 Event reporting from the PCF

The AF may subscribe/unsubscribe to notifications of events from the PCF for the PDU Session to which the AF session is bound.

The events that can be subscribed by the AF are listed in 3GPP TS 23.503), some example events include:

- PLMN Identifier Notification
- Change of Access Type
- Signaling path status
- Access Network Charging Correlation Information
- Access Network Information Notification
- Reporting Usage for Sponsored Data Connectivity
- Resource allocation status
- QoS targets can no longer (or can again) be fulfilled
- Out of credit

As can be seen from some of these events themselves, AF can use them to adjust the application behavior, apply credit/charging action toward the user or even terminate access to the application, as may deem appropriate.

10.10 Charging

As operators invest in new infrastructure and persuade end-users to enjoy the benefits of the newly deployed networks, the revenue generating options become a key factor for the business cycle. How the end-users/subscribers are actually charged and how billing information is packaged toward them is very much according to individual operator's business model and competitive environment they are operating in. From the 5GS point of view, same as has been for its predecessors in EPS, the system needs to enable collection of enough information related to different aspects of the usage for individual users, so that the operator has the flexibility to determine his own variant of billing as well as packaging toward the end-users. It has become increasingly important in today's competitive business environment for the operators to be able to provide lucrative and competitive option packages toward their potential customers while competing against free downloadable services. In addition, with new emerging partnerships with potential industry partners like in case of factory automation, Industrial IoTs, energy sectors etc. the single type of billing model or volume-based charging models may no longer be appropriate nor sustainable. The process of collecting information related to charging can provide tools and means for the operators to make flexible business driven charging models toward its varied customers.

The two main charging mechanisms provided by the model are still Offline and Online charging, though the terms Online and Offline are not necessarily related to how the end-users are billed at all. These two mechanisms are the means of how the charging-related data are collected and transported to the billing system for further processing as per individual customer's billing options and for settling accounting relations between operators and between operators and subscribers.

Offline charging facilitates collection of charging-related data concurrently as the resources are being used. The offline charging data is collected in the various elements provisioned to support such collection. The data is collected on an individual basis and then it may be sent toward the billing domain according to the operator's configuration.

Note that even though it is not required for the various types of information to be sent in a synchronous manner, the overall charging event must be able to receive and process all relevant data for a specific service/session in real-time in order to provide accurate, billable data toward end-users. Therefore, all offline processing of Charging Data Records (CDR) toward the end-user's billing is performed after the usage of the network resources is complete. The billing domain is responsible for generation and handling of the settlement/billing process offline.

In the case of Online charging, the network resource usage must be authorized and thus a subscriber must have a pre-paid account in the CHF in order for the Online pre-network resource usage authorization to be performed. The two methods used to achieve this are known as Direct Debiting and Unit Reservation. As their names imply, in case of Direct Debiting, the user is immediately debited the amount of resource usage needed for that specific service/session, where as in case of Unit Reservation a predetermined unit is reserved for the usage and the user is then allowed to use that amount, or less, for that service/session. When resource usage has been completed (i.e. session terminated, or the service is completed, etc.), the actual amount of resource usage (i.e. the used units) must be returned by the network entity responsible for monitoring the usage, to the CHF so that over-reserved amounts can be re-credited to the subscriber account, ensuring that the correct amount gets debited.

PCC makes it possible to have quite detailed charging mechanisms and allows for the possibility of operators having granular control over the subscriber's usage of the network resources. PCC also allows operators to offer various flexible charging and policy schemes toward their subscribers. More details can be found in Sections 10.8 and 10.9 on policy and charging control enabling better charging management and billing options. Features like Application Detection and Control, Usage Monitoring Control, Policy control based on subscriber spending limits, sponsored connectivity, all contributes directly in enhanced capability for more complex and dynamic charging capability in a standardized manner. The tools available allows for customized billing to be developed by the vendors catering to specific market/customer needs that allows operators the ability to distinguish themselves from each other and apply these in creative marketing campaign.

For 5G with EPC (e.g. EN-DC), Chapter 12 describes additional data volume collection associated with secondary RAT in use (i.e. NR), that is then provided to the Charging System. These data, together with the EPS charging support, can provide operators additional tools toward their end customers.

5GS Charging architecture for Release 15 has been quite simplified using Service Based architecture model, for more details on SBA see Chapter 13. The Converged Charging System (CCS) in 5G uses service based interface as shown in Fig. 10.13 as

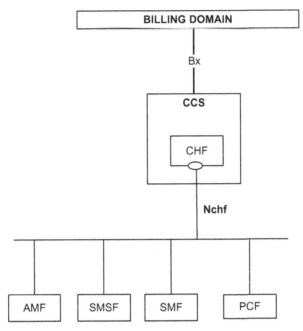

Fig. 10.13 Converged Charging System using SBI.

described in 3GPP TS 32.240. There is also a single NF (CHF) offering services for both online and offline charging.

The Nchf_SpendingLimitControl service exposed by CHF and consumed by the PCF is specified in 3GPP TS 23.502.

The main 3GPP specification that captures the mapping and details of incorporating 5G data connectivity into the overall charging system is 3GPP TS 32.255 and 3GPP TS 32.240.

The main components for the converged charging architecture for 5GS include (Fig. 10.14):

- The Charging Trigger Function (CTF) is embedded in the SMF and is responsible for generating charging events toward the CHF for PDU session connectivity converged online and offline charging.
- The CDRs generation is performed by the CHF acting as a Charging Data Function (CDF), which transfers them to the Charging Gateway Function (CGF).
- The CGF creates CDR files and forwards them to the Billing Domain (BD).

For Offline charging, the scenarios remain the same as for EPS, generating CDRs and reporting to the BD for subscriber's billing as well as inter-operator settlements in scenarios supported by:

- Event based charging;
- Session based charging.

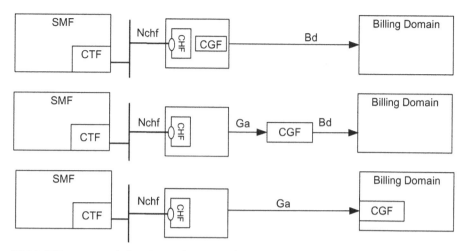

Fig. 10.14 5GS converged charging architecture, functional mapping to 5GC components.

In case of Online charging, the high-level scenarios supported are:
• Immediate Event Charging;
• Event charging with Unit Reservation;
• Session charging with Unit Reservation.
In addition, a converged charging scenario is enabled when offline charging and online charging are both applicable to a service delivery, the charging information of both offline charging and online charging can be provided in a single command, upon any triggers of the offline charging or online charging.

For more details for understanding the events and session based charging see 3GPP TS 32.290.

Subscriber charging provides the means to configure the end-user's charging information into the network. Charging data collected by the different PLMNs involved (e.g. HPLMN and VPLMN) and may be used by the subscriber's home operator, dependent upon the deployment and user's roaming status, to determine the network usage and the services, either basic or supplemental. There may also be the possibility to use external Service Providers for billing.

For those subscribers handled by Service Providers, the billing information is utilized for both wholesale (Network Operator to Service Provider) and retail (Service Provider to subscriber) billing. In such cases, the charging data collected from the network entities may also be sent to the Service Provider for further processing after the Home PLMN operator has processed the information as may be desired.

CHAPTER 11

Network slicing

11.1 Introduction

Traditional networks and their one-size-fits-all approach needs to be adapted so that the expected large number of network deployment use cases, many different subscriber types with diverse and sometimes contradictory requirements, and varying application usage can be supported. So, instead of using a single monolithic network serving multiple purposes, technology advancements such as Virtualization and SDN allows us to build logical networks on top of a common and shared infrastructure layer. These logical networks are then called Network Slices.

As mentioned in Chapter 3, the meaning of the term Network Slice vary in the industry, but in general a Network Slice is a logical network serving a defined business purpose or customer, consisting of all required network resources configured together. A Network Slice is realizing a complete network for any type of access and is an enabler for providing services. The used physical or virtual infrastructure resources may be dedicated to the Network Slice or shared with other Network Slices.

As the network slicing concept allows multiple logical networks to be created, they can then be accommodated to realize a wanted network characteristic and provide specific network capabilities to address a specific customer need. The customer here is not directly the end-user, but a business entity that has requested specific services from the network operator, e.g., an enterprise, another service provider or the network operator itself. The Network Slices are orchestrated and managed by management functions. The concept of network slicing and one definition is summarized in Fig. 11.1.

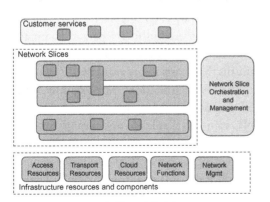

Network Slice is a **logical network** serving a defined **business purpose** or **customer**, consisting of **all** required network resources **configured** together.

- Complete network within a provider
- Enabler for services
- All access types
- Resources may be physical or virtual, dedicated or shared
- Independent/Isolated but may share resources

Fig. 11.1 A Network Slice definition.

5G Core Networks
https://doi.org/10.1016/B978-0-08-103009-7.00011-9

What, then, is the benefit with network slicing? The network slicing concept assumes virtualization and automated orchestration and management, and the expectations is that when these are used together they provide:

• Better customer experience by per customer adaptations and optimizations
• Shorter time-to-market and time-to-customer
• Simpler resource management
• Increased automation
• Flexibility and agility
• Reduced risks by separation of concerns.

Depending on the service type, e.g., eMBB, URLLC, mIoT, and customer expectations, there may be different requirements to be addressed by a Network Slice, for example:

– Traffic capacity requirements per geographical area
– Charging requirement
– Coverage area requirement
– Degree of isolation requirement
– End-to-end latency requirement
– Mobility requirement
– Overall user density requirement
– Priority requirement
– Service availability requirement
– Service reliability requirement
– Security requirement
– UE speed requirement.

To address the various and possibly diverse requirements when designing Network Slices, the various resources, and logical functions may be placed in different parts of the network. Fig. 11.2 provides example realizations for some type of Network Slices.

To separate and "Slice" parts of the network is not a new concept. It is supported already by different mechanisms and for different purposes, e.g., the ability to share a radio network between different operators is supported by each operator using separate PLMN identities, or separating PS data can be done by establishing separate data paths (i.e., PDP Contexts in 3G, PDN Connections in 4G and PDU Sessions in 5G), see Chapter 4 for more information of the mechanisms that exists in previous 3GPP systems. These techniques are also available in 5GS, but they have limitations. For example, even if it is possible for an operator to get more than one Mobile Network Code (i.e., MNC part of the PLMN identity) the number of MNC values available are not enough for each Network Slice to be given a separate MNC. Also, the separation of PS data paths using separate DNNs would only enable separation of part of the network and would not meet the expectation of complete logical networks dedicated for the customer needs. Therefore, it was not possible to re-use any of the existing mechanisms for addressing the

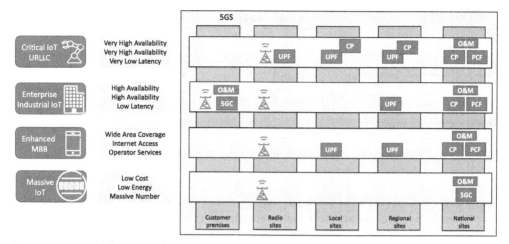

Fig. 11.2 Network Slice examples.

complete network slicing concept. However, of course, the existing means can also be used within a Network Slice, to achieve a limited separation between resources.

An automated management process is important to realize the expectations from the operators' customers and to enable the possibly large number of Network Slices in an operator's network. Some aspects of the Management and Orchestration of Network Slices are described in Section 11.2.

As to allow any type of Network Slices to be established and used it was agreed to develop a generic framework for the Network Slice selection. This framework is described in Section 11.3.

11.2 Management and orchestration

During the preparation and whole Lifecycle management process, the customer is able to provide its requirements using APIs from which the customer gets information of how the Network Slices perform, and is able to modify its requirements as to adapt to the needs of the customer. Fig. 11.3 provides a high-level view of the process in the preparation and the Lifecycle management of a network Slice Instance (NSI). In each of the steps the nature of isolated Network Slices aids to increase the speed in the process as there are less dependencies to consider.

The process of preparation and Lifecycle management of an NSI can be described as follows:

11.2.1 Preparation

Network Slice "blueprints" or "templates" are used to simplify the process. If a Network Slice template exists that meets the customer requirements, then the preparation process

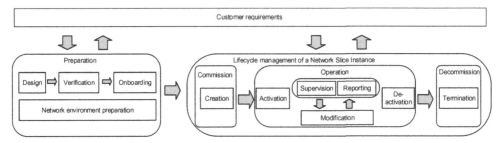

Fig. 11.3 Preparation and Lifecycle management of a Network Slice Instance.

can be shortened, as either the customer may be able to use an existing NSI, i.e., an existing NSI is then scaled to also meet the requirements from the new customer, or a new NSI is to be created using an existing Network Slice template. If that is the case, then the preparation phase can be excluded, and the ordering of the creation and activation can begin.

If there is no suitable Network Slice template, then a new one is designed using the customer requirements. Once a new Network Slice template is designed it is normally added to a catalogue of services and Network Slice templates that allow the preparation phase to be skipped or shortened for the next customer with the same or similar requirements.

The verification is simplified when done for dedicated Network Slices as there are fewer dependencies to consider compared to when using one network for a large range of customers, applications and services.

The onboarding includes uploading required information, e.g., the designed templates into the production system, validation of, e.g., templates and virtual machines (VM) images, and everything that is needed by the orchestration system in the next step.

During the preparation phase the network environment is prepared and other necessary preparations are done as required for the creation of an NSI. Then the ordering of the creation and activation can be done.

11.2.2 Commissioning

NSI provisioning in the commissioning phase includes creation of the NSI. During NSI creation all needed resources are allocated and configured to satisfy the Network Slice requirements.

11.2.3 Operation

The Operation phase includes the activation, supervision, performance reporting (e.g., for KPI monitoring), resource capacity planning, modification, and de-activation of an NSI.

Activation makes the NSI ready to support communication services.

Resource capacity planning includes any actions that calculates resource usage based on an NSI provisioning, and performance monitoring and generates modification polices as a result of the calculation.

The supervision and performance reporting include, e.g., monitoring, assurance and reporting of the performance according to the KPIs agreed as part of the Service Level Agreements (SLAs) for the NSI.

NSI modification could include, e.g., capacity or topology changes. The modification can include creation or modification of NSI resources. NSI modification can be triggered by receiving new Network Slice requirements or as the result of supervision/reporting.

The deactivation includes actions that make the NSI inactive and stops the communication services.

11.2.4 Decommissioning

NSI provisioning in the decommissioning phase includes decommissioning of non-shared resources if required and removing the NSI specific configuration from the shared resources. After the decommissioning phase, the NSI is terminated and does not exist anymore.

11.2.5 Summary

Each Network Slice can be used as a fully working network with all the functions and resources required for independent service. With a "Network as a Service" (NaaS) business model, customers can be granted visibility of *their* Network Slice, and then modify it to suit their changing needs, or create new Network Slices for new business opportunities.

Further details and additional references of the management and orchestration of Network Slices can be found in 3GPP TS 28.530.

11.3 Network Slice selection framework
11.3.1 Introduction

A flexible framework was developed for Network Slice selection purposes. This section describes the used identifiers and the selection mechanism.

11.3.2 Identifiers

As mentioned in Chapter 3, a Network Slice, or a Network Slice instance, is identified by a parameter called Single Network Slice Selection Assistance Information (S-NSSAI). The format of the S-NSSAI is shown in Fig. 11.4.

The SST part of the S-NSSAI is mandatory and indicates the type of characteristics of the Network Slice. The SST value range includes a standardized part, see Fig. 11.4 and

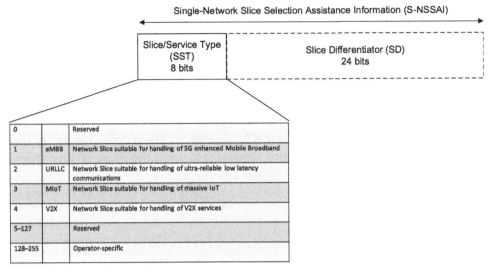

Single-Network Slice Selection Assistance Information (S-NSSAI)

| Slice/Service Type (SST) 8 bits | Slice Differentiator (SD) 24 bits |

0		Reserved
1	eMBB	Network Slice suitable for handling of 5G enhanced Mobile Broadband
2	URLLC	Network Slice suitable for handling of ultra-reliable low latency communications
3	MIoT	Network Slice suitable for handling of massive IoT
4	V2X	Network Slice suitable for handling of V2X services
5–127		Reserved
128–255		Operator-specific

Fig. 11.4 Format of the S-NSSAI.

3GPP TS 23.501 Table 5.15.2.2-1 for latest standardized values, and an operator specific part, i.e., non-standardized value range. The SD part is optional and can be used to differentiate between different Network Slices used for different customers, e.g., enterprises, or to differentiate between different Network Slice instances.

The S-NSSAI is defined in the scope of a PLMN, but, e.g., an S-NSSAI with solely a standardized SST part can be understood by any PLMN.

Before a UE can access a Network Slice, the UE needs to register it with the network and that is done using the Registration procedure. As to enable support for multiple Network Slices for the same UE, there is a need to be able to send one or more S-NSSAIs at the same time from the UE to the network and from the network to the UE. Therefore, one or more S-NSSAIs can be provided in an NSSAI. While sometimes "S-NSSAI(s)" or "S-NSSAIs" are used to describe that there may be one or more S-NSSAIs. The S-NSSAI and network slicing related information is used in various information elements in the system and their purpose is summarized in Table 11.1. Note, when Serving PLMN is mentioned in the table it can be either the HPLMN (i.e., non-roaming) or a VPLMN.

11.3.3 Network Slice availability

A Network Slice may be available in the whole PLMN or in one or more Tracking Areas of the PLMN. The availability means where the S-NSSAIs are to be supported by the involved NFs. A basic availability applicable for all UEs are configured as UE independent information at a configuration phase and updated when changed.

Table 11.1 Overview of Network Slice information used in the selection framework

Network Slice information	Purpose
Requested NSSAI	Contains up to eight S-NSSAIs that the UE wants to register, and that the UE provides to the Serving PLMN during Registration procedures. The UE needs to register, separately for each Access Type, the S-NSSAIs before the UE is allowed to use them, e.g., for establishing PDU Sessions
	The Requested NSSAI signaled by the UE in 5G-AN signaling allows the 5G-AN to select an AMF, and the Requested NSSAI sent in NAS is used as input to 5GC's selection of Network Slice(s) for the UE
	The UE may also send mapping of S-NSSAIs of the Requested NSSAI to HPLMN S-NSSAI values that is used by the Serving PLMN to understand which Subscribed S-NSSAIs the UE's request is referring to
Allowed NSSAI	NSSAI (one up to eight) provided by the Serving PLMN during, e.g., a Registration procedure, indicating the S-NSSAI values the UE can use in the Serving PLMN for the current RA. All the S-NSSAIs part of the Allowed NSSAI are available in the TAs of the RA
	Used as input to set the RFSP Index that 5GC provides to NG-RAN for steering the NG-RAN RRM strategies
	Can be sent by 5GC to 5G-AN over N2, per Access Type, when a UE is successfully registered, and used for 5G-AN specific policies
	AMF or NSSF determines and provides mapping of the Allowed NSSAI to HPLMN S-NSSAI values in case the S-NSSAI values differs from the Subscribed S-NSSAI values. This mapping information allows the UE to associate Applications to S-NSSAIs of the HPLMN as per the URSP rules or as per the UE Local Configuration
	Stored by the UE per PLMN and Access Type, also when switched off, until a new Allowed NSSAI is received by the PLMN. Used by the UE as input to set the Requested NSSAI
Configured NSSAI	NSSAI provisioned in the UE by a Serving PLMN, during Registration procedure or using UE Configuration Update procedure, and is applicable to one or more PLMNs
	Stored by the UE per PLMN until a new Configured NSSAI is received for the same PLMN. Used by the UE as input to set the Requested NSSAI Includes S-NSSAI values that can be used in the PLMN that provisioned it and may be associated with mapping of each S-NSSAI of the Configured NSSAI to one or more corresponding HPLMN S-NSSAI values

Continued

Table 11.1 Overview of Network Slice information used in the selection framework—cont'd

Network Slice information	Purpose
Default configured NSSAI	May be provided to the UE by the UDM in the HPLMN, via the AMF
	If no Configured NSSAI and Allowed NSSAI for the Serving PLMN are available, the UE sets the S-NSSAIs in the Requested NSSAI corresponding to the Default Configured NSSAI, if it is configured in the UE
	May be associated with mapping of each S-NSSAI of the Default Configured NSSAI to one or more corresponding HPLMN S-NSSAI values
Subscribed S-NSSAI(s)	S-NSSAIs (one or more) that a UE is subscribed to use in a PLMN, i.e., during roaming the UDM only sends the Subscribed S-NSSAIs to AMF in the VPLMN that the UE is allowed to use in the VPLMN
	The values correspond to the S-NSSAI values provided to the UE as part of the URSP rules and to the mapping of Serving PLMN S-NSSAI values to HPLMN S-NSSAI values
	HPLMN may set the RFSP Index taking into account the Subscribed S-NSSAIs
	Some subscription information is set per Subscribed S-NSSAI, e.g., DNN
Default S-NSSAIs	Subscribed S-NSSAIs marked as default S-NSSAI S-NSSAIs to be used in case the UE does not provide any permitted S-NSSAIs in the Requested NSSAI
S-NSSAI	Used for Network Slice selection, e.g., provided by the UE at PDU Session Establishment Can be used for NF discovery, i.e., NF producer registers with the NRF the S-NSSAI(s) it supports such that NF Consumers can discover NFs within a certain Network Slice
Rejected S-NSSAI(s)	An S-NSSAI may be rejected for the entire PLMN, or for the current Registration Area. While the condition is valid the UE does not try to register a rejected S-NSSAI Contain only values for the Serving PLMN
Network Slice instance (NSI) ID	Identifier for identifying the Core Network part of a Network Slice instance when multiple Network Slice instances of the same Network Slice are deployed, and there is a need to differentiate between them in the 5GC
	Optionally used within the 5GC, i.e., not provided to the 5G-AN nor to the UE
NSSAI inclusion mode	Access Stratum Connection Establishment NSSAI Inclusion Mode parameter, indicating whether and when the UE shall include NSSAI information in the 5G-AN signaling, e.g., RRC

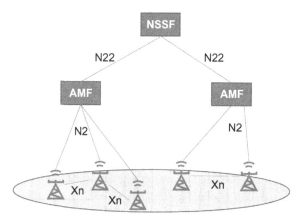

Fig. 11.5 Signalling of Network Slice availability information at configuration phase.

When a specific UE registers, UE specific policies may be applied on a per UE basis, e.g., dependent on the HPLMN of the UE. Such per PLMN specific policies may be configured in the AMF or in the NSSF.

The knowledge of where the Network Slices are available, from a UE independent perspective, can be configured by O&M and also be spread by signaling between the entities connected to each other as shown in Fig. 11.5.

The NSSF is configured by O&M on where Network Slices are to be made available, and also the 5G-AN is configured with Network Slice availability by O&M on a per TA level. Then the Network Slice availability information is spread using signaling over N22, N2 and Xn as follows:

- Over N2, see 3GPP TS 38.413, when the 5G-AN nodes establish the N2 connection, using N2 Setup, or when the N2 connection is updated, using RAN Configuration Update or AMF Configuration Update:
    - AMF learns the S-NSSAIs supported per TA by the 5G-AN.
    - 5G-AN learns the S-NSSAIs per PLMN ID the AMF to support.
- One or all AMFs per AMF Set provides and updates NSSF with the S-NSSAIs support per TA.
- The NSSF provides to AMF the per TA restricted S-NSSAIs at setup of the network and whenever changed. For the NSSF services, see 3GPP TS 23.502.
- At Xn Setup and NG-RAN node Configuration Update the 5G-AN nodes exchange the supported S-NSSAIs per TA with each other. For Xn procedures, see 3GPP TS 38.423.

11.3.4 Network Slice selection

A UE can be configured either by URSP rules (see Chapter 10) or by local configuration, the selection of a Network Slice or Network Slices is based on the UE request using this configured information. The Network Slice or Network Slices to be used is then decided by the network using subscription information, network policies, Service Level

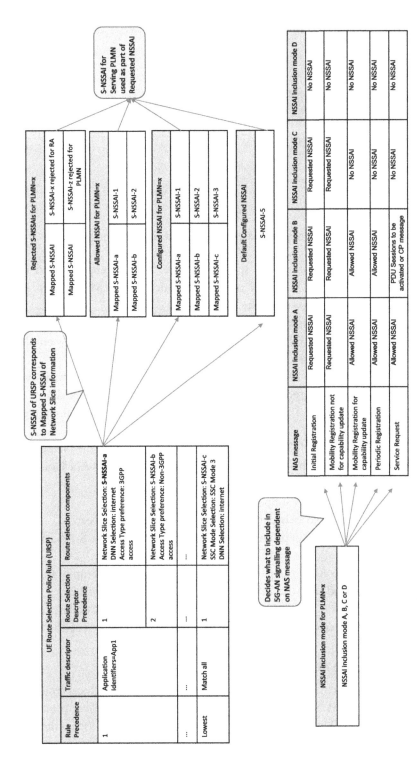

Fig. 11.6 Selecting S-NSSAIs for Requested NSSAI.

Agreements, Network Slice availability and what the UE requests as part of the Requested NSSAI.

The UE selects the S-NSSAIs to be used in a Requested NSSAI as follows:

- The UE decides which applications it wants to enable. How and when this is done is UE implementation specific.
- When selecting S-NSSAIs to be included in the Requested NSSAI the UE uses URSP rules, if available. The UE matches the applications the UE wants to enable with the URSP rules in precedence order and may get a matching rule containing S-NSSAI (e.g., S-NSSAI-a) from the Route selection components, see Fig. 11.6.
- The S-NSSAI from the matched URSP rule's Route selection component corresponds to the HPLMN S-NSSAI values as part of the UEs subscribed S-NSSAIs in the UDM and to the HPLMN values of the Network Slice information that the 5GC provides to the UE, e.g., Allowed NSSAI. When available the Mapped S-NSSAIs from each of the Rejected S-NSSAIs, Allowed NSSAI, Configured NSSAI includes the S-NSSAI with HPLMN values. If there is no Mapped S-NSSAI part, then the Serving PLMN S-NSSAI value is the same as the HPLMN value.
- The UE retrieves the S-NSSAI for the Serving PLMN to be used in the Requested NSSAI by first checking if the S-NSSAI from the URSP rules matches any of the Mapped S-NSSAIs of the Rejected S-NSSAIs, Allowed NSSAI, and Configured NSSAI respectively. If there is a match then the UE uses the S-NSSAI value in the right column corresponding to the Serving PLMN value.
- If the S-NSSAI matches an S-NSSAI of the Rejected S-NSSAIs, then the UE is not allowed to use the S-NSSAI either while the UE is in the current RA (if the S-NSSAI was rejected for the RA) or while the UE is using registered in the PLMN (if the S-NSSAI was rejected for the PLMN). In such case the UE tries to find another matching URSP rule for the application.
- If the S-NSSAI, from the URSP, matched any HPLMN value of the Allowed NSSAI or the Configured NSSAI, the UE uses the corresponding S-NSSAI for the Serving PLMN as input to the Requested NSSAI.
- If the UE did not find such match, and the UE has a Default Configured NSSAI, then the UE uses the S-NSSAI(s) as part of the Default Configured NSSAI as input to the Requested NSSI.
- Finally, if no Default Configured NSSAI is available then the UE does not include any S-NSSAI in the Requested NSSAI for the application. However, there is an exception to this and that is upon mobility from EPS, see Section 11.3.5.
- The UE matches all the applications the UE wants to enable with the URSP rules, and if there is no URSP rule match for an application except the URSP rule with the "match all" Traffic descriptor, then the UE can use local configuration, if available, to decide an S-NSSAI for the application.
- If no local configuration is available, then the UE uses the match-all URSP rule, if available.

• If no information is available, then the UE does not include any Requested NSSAI. When the UE has matched all applications with URSP rules and retrieved the S-NSSAIs for the Serving PLMN out of the Network Slice information as input to the Requested NSSAI, the UE checks the NSSAI inclusion mode for the PLMN, if available.

If the UE does not have NSSAI inclusion mode stored for the current PLMN and the Access Type, then the UE operates in NSSAI inclusion mode D over 3GPP access in the current PLMN and for non-3GPP access the UE operates in NSSAI inclusion mode C. The NSSAI inclusion mode controls how the UE provides Requested NSSAI in lower layers, i.e., in RRC for 3GPP access and in an EAP message for non-3GPP access. The NSSAI inclusion mode was introduced to protect the user privacy, i.e., as RRC messages from RRC-IDLE are not encrypted then anyone able to read the RRC message would be able to derive to which Network Slices the UE wants to register. As for non-3GPP access the access signaling is encrypted the UE can include the Requested NSSAI which enables the 5G-AN to select an AMF from an AMF Set supporting the Requested NSSAI.

Primarily the 5G-AN uses the UE provided GUAMI (or 5G-S-TMSI) to select an AMF, but for the cases when the UE has no GUAMI or the 5G-AN cannot use them, then the 5G-AN uses the Requested NSSAI for selecting the AMF. When the 5G-AN cannot use the GUAMI and does not get any Requested NSSAI from the UE, the 5G-AN selects a default AMF, which then selects appropriate Network Slice(s) and an AMF for the UE, usually by the help of an NSSF.

A high-level description of the Network Slice selection during a Registration procedure is shown in Fig. 11.7.

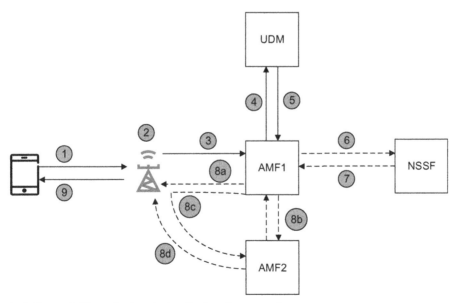

Fig. 11.7 Network Slice selection during Registration procedure.

(1) The UE provides a Requested NSSAI in 5G-AN signaling, e.g., RRC, and as part of the Registration message. The UE provides mapping of Requested NSSAI, if available, in the Registration message.

(2) If the UE is in CM-CONNECTED state the 5G-AN forwards the Registration message to the AMF based on the N2 connection of the UE.

If the UE is in CM-IDLE, the 5G-AN selects an AMF based on the GUAMI or 5G-S-TMSI, if the UE 5G-S-TMSI or the GUAMI cannot be used, then the 5G-AN uses the Requested NSSAI to select an AMF that during N2 Setup or AMF Configuration Update has indicated that it supports the S-NSSAIs in the Requested NSSAI.

If the 5G-AN cannot select an appropriate AMF, it forwards the Registration message to a default AMF.

(3) The 5G-AN forwards the Registration message to the selected AMF.

(4) If the AMF does not have subscription data for the UE, and the AMF cannot retrieve it from another AMF or UDSF, the AMF queries the UDM to retrieve UE subscription information for Network Slice selection, i.e., the Subscribed S-NSSAIs, or AMF retrieves the Access and Mobility Subscription data, including the Subscribed S-NSSAIs.

(5) The UDM provides the subscription information requested, including the Subscribed S-NSSAIs applicable for the Serving PLMN with an indication per S-NSSAI whether it is also a default S-NSSAI.

The UDM may provide an indication that the subscription data for network slicing has been updated for the UE.

The AMF may check whether the AMF can serve the UE, by checking if the AMF can serve all the S-NSSAI(s) from the Requested NSSAI present in the Subscribed S-NSSAIs, or all the S-NSSAI(s) marked as default in the Subscribed S-NSSAIs in the case that no Requested NSSAI was provided or none were part of the Subscribed S-NSSAIs. If mapping of Requested NSSAI was provided the AMF uses it to associate to the Subscribed S-NSSAIs.

If the AMF can serve the S-NSSAIs in the Requested NSSAI, the AMF remains the serving AMF for the UE and steps 6 and 7 are skipped.

(6) If the AMF cannot serve the UE or the AMF is configured to let NSSF perform the Network Slice selection, the AMF queries the NSSF by providing the available S-NSSAI information to the NSSF and in addition, PLMN ID of the SUPI and UE's current Tracking Area.

(7) The NSSF performs the Network Slice selection based on the received information, operator policies including SLA with the HPLMN in case of roaming, the availability of the Network Slice instances in the current TA, and the NSSF may return the following to the AMF:

    a. Allowed NSSAI, the mapping of each S-NSSAI of the Allowed NSSAI to HPLMN S-NSSAIs.

    b. Target AMF Set, or, based on configuration, the list of candidate AMF(s).

    c. NRF(s) to be used to select NFs/services within the selected Network Slice instance(s), and the NRF to be used to determine the list of candidate AMF(s) from the AMF Set.

    d. NSI ID(s) to be associated to the Network Slice instance(s) corresponding to certain S-NSSAIs.

    e. Rejected S-NSSAI(s).

    f. Configured NSSAI for the Serving PLMN and the associated mapping of the Configured NSSAI to HPLMN S-NSSAIs.

(8) The AMF may perform one of the following options:

    a. If the AMF is to serve the UE the AMF accepts the UE's Registration message and provides the following information to the UE:

        i. Allowed NSSAI, and the mapping of each S-NSSAI of the Allowed NSSAI to HPLMN S-NSSAIs, if any.

        ii. Rejected S-NSSAI(s), if any.

        iii. Configured NSSAI for the Serving PLMN and the associated mapping of the Configured NSSAI to HPLMN S-NSSAIs, if any.

        iv. RA by using the current TA and possibly adding TAs based on Network Slicing availability while keeping homogenous support of the S-NSSAIs for the whole RA.

        v. An indication that the subscription data for network slicing has been updated for the UE.

    b. If another AMF is to be re-allocated, by direct forwarding, the AMF forwards the NAS message from the UE to the target AMF, and the UE's SUPI and MM Context if available, and the information provided by the NSSF as described at step 7, except the AMF Set or list of AMF addresses.

    c. If the AMF decides to forward the NAS message to the target AMF via 5G-AN, the AMF sends a Reroute NAS message to the 5G-AN. The Reroute NAS message includes the information about the target AMF and the Registration Request message. If the AMF has obtained the information from NSSF, that information is included. The 5G-AN sends the information to the target AMF and indicates reroute due to network slicing such that the target AMF does not do yet another Network Slice selection.

    d. The target AMF then updates the 5G-AN with a new updated N2 termination point for the UE and provides the UE with the same information as described in step 8a. The AMF also includes Allowed NSSAI to the 5G-AN.

(9) The 5G-AN forwards the Registration acceptance to the UE.

    The UE stores the received information. If the AMF provided an indication that the subscription data for network slicing has been updated for the UE, the UE removes stored network slicing information for other PLMNs except the Default Configured NSSAI.

The UE first need to register S-NSSAIs before the UE can use S-NSSAIs in the 5GS services. Therefore, after the Registration procedure, the UE can use the S-NSSAIs that the UE received in the Allowed NSSAI, e.g., when establishing a PDU Session. The selection of 5GC parts during a PDU Session Establishment procedure is shown in Fig. 11.8.

(1) The UE sends a PDU Session Establishment request message with:

    a. An S-NSSAI from the Allowed NSSAI of the current Access Type that the UE derived using the URSP rules or by local configuration.

        i. If the UE cannot determine any S-NSSAI after performing the association of the application to a PDU Session using URSP rules and local configuration, the UE does not indicate any S-NSSAI in the PDU Session Establishment procedure.

    b. If the Mapping of Allowed NSSAI was provided to the UE, the UE provides both the S-NSSAI from the Allowed NSSAI and the corresponding S-NSSAI from the Mapping Of Allowed NSSAI, i.e., the S-NSSAI of the HPLMN.

(2) If the PDU Session Establishment request message does not contain an S-NSSAI, the AMF determines a default S-NSSAI for the requested PDU Session either according to the UE subscription, if it contains only one default S-NSSAI, or based on operator policy.

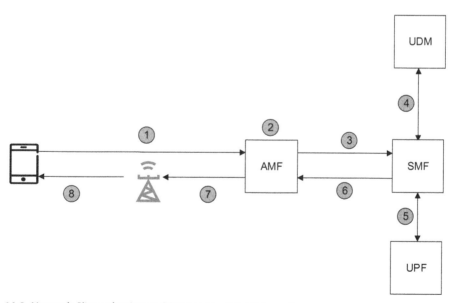

Fig. 11.8 Network Slice selection at PDU Session Establishment.

The AMF selects an SMF using the S-NSSAI for the Serving PLMN, and in the case of roaming with Home Routed PDU Session the AMF also selects an H-SMF using the S-NSSAI for the HPLMN.

(3) The AMF forwards the PDU Session Establishment request message to the SMF and includes:

a. the S-NSSAI from the Allowed NSSAI to the SMF, and

b. the S-NSSAI for the HPLMN in the case of roaming.

(4) If Session Management Subscription data for corresponding SUPI, DNN and S-NSSAI is not available, then SMF retrieves it from the UDM. The S-NSSAI used with the UDM is the S-NSSAI value for the HPLMN.

(5) At this point, the SMF creates an SM context and replies to the AMF and Secondary authentication/authorization may be executed as well as PCF selection, but that is not shown in the sequence.

The SMF then selects one or more UPFs using a wide range of information including S-NSSAI. See Chapter 6 for information regarding UPF selection.

(6) The SMF provides the PDU Session Establishment Accept message for the UE, and also N2 SM information for 5G-AN including the S-NSSAI, with the value for the Serving PLMN, for the PDU Session.

(7) The AMF forwards the information towards 5G-AN and the UE.

(8) The 5G-AN stores the S-NSSAI as associated to the PDU Session and provides the PDU Session Establishment Accept message to the UE and establishes the necessary User Plane resources for the PDU Session.

11.3.5 Network Slice selection at interworking with EPS

The principles for interworking with EPS apply as described in Chapter 7, but some additional functionality is needed to better support network slicing.

3GPP Release-15 supports EPS and 5GS interworking with network slicing, but with some limitations, e.g., when default AMF and default V-SMF were selected at mobility from EPS to 5GS there is no defined procedure to reallocate them during CM-CONNECTED state. However, 3GPP Release-16 allows reallocation of AMF and V-SMF after mobility to 5GS during CM-CONNECTED state, as well as AMF reallocation of the AMF selected by the MME at idle mode mobility from EPS based on the S-NSSAIs associated with the established PDU Sessions.

The additions to the interworking procedures to support network slicing can be summarized for each type of mobility procedure as follows.

The Fig. 11.9 provides an overview of the Network Slice selection aspects during idle mode mobility from EPS to 5GS.

For idle mode mobility from 5GS to EPS, the following additional principles applies for the Network Slice selection aspects:

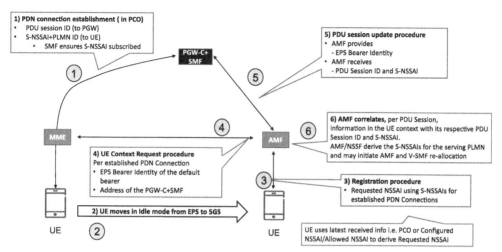

Fig. 11.9 Idle mode mobility from EPS to 5GS.

- In case of Dedicated Core Network are used in EPS, see 3GPP TS 23.401, the UE Usage type may be retrieved by the new MME from the old AMF as part of the UE MM context transfer procedure, and may be used to decide if a different MME is to be selected by invoking redirection.
- Since, in EPS, the same PGW shall be used for all PDN Connections using the same APN.
  - o At EBI bearer allocation, if PGW-C + SMF serves multiple PDU Sessions for the same DNN but different S-NSSAIs for a UE then SMF only request EBIs for PDU Sessions served by a common UPF (PSA). In case different UPF (PSA) are serving those PDU Sessions, then SMF chooses one UPF (PSA).
  - o If a PDU Session from another SMF already exists towards same DNN, AMF either rejects the EBI assignment request, or revokes EBI(s) from existing PDU Session(s) to same DNN but different SMF. The AMF makes the decision based on the operator policy.
  - o The AMF stores DNN and PGW-C + SMF for PDU Session(s) supporting EPS IW to UDM
  - o The above applies when S-NSSAI(s) for PDU Sessions are different, otherwise same SMF is selected for PDU Sessions to same DNN.

For connected mode mobility from 5GS to EPS the following additional principles applies for the Network Slice selection aspects:

- At connected mode mobility from 5GS to EPS the source AMF selects the target MME based on the AMF Region ID, AMF Set ID and target location information and forwards the UE context (including UE Usage type, if available) to the selected MME.

- When the handover completes the UE performs a Tracking Area Update. This completes the UE registration in the target EPS.
- After concluded handover the initially selected MME may select a different target MME, e.g., based on the UE Usage type, by invoking redirection. This is achieved via a Tracking Area Update from the UE after being forced to Idle mode.
- Note that only PDU Sessions which have EBI allocated are candidates for being moved to EPS.

For connected mode mobility from EPS to 5GS the following additional principles applies for the Network Slice selection aspects:

- At connected mode mobility from EPS to 5GS the source MME selects the target AMF based on target TAI and any other available local information (including the UE Usage Type if one is available for the UE in the subscription data) and forwards the UE context to the selected AMF.
- In the home-routed roaming case, the AMF selects default V-SMFs (using default/ special S-NSSAI used for interworking).
- During the preparation phase the PGW-C + SMF sends the PDU Session IDs and the related S-NSSAIs to AMF.
- If S-NSSAI for interworking is different from S-NSSAI to be used in VPLMN, AMF updates default V-SMF which updates NG-RAN with appropriate VPLMN S-NSSAI value(s).
- When the Handover completes the UE performs a Registration procedure. PGW-C + SMF has provided the UE with corresponding S-NSSAIs per PDU Session ID at PDN Connection establishment thus enabling a Requested NSSAI. As part of the Registration procedure the UE obtains an Allowed NSSAI.

CHAPTER 12

Dual connectivity

12.1 Introduction

Dual Connectivity (DC) is a feature where the UE and RAN can receive and transmit data over two base stations simultaneously. Or, in more technical terms, it provides the ability to utilize radio resources provided by two independently operating Cell Groups, where the two Cell Groups may be controlled by two different radio nodes, connected to a single core network. DC aims, e.g., at increasing user throughput, providing improved mobility robustness, and supporting load-balancing between RAN nodes.

A Cell Group in DC is defined as a group of serving cells associated with either of the RAN nodes forming the DC. Dual Connectivity is specified for both EPS and 5GS. In this chapter we focus on the DC architecture and features associated with an NR access Cell Group and an E-UTRA Cell groups, specified in both variants and controlled by radio nodes connected to 5GC. Chapter 4 described the concepts and functions of EPC for 5G; the principle 5G component is the use of NR access with the E-UTRA.

The concept of Dual Connectivity (DC) was first introduced for EPS with two Cell Groups, both providing E-UTRA resources, for UEs capable of multiple Receive/Transmit (Rx/Tx) in connected state. The eNB the UE first connects to and from which all signaling toward Core Network is performed is known as the Master node and this is the access node that UE's location is associated with. Once the UE reaches Connected state this Master eNB can request another, Secondary eNB to offload data traffic. Dual Connectivity is enabled by two independently operating schedulers located in the involved eNBs. One scheduler, located in the Master node, controls access to radio resources provided by a Master Cell Group (MCG), the other scheduler, located in the Secondary node, controls access to radio resources provided by a Secondary Cell Group (SCG), These eNBs connected via non-ideal backhauled X2 interface, as described in 3GPP EUTRA architecture specification 3GPP TS 36.300.

While the first DC variant that was specified operates on different frequency bands of the same Radio Access technology (E-UTRA), the Multi-Radio DC, MR-DC, an evolution of the first DC concept, allows dual connectivity via two different Radio Access technologies, NR access and E-UTRA, provided by two radio nodes. The two radio nodes are in general connected via non-ideal backhaul, operates also in this case with two independent schedulers. The MR DC concept is developed by 3GPP and specified in 3GPP TS 37.340. One example of a MR-DC configuration is NR connected to 4G core (aka EPC) via EUTRA (4G Radio). This configuration is an attractive option for operators as it enables fast market roll-out of the higher bandwidth/data of NR, while re-using the EPC.

The MR-DC concept can be deployed in different configurations and connectivity options. The 3GPP RAN structured standardization discussions along various configurations and connectivity options for 5G Radio to 5G as well as 4G core, these were initially called "Options 1 to 8."

These "Options 1 to 8" depicted variants permuting the following variables:

1) Core Network to which the RAN is connected (EPC or 5GC).

2) Radio Access Technology used for user plane data (NR access or E-UTRA or both). We remind the readers again here these options (detailed analysis of which is provided in Chapter 3) in the context of and for DC we are dealing with Options 3, 4 and 7 exclusively. Options 1, 2, 5 and option 6 do not use DC (Fig. 12.1).

3GPP specified various connectivity variants for each of the options. The terminology can be depicted as follows (Fig. 12.2).

For example, if E-UTRA provides the Master Cell Group (MCG), this Cell Group is controlled by a Master Node (MN), which is also associated with a user plane termination toward the Core Network and a higher radio layer part associated with lower radio layer parts of the E-UTRA Cell Group:

- A data radio bearer is said to be Master-Node-terminated (MN-terminated), if the user plane termination resources are associated with Master-Node owned UP termination resources.
- If, in the above example the radio resources are provided by an E-UTRA Cell Group only, the data radio bearer is said to be an *MN-terminated MCG bearer*, if only be the NR access Cell Group, it is an *MN-terminated SCG bearer*, if both Cell Groups provide radio resources, it is an *MN-terminated split bearer*.

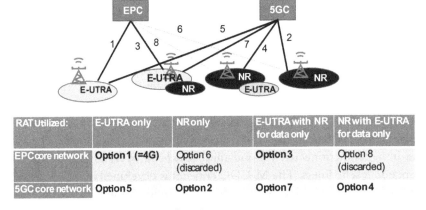

RAT Utilized:	E-UTRA only	NR only	E-UTRA with NR for data only	NR with E-UTRA for data only
EPC core network	Option 1 (=4G)	Option 6 (discarded)	Option 3	Option 8 (discarded)
5GC core network	Option 5	Option 2	Option 7	Option 4

Fig. 12.1 DC related architecture options for 5G.

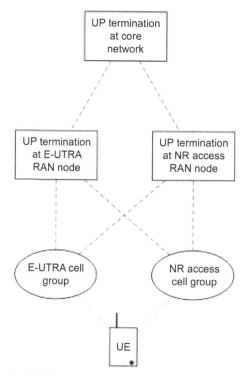

Fig. 12.2 UP Termination for MR-DC.

Table 12.1 outlines the UP Termination across Master Cell Groups and Slave Cell Groups.

Fig. 12.3 illustrates the DC variant of these options with option 3 is represented as EN-DC (which was at the end replaced by the term MR-DC with EPC) and options 4 and 7 are represented by NE-DC and NGEN-DC. In addition, a new DC variant has been developed, similar to the original DC architecture for 4G, but with the combination of NR as both Master and Secondary RAN nodes connected to 5GC.

So, for MR-DC with EPC, DC is an optional feature where the Master Node (MN) always provides E-UTRA and NR access may be added as Secondary RAT.

Table 12.1 UP termination

UP termination/cell group involved	MCG	SCG	MCG and SCG
MN terminated	MN terminated MCG bearer	MN terminated SCG bearer	MN terminated Split bearer
SN terminated	SN terminated MCG bearer	SN terminated SCG bearer	SN terminated Split bearer

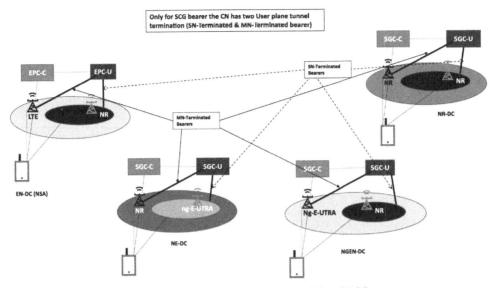

Fig. 12.3 High level architecture combination for MR-DC via EPC and 5GC.

The 5G specification process in 3GPP adopted Multi-RAT DC from the beginning, enabling an interchangeable and flexible DC architecture where radio resources are provided by either Cell Group (MCG or SCG) or both. Each Cell Group may assume either RAT, i.e., NR access or E-UTRA, and the termination of UP connectivity toward the Core Network is provided by either the Master Node or the Secondary Node (denoted as MN-terminated or SN-terminated bearers).

We focus on MR-DC with 5GC in this chapter since this book's primary scope is the 5G Core Network with 5G Radios. MR-DC with EPC architecture, from an overall architecture and core network perspective, follows the same functional evolution so readers should be able to deduce the MR-DC with EPC concept easily from MR-DC with 5GC.

12.2 Multi-RAT Dual Connectivity overall architecture

Multi-RAT Dual Connectivity allows E-UTRA and NR access to be connected in different combinations enabling a multiple Rx/Tx capable UE with active radio bearers to benefit from radio resources that can be made available to the UE simultaneously, while connectivity toward the core network is under one of the RAN nodes' control. Depending on the configuration, which is described later, the core network may either be EPC or 5GC.

The Radio Access Network Architecture is defined to consist of the following nodes listed in Table 12.2.

The general principle for DC remains the same, the RAN node controlling the UE's control plane connectivity toward the Core Network, i.e., the Master Node (MN), may

Table 12.2 Radio access network architecture nodes

Core network connectivity	Radio access network	RAN node providing E-UTRA	RAN node providing NR access
EPC	E-UTRAN	eNB	en-gNB
5GC	NG-RAN	ng-eNB	gNB

provide the UE (multiple Rx/Tx capable) with user plane radio resources controlled by either the MN or the Secondary Node (SN) or both. The Master RAN node is the node that provides the control plane connection to the Core network and thus manages all the NAS signaling and core network/UE related signaling toward the Core network. The SN connects to the Core network via the user plane only. Depending on the type of MR-DC bearer, the user plane from the SN may be directly connected to the core network (SN terminated) or transfer user data via the MN user plane using X2/Xn interface. The MN and SN are connected to each other via both control and user plane interfaces known as X2/Xn. Depending on the UE's Cell Group configuration, the UE to RAN connections vary for the user plane usage of DC. Whereas, from the Core Network perspective, the MR-DC Cell Group configuration is transparent, the Core Network is only affected by providing connectivity to either the SN or MN UP termination, in a sufficiently transparent manner.

Fig. 12.4 shows a high-level view of the architecture for MR-DC with CP and UP connectivity to CN. The left diagram shows a two user plane tunnel termination configuration where both MN and SN have GTP-U tunnels toward the Core Network GW (i.e., UPF in case of 5GC) and the right diagram show a single user plane tunnel configuration where MN terminates the GTP-U tunnel to the Core Network GW and

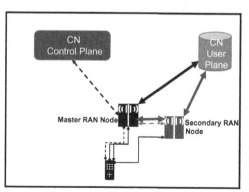
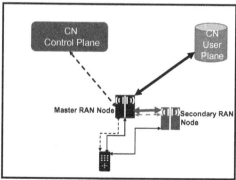

Fig. 12.4 High level generic overall architecture for MR-DC with CN connectivity.

SN forwards user data to MN via Xn-U which is the user plane connection between the two RAN nodes.

The MR-DC architecture described above has four different variants, with one connected to EPC and the other 3 connected to 5GC as described in Section 12.1. These variants are further described in detail below.

MR-DC with EPC (aka EN-DC) contains one eNB and one gNB (also referred to as en-gNB to separate from gNB where the gNB also has control plane interface toward 5GC). With EN-DC the UE is connected to one eNB that acts as a MN and one gNB that acts as a SN. The MN eNB is connected to the EPC via the S1-C/S1-U interfaces and to the gNB via the X2 interface. The SN gNB may be connected to the EPC via the S1-U interface and to other gNBs via the X2-U interface, depending on the type of MR-DC bearers supported, depending on the combination of MCG, SCG, Split bearers used for a PDN connection.

Fig. 12.5 illustrates MR-DC with EPC architecture where two S1-U user plane tunnels have been established.

MR-DC with 5GC may have any of the following configuration and their combination:

1. **NG-EUTRA—NR dual connectivity (aka NGEN-DC)** where a UE is connected to one ng-eNB that acts as a MN and one gNB that acts as a SN. The ng-eNB is connected to the 5GC via NG-C/NG-U (N2/N3) interface and the ng-eNB is connected to the gNB via the Xn interface.
2. **NR—NG-EUTRA Dual Connectivity (aka NE-DC)** where a UE is connected to one gNB that acts as a MN and one ng-eNB that acts as a SN. The gNB is

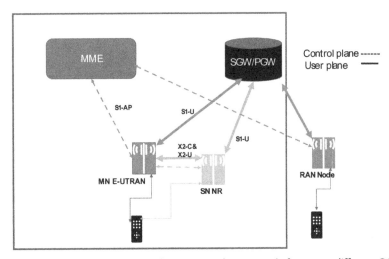

Fig. 12.5 MR-DC with EPC architecture with two user plane tunnels from two different RAN nodes.

connected to 5GC via NG-C/NG-U (N2/N3) interface and the ng-eNB is connected to the gNB via the Xn interface.

3. **NR—NR Dual Connectivity (aka NR-DC),** where a UE is connected to one gNB that acts as a MN and another gNB that acts as a SN. The master gNB is connected to the 5GC via NG-C/NG-U (N2/N3) interface and to the secondary gNB via the Xn interface. The secondary gNB may be connected to the 5GC via the NG-U interface.

Fig. 12.6 illustrates the high-level architecture for MR-DC with 5GC where either NG-RAT may be the MN or SN.

In addition, NR-DC can also be used when a UE is connected to two gNB-DUs, one serving the MCG and the other serving the SCG, connected to the same gNB-CU, acting both as a MN and as a SN. This characteristic is unique to NR-DC and comes from the Control and User plane functions within gNB architecture as defined in 3GPP TS 38.401. A simplified description of the NR architecture using CU and DU split is illustrated in Fig. 12.4. Fig. 12.7. shows overall NR architecture as defined in 3GPP TS 38.401, shown here for reference merely to illustrate the gNB-CU/gNB-DU configuration. A gNB which may be further decomposed into a gNB-CU-CP, multiple gNB-CU-UPs and multiple gNB-DUs has the following association:

– One gNB-DU is connected to only one gNB-CU-CP;
– One gNB-CU-UP is connected to only one gNB-CU-CP;
– One gNB-DU can be connected to multiple gNB-CU-UPs under the control of the same gNB-CU-CP;
– One gNB-CU-UP can be connected to multiple gNB- DUs under the control of the same gNB-CU-CP.

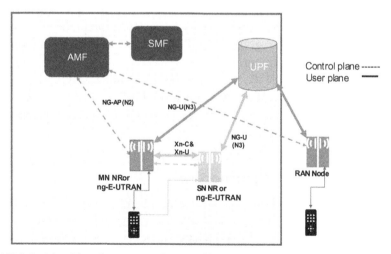

Fig. 12.6 MR-DC with 5GC architecture with two different NG-RAN nodes.

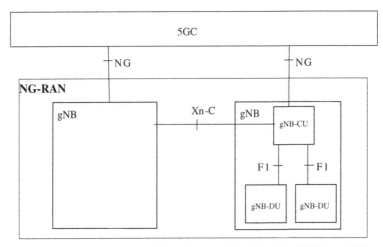

Fig. 12.7 Overall NG-RAN connected to 5GC architecture for NR with gNB CU and DU separation.

On a high level, the functionality of these entities can be described as follows:

For NG-RAN, the NG and Xn-C interfaces terminate in the gNB-CU when a gNB comprises of a gNB-CU and gNB-DU(s). For EN-DC, the S1-U and X2-C interfaces terminate in the gNB-CU when a gNB comprises of a gNB-CU and gNB-DU(s). The gNB-CU and connected gNB-DUs are seen only as a gNB to other gNBs and the 5GC.

The node hosting user plane part of NR PDCP (e.g., gNB-CU, gNB-CU-UP, and for EN-DC, MeNB or SgNB depending on the bearer split) shall perform user inactivity monitoring and further informs its inactivity or (re)activation to the node having C-plane connection toward the core network (e.g., over E1 for 5GS, X2 for EPS). The node hosting NR RLC (e.g., gNB-DU) may perform user inactivity monitoring and further inform its inactivity or (re)activation to the node hosting control plane, e.g., gNB-CU or gNB-CU-CP. The details of the gNB internal architecture is not further discussed here.

In the case of MR-DC with 5GC, irrespective of the number of QoS Flows, there are two N3 tunnel terminations at the RAN and at the UPF for PDU Session(s) in case of SCG bearer. In case of MR-DC with EPC the two S1-U tunnel termination points are between RAN and SGW node.

12.3 MR-DC: UE and RAN perspective

In order for MR-DC to work, it requires that two RATs serving a single UE in connected state via a single CN. The serving cells are configured with Master Cell Group (MCG) containing the serving cells of the MN and Secondary Cell Group (SCG) containing the serving cells of the SN. There are three types of bearers known as MCG, SCG and Split bearers. From a network perspective, each bearer (MCG, SCG and split bearer) can be terminated either in MN or in SN.

When SCG is configured, there is always at least one SCG bearer or one Split bearer. In MR–DC, MCG bearer is typically defined as *a radio bearer with an RLC bearer only in the MCG*. For the User plane:

- For MCG bearers, the SN is not involved in the transport of user plane data for this type of bearer(s) over the Uu.
- For Split bearers, the PDCP data is transferred between the MN and the SN via X2-U/Xn-U. The SN and MN are involved in transmitting data of this bearer type over the Uu.
- For SCG bearers, the MN is not involved in the transport of user plane data for this type of bearer(s) over the Uu.

The core network is agnostic to the mapping of DC bearers between the UE and the Radio network. In case the addition of bearers requires the addition or modification of the SN user plane tunnel toward the CN, CN procedures are invoked to trigger addition of a new user plane tunnel. But all control signaling between the UE and CN are handled via the MN itself (e.g., NAS signaling) and the MN only has control plane signaling to the CN (e.g., S1-AP or NG-AP signaling). For split bearers involving *SN terminated SCG bearers* and *MN terminated MCG bearers*, PDCP data is transferred between the MN and the SN via the MN–SN user plane interface and then transferred to the CN via MN user plane interface.

Fig. 12.8 explains how the user plane protocol structure works in the UE as defined in 3GPP TS 37.340.

In the case of the Control Plane, a UE has a single RRC state, based on the MN RRC and corresponding single Control plane connection via the MN toward the Core

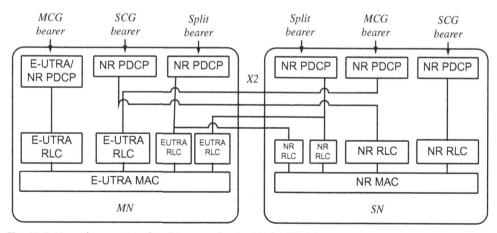

Fig. 12.8 User plane protocol architecture for the UE for MR-DC.

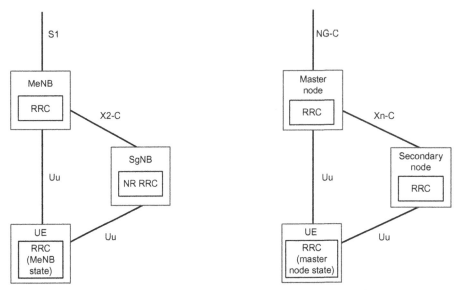

Fig. 12.9 UE control plane Radio protocol architecture for MR-DC for EPC and 5GC.

Network. Each RAN node has its own RRC entity, i.e., E-UTRA version if the node is an eNB or NR version if the node is a gNB, which generates the RRC PDUs to be sent to the UE accordingly. Fig. 12.9 (from 3GPP TS 37.340) illustrates the UE control plane connectivity protocol structure.

Since the MN node has the control plane connection to the Core Network, all associated functions such as mobility management, session management, location reporting, access restriction, user plane management are handled by the MN toward the UE and the core network as applicable.

A UE that supports MR-DC uses different capabilities (e.g., specific radio related information) and these capabilities are transported in the system via what is known as UE capability containers. These sets of capability information are required to establish appropriate DC support between the UE and the MN and the SN. These capability containers contain a common set which is related to MR-DC and accessible to both MN and SN but there are also capability containers that are specific to the radio access technology only relevant to the specific node supporting the RAT in question. These different containers are known as MR-DC capability container, E-UTRAN capability container and NR capability container.

12.4 MR-DC: Subscription, QoS flows and E-RABs, MR-DC bearers

To better understand how MR-DC works, we need to get into some details of the radio network and UE aspects first. Here we provide some basic examples of such functions required by MR-DC feature without going into details about the actual Radio system itself. As indicated in Section 12.1, the UE must be able to connect to two RATs simultaneously and be able to receive and transmit on both RATs simultaneously. This requires careful configuration of radio layer 1 (physical layer) between the two cells that are enabling the MR-DC operation. In addition, the UE must be able to interpret how to operate in MR-DC environment based on information made available by the MN via dedicated RRC signaling which includes Radio related information such as radio frame timing, system information for initial configuration required by the UE. On the RRC layer, for example, depending on the reason for RRC configuration, either MN itself or MN assisted by information provided by SN, or SN, may (re)configure the UE with required parameters. In case combined MN/SN RRC messages are required for both MCG and SCG reconfiguration, MN is responsible for the coordination between MN and SN and the SN reconfiguration is encapsulated in an MN RRC message so that the UE has combined configuration and UE can process them jointly.

Similarly, the UE needs to be configured with two MAC entities: one MAC for the MCG and one MAC for the SCG. The PDCP and SDAP layers, as applicable, also needs to adhere to the specific requirements for MR-DC. For example, for NR-DC, the UE has single SDAP layer per PDU Session whereas in the network both the MN and SN may have their own SDAP layer for the same PDU session resulting in two SDAP layers per PDU session.

Some of the important aspects specific to a MR-DC configuration are handling of bearers, QoS, SN addition/modification/removal, Inter MN Handover with or without SN, MN to/from ng-eNB/eNB/gNB change, User Data usage reporting in relation to SN. We discuss very briefly some of these functions later, but firstly some of the additional core network managed functions added to facilitate better utilization of the DC are discussed first.

From an operator's perspective, MR-DC provides an opportunity to enhance end users' experience in busy/crowded areas via SN support. Enabling the feature specially in EN-DC configuration (aka EPC with 5G) allows for the possibility of immediately visible performance improvements and therefore the opportunity to add new services to attract prospective customers. With that in mind, requirements to manage and control which users may receive such services as well as knowledge of the usage of the SN by a user become important differentiators and requirements.

The addition of subscription control determining which user can receive DC service or not based on mobility restriction from a specific RAT was therefore developed. For a user in EPS, this is delivered by the HSS having additional subscription information related to permitting the use of NR as SN. Based on this subscription information and the UE indicating that it supports DC, the MME indicates to the MN (i.e., E-UTRAN) to activate the MR-DC feature in RAN. The AMF provides the Mobility Restriction list from UDM in case of 5GS and the MME provides the Handover

Restriction List with an indication of whether Secondary RAT NR is permitted or not in case of EPS. The MME/AMF may also have local configuration parameters which prevents DC (i.e., setting up NR as Secondary RAT) for roaming users. The MN has the final decision regarding activation of SN and on the type of bearers (MCG, SCG or Split) to activate based on information provided by the MME/AMF, as well as the information that the UE has provided. The UE determines whether DC is supported in a specific cell based on the broadcast information indicating DC support for MR-DC.

The MME may also select specific SGW/PGW nodes for DC, based on the knowledge the UE provides about its capability for DC support in case of MR-DC, in addition to the subscription information. In the case of MR-DC, the MME also indicates to the UE when it is not permitted to use DC. If available on the device in question, the end-user can receive visual indication of the availability of NR while it is connected to EPS.

In the case of MR-DC connected to 5GC, there is no restriction about which RAT is the MN and which is the SN. The availability of DC is enabled in the specifications from Release 15, eliminating the need for the UE to be informed of the system status of DC availability explicitly.

In the case of roaming users, if DC is activated in the VPLMN, then transfer of User data volume reporting is performed based on the HPLMN-VPLMN roaming agreement and on the indication from HPLMN to the VPLMN. For EPC this information is transferred between SGW and PGW and for 5GC this information is transferred between the V-SMF and H-SMF.

In case of MR-DC with EPC, the MN determines the Radio bearers handling and manages its allocation to the SN. That means that the MN determines the PDCP location as well as which cell group(s) the radio resources are to be allocated to. When there is a Split bearer on SN, the SN may remove any SCG resources for that specific E-RAB, ensuring that the QoS requirements are maintained.

When in MR-DC, the MN is responsible same as for non-DC operation for QoS framework enforcement, as discussed in Chapter 9, QoS. MN is responsible for appropriately guiding the SN with relevant information in order to manage the QoS operation.

In order to be able to map PDU sessions to different bearer types in MR-DC, the MN can request the core network to:

- Direct the User Plane traffic of the whole PDU session either to the MN or to the SN. In that case, there is a single user plane tunnel termination at the NG-RAN for such PDU session.
- Direct the User Plane traffic of a subset of the QoS flows of the PDU session to the SN (MN) while the rest of the QoS flows of the PDU session is directed to the MN (SN). In that case, there are two user plane tunnel terminations at the NG-RAN for such PDU session.
- Regardless of the type of setup, the MN can request to change this assignment during the life time of the PDU session. For MR-DC, NG-RAN may initiate moving QoS Flows from one RAN node to another (i.e., between MN and SN). This procedure

works when there is connectivity between the User plane GW terminating the N3 tunnel from both MN and SN RAN nodes (i.e., UPF with N3 termination) and there is no change of SMF and UPF with N3 termination during this procedure.

In MR-DC with 5GC, the MN and SN can support any bearer type and thus makes it possible to change the bearer types accordingly enabling changing of:
- MCG bearer to/from Split bearer;
- MCG bearer to/from SCG bearer;
- SCG bearer to/from Split bearer.

For MR-DC with 5GC, Fig. 12.10 (from 3GPP TS 37.340) illustrates how these bearers are defined on the Radio side:
- The MN is responsible for the location of the SDAP function per PDU session, i.e., whether it shall be hosted by the MN or the SN or by both (split PDU session);
- When the MN itself hosts an SDAP function, it makes the decision on how some of the related QoS flows to be realized (i.e., some as MCG bearer, some as SCG bearer, and others to be realized as Split bearer);
- When the MN designates the SN to host an SDAP function, some of the related QoS flows may be realized as SCG bearer, some as MCG bearer, while others may be realized as Split bearer. The SN assigns the corresponding DRB IDs, based on the DRB IDs indicated by the MN. The SN may remove or add SCG resources for the respective QoS flows, if the QoS for the respective QoS flow is guaranteed;
- For each PDU session, including split PDU sessions, at most one default DRB may be configured.

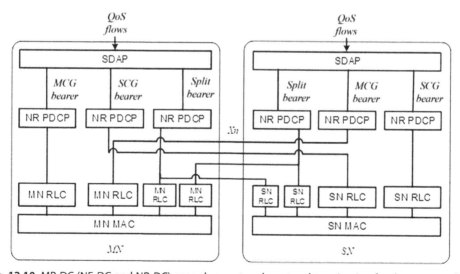

Fig. 12.10 MR-DC (NE-DC and NR-DC) user plane network protocol termination for three types of DC bearers.

12.5 Managing secondary RAN node handling for mobility and session management

This section deals with procedures related to addition, modification, release of SN, Inter MN change with or without SN and handover from MN to non-DC RAT and vice versa, QoS Flows move between MN and SN. From the CN perspective, the procedures are consistent with non-DC related procedures for Path Switch and Handover flows as described Chapter 15 but the trigger for change is due to MN or SN triggered modification of the ongoing PDU session(s).

We illustrate some example high level flows for few of the procedures to give the reader an understanding of how SN node is managed toward the CN.

The establishment of an SN node addition/modification procedure toward the CN is illustrated in Fig. 12.11.

The addition of the SN can only be initiated by the MN but a modification procedure may be initiated either by the MN or the SN. Since all CN signaling is performed by the MN only, the MN triggers addition/modification of an SN toward CN by initiating S1 user plane E-RAB Modification procedure for EPC and N3 PDU Session Modification procedure for 5GC. This procedure is used for bearers requiring SCG radio resources, to add at least the first cell of the SCG and can also be used to configure an SN terminated MCG bearer (where no SCG configuration is needed).

The 5GS detailed procedure for moving QoS flows between MN and SN per PDU session (repeated for each PDU session) is illustrated in Fig. 12.12. This procedure shows the QoS flow being moved from the MN to the SN but the same procedure applies when a QoS flow is moved from the SN to the MN as well.

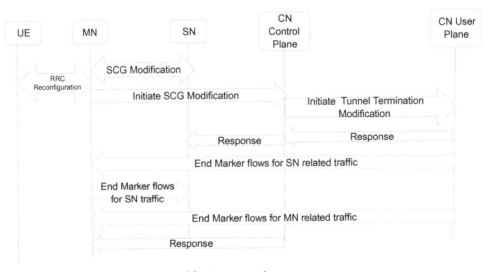

Fig. 12.11 High level SN addition/modification procedure.

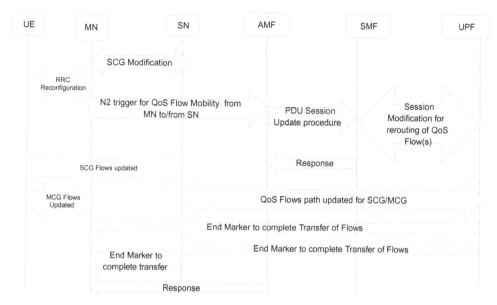

Fig. 12.12 SN addition for MR-DC with 5GC.

The steps show the trigger for SCG bearer QoS flow trigger causing message sent to CN (i.e., AMF) to modify the flow termination. AMF indicates the appropriate SMF to update the SM Context which leads SMF to request UPF to modify the N3 tunnel termination end-point for the specific QoS Flow. Once the QoS Flows have been switched between UPF, SN and MN, UPF ensures that End Marker is sent to each RAN node to complete the transaction.

In the case of MR-DC with EPC, similar procedure takes place for E-UTRAN initiated E-RAB modification procedure between MN, MME and SGW where the SGW changes the S1 tunnel termination end-point for that specific bearer.

At the end of these procedures, the SGW/UPF is connected to two RAN nodes for user plane connectivity.

In case the SN chooses to initiate SN modification of the SN configuration that do not require any MN coordination, the procedure then only involves the SN and the UE and since MN is not involved, the procedure also is not visible in the CN. This procedure may be used to for example security key handling, PDCP recovery etc. Fig. 12.13 illustrates this procedure.

The UE may require Random Access procedure to perform appropriate synchronization as may be needed due to the changes performed, otherwise the UE may start UL data transmission according to the new configuration.

SN release procedure may be used either by the MN or by the SN to release the SN UE context. This procedure may be triggered either by the MN or by the SN itself. Not

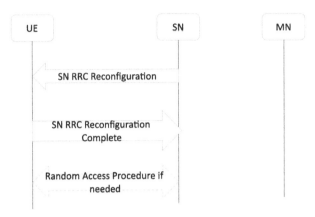

Fig. 12.13 SN Initiated SN modification without MN toward the UE.

all SN releases require signaling toward the UE, such as in certain radio failure conditions. The High-level procedure for SN release is shown in Fig. 12.14.

When the procedure is executed to remove SN, then the UE is triggered to release the SN configuration if needed. At the end of the procedure, the SN is notified by the MN so that SN can release all the information related to the UE in question.

For an SN change, either the MN or the source SN (currently serving SN) may initiate the change as illustrated in Fig. 12.15 and used to transfer a UE context from a source SN to a target SN and to change the SCG configuration in UE from one SN to another. The SN Change procedure always involves signaling over MCG SRB toward the UE.

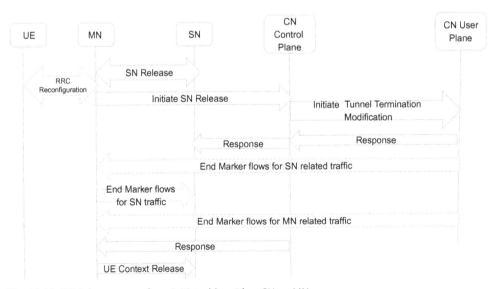

Fig. 12.14 SN Release procedure, initiated by either SN or MN.

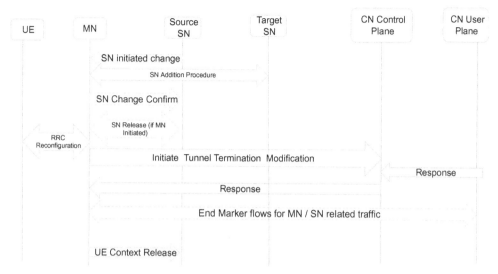

Fig. 12.15 SN Change initiated by MN or Source SN.

The steps show that an SN change triggered either by the SN or by the MN. For an MN initiated procedure, the MN is responsible for triggering the Source SN release procedure and the UE reconfiguration procedure. At the end of the procedure, the UE context is released in the Source SN. From the CN perspective, the procedure is similar to the SN addition/modification mechanism.

In case of Inter MN handover (with or without SN), currently Inter-RAT Inter-MN handover is not supported for EN-DC (i.e., EN-DC to NR-NR DC). In all other cases it is the Source MN that initiates the Handover request and Target MN initiates any SN change and selection of the target SN. An example procedure for Inter MN change with SN change is shown as follows, more details can be found in 3GPP TS 37.340.

Inter MN handover with SN change illustration as described in Fig. 12.16, follows similar procedure as handover except for possible SN change that may occur .

At request for handover from Source MN, the Target MN determines that an SN change is required and selects the target SN and informs Source MN. Source MN releases the Source SN and asks UE for RRC reconfiguration. The CN is notified of the change of MN (and SN, i.e., the new RAN tunnel termination end points) and UE context is released in the Source SN and Source MN.

In case of MN to ng-eNB/g-NB/eNB handover, the MN initiates handover to target RAN node and releases the SN first. Then the Source MN continues with normal handover procedure as applicable for RAN node to RAN node handover.

In case of RAN node to MN node handover, the MN is responsible for selecting the SN first before responding to the handover request. Target MN responds with all necessary information in order for the source RAN node to trigger RRC reconfiguration

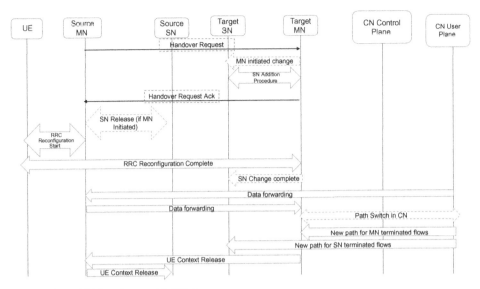

Fig. 12.16 Inter MN handover with SN change.

including SCG information, so the UE can connect to the target MN and SN. The target MN continues to complete the handover procedure toward the core network as described in Chapter 15?.

In case of MR-DC with EPC, user data forwarding may need to be performed for E-RABs for which the bearer type change from/to MN terminated bearer to/from SN terminated bearer is performed. The data forwarding node behaves as the "source eNB" for handover and node where the data is forwarded to behaves as the "target eNB" for handover.

For MR-DC with 5GC, user data forwarding may be performed between NG-RAN nodes whenever the logical node hosting the PDCP entity changes. The data forwarding node behaves as the "source NG-RAN node" for handover, the node to which data is forwarded to behaves as the "target NG-RAN node" for handover.

12.6 Security

To enable security, MR-DC can only be configured after security activation in the MN. The following security parameter needs to be configured in the UE for each type of MR-DC:

- In EN-DC and NGEN-DC, for bearers terminated in the MN the network configures the UE with K_{eNB}; for bearers terminated in the SN the network configures the UE with $S-K_{gNB}$.

- In NE-DC, for bearers terminated in the MN the network configures the UE with K_{gNB}; for bearers terminated in the SN the network configures the UE with S-K_{eNB}.
- In NR-DC, for bearers terminated in the MN the network configures the UE with K_{gNB}; for bearers terminated in the SN the network configures the UE with S-K_{gNB}.

The details of 5G Security, including a description of the key hierarchy, is available in Chapter 8.

12.7 Reporting User Data Volume traversing via SN

To enable the Core Network and the operator to collect User Data Volumes that have traversed over the SN, the MN may request the SN to count the user data transported via the SN. The MN then reports that information to the CN. This secondary RAT data volume reporting function is used to report the data volume of the secondary RAT to the 5GC or EPC depending on the MR-DC option. If configured, the MN collects and reports both the uplink and downlink data volumes of used resources to the CN. Configuration for reporting of secondary RAT data volume may happen separately for NR and E-UTRA in case of 5GS. Secondary RAT data volume reporting also indicates the secondary RAT type used during the data volume collection in RAN.

For MR-DC, the data volume is counted by the node hosting PDCP.

In case of MR-DC with 5GC:

- Downlink data volume is counted in bytes of SDAP SDUs successfully delivered to the UE or transmitted to the UE.
- Uplink data volume is counted in bytes of SDAP SDUs received by the node hosting PDCP.

In case of MR-DC with EPC:

- Downlink data volume is counted in bytes of PDCP SDUs successfully delivered to the UE over NR or transmitted to the UE over NR.
- Uplink data volume is counted in bytes of PDCP SDUs received by the node hosting PDCP over NR.

Forwarded packets are not counted when PDCP is relocated. When PDCP duplication is activated, packets are to be counted only once. This allows the reporting to be as accurate as possible, but it does not guarantee that all data has been delivered.

The conditions when RAN should provide the usage data volume to the CN are related to the events that may cause SN node to be changed, released, reconfigured or the status of the session changed (e.g., release of a bearer or QoS flow), handover (Xn/X2, N2/S1) or release of the N2/S1 connection.

There are two mechanism for reporting the data volume from the SN: (1) Piggybacking on messages related to ongoing events that anyway causes signaling toward the CN and can also cause reset/stop of the user data collection or (2) dedicated signaling to report the user data volume at certain internal. It is possible to configure the RAN to provide both mechanisms.

The User data volume reporting may be provided to the CN using existing events related triggers such as RAN release, PDU session Modification/Release, Selective deactivation of User plane connection for an existing PDU session procedures. This data reporting traverses from NG-RAN to AMF, to SMF and may be reported to the charging system during existing reporting events trigger. This information may provide the operators with sufficient level of accuracy about the amount of data transferred via the Secondary RAT. During Xn/X2 handover and S1/N2 handover, the source RAN node reports the data volume to the CN. The reported data volume excludes data forwarded to the target RAN node.

The RAN may also be configured to report the user data volume on a predefined interval, MN then periodically reports to the CN according to the time interval configured in the MN. So, if this periodic reporting is desired, then the internal timer is configured in the MN taking care that the reporting is not too frequent in order to not incur excessive signaling in the core network in an unpredictable manner. This is because in certain mobility conditions, the SN may be changed more frequently and thus closing report gathering and starting new reporting is required. In this case no existing signaling may be used (e.g., bearer release, PDU session release etc.) and dedicated signaling is used as illustrated in Fig. 12.17 to report the user data volumes accumulated by the RAN.

In order to get consistent data volume reported, the PLMN operator needs to ensure that all RAN nodes are configured to report the data to CN.

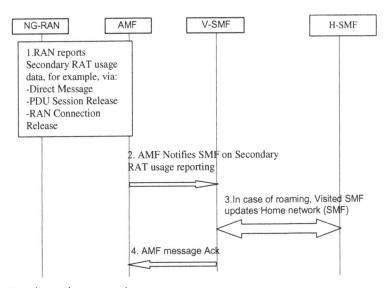

Fig. 12.17 User data volume reporting.

In roaming cases, if VPLMN and HPLMN agreement exists, then the user data volume is reported from SGW to PGW. The reporting of this information toward the charging system is performed both in the VPLMN and in the HPLMN, as per local operators' configuration. It is important that the data volume reporting does not cause excessive signaling toward the core network and as such care is taken to report the volume using existing procedures whenever possible.

For 5GC, if VPLMN operator has agreement to receive the information for the roaming users, then V–SMF in the VPLMN will need to provide that information to the H–SMF in the HPLMN. The reporting of this information toward the charging system is performed both in the VPLMN and in the HPLMN, as per local operators' configuration.

Some of the conditions for reporting the data allow for the operators to estimate the usage on a granularity that provides sufficient information to provide additional value toward their customers and as such the data is reported to the charging system. The RAN reports uplink and downlink data volumes to the 5GC for the Secondary RAT on a per QoS flow and per time interval when it is MR–DC connected to MGC, whereas per bearer and per time interval level reporting is performed for the MR–DC connected to EPC.

To summarize the user data volume reporting for dual connectivity (which is the main addition to CN for DC support), at the time of RAN connection release, Secondary Node change/release, deactivation of UP connection for a PDU Session, the RAN node reports the data volumes to the CN. An operator may want to utilize the volume reporting on a periodic manner, for example, to align sufficiently to assist with partial charging record generation. In such cases, if RAN node is configured with the time interval then dedicated procedure is used to report the user data volume to the core network (e.g., MME or AMF) and then the information is further forwarded to SGW and PGW in EPC and to SMF in 5GC. This information assists operators directly regarding an end user using DC feature while connected to the network and specifically for EPS, that the user is using NR (and as such 5G) radio.

CHAPTER 13

Network functions and services

This chapter describes in more detail the different network functions, reference points and services that are developed for 5GC. Before jumping into the actual descriptions, it may be useful to reiterate what we actually mean by a Network Function, a service and service based interfaces, as introduced in previous chapters.

With the adoption of cloud technologies and service based interfaces 3GPP has evolved the way the logical architecture is described. What was called a node or logical entity in EPS, e.g., the MME, is now called a Network Function e.g., the AMF. The reason for this change in terminology is to indicate that the Network Function is typically a set of software running on a cloud platform rather than an integrated product with dedicated HW.

The 5G Core Network Functions supports or hosts a collection of services and each Network Function offers one or more services to other Network Functions in the network. These services are made available over Service Based Interfaces in the Service Based Architecture (SBA). In practice this means that functionality supported in a specific Network Function is made available and accessible over an API.

The Network Functions that includes logic and functionality for processing of signaling are exposing and services making them available to other network functions. For each interaction between network functions, one of these acts as a "Service Consumer," and the other as a "Service Producer."

13.1 5G core network functions

13.1.1 AMF—Access and mobility management function

The AMF interacts with the access network over the N2 interface and with UEs over the N1 interface. Interactions with all other Network Functions are done via service-based interfaces. The AMF supports establishing encrypted signaling connections towards UEs, allowing these to register, to be authenticated, and to move between different radio cells in the network. The AMF also supports paging devices in idle mode.

When a UE is connected via one access network e.g., NG-RAN there is a single AMF that handles all the signaling interactions with the UE and via one N1 interface.

The AMF relays all session management-related signaling messages between the devices and the SMF Network Function. The AMF also relays SMS messages between the UEs and SMSF and it also relays location Services messages between UE and LMF as well as between RAN and LMF. Furthermore, the AMF relays UE policy messages between the PCF and the UE.

The AMF includes security anchor functionality supports the authentication and authorization of UEs (in cooperation with AUSF and UDM). After successful authentication the AMF derives separate sets of keys for integrity protection for the:

- N1 NAS signaling between UE and AMF.
- N2 RRC signaling between UE and NG-RAN.
- User Plane traffic between UE and eNB.

13.1.2 SMF—Session management function

The Session Management functionality of the 5G system has the responsibility for the setup of the connectivity for the UE towards Data Networks as well as managing the User Plane for that connectivity. The SMF is the control function that manages the user sessions including establishment, modification and release of sessions, and it can allocate IP addresses for IP PDU sessions. The SMF communicates indirectly with the UE through the AMF that relays session-related messages between the devices and the SMFs.

Compared to EPS the session management has added flexibility e.g., there are new options for end-user protocol types, different options for how to handle service and session continuity, as well as a flexible User Plane architecture.

The SMF interacts with other Network Functions through service based interfaces, and it also selects and controls the different UPF network functions in the network over the N4 network interface.

The SMF interacts with the PCF Network Function to retrieve policies which are used by the SMF to configure the UPF for the PDU session. Including configuration of the traffic steering in the UPF for individual sessions.

The SMF also collects charging data, and also controls the charging functionality in the UPF. The SMF supports both offline and online charging.

13.1.3 UPF—User plane function

The UPF processes and forwards user data. The functionality of the UPF is controlled by the SMF. It interconnects with external IP networks and acts as an anchor point for the UEs towards external networks, hiding the mobility. This means that an IP address of a specific UE PDU session is routable to the UPF that is serving this UE and Session.

The UPF performs various types of processing of the forwarded data. It generates charging data records and traffic usage reports. It can apply "packet inspection," analyzing the content of the user data packets for usage either as input to policy decisions, or as basis for the traffic reporting.

It also executes on various network or user policies, for example enforcing gating, redirection of traffic, or applying different data rate limitations.

When a device is in idle state and not immediately reachable from the network, any traffic sent towards this device is buffered by the UPF which triggers a page from the network to force the device back to go back to connected state and receive its data.

5G Core UPF can be deployed in series, e.g. one UPF distributed towards the edge of the network, and one UPF located in a more central network site. Network rules can then be used to control the traffic forwarding of the distributed UPF. Classification of data packets coming from the UE (uplink packets) can be applied to determine if the data should be sent out onto a local, distributed IP network, or if the packets are to be forwarded to the centralized UPF.

The UPF can also apply Quality-of-Service (QoS) marking of packets towards the radio network or towards external networks. This can be used by the transport network to handle each packet with the right priority in case of congestion in the network.

13.1.4 NRF—Network repository function

The NRF is a repository of the profiles of the network functions that are available in the network. The purpose of the NRF is to allow service consumer (e.g., an NF) to discover and select suitable service producers i.e., NFs and NF services without having to be configured beforehand.

When a new instance of a network function deployed or changed, e.g., due to scaling, the NRF is updated with the new profile information. The NRF profile can be updated by the network function itself or by another entity on behalf of the network function. There is also a keep alive mechanism that allows the NRF to maintain the repository and remove the profiles of missing or dormant network functions.

The NF profile in the NRF contains information like NF Type, address, capacity, Supported NF services and addresses for each NF service instance. The information is provided to the NF service consumer in the discovery procedure and provides to enough for information for the service consumer to use the service based interface of the selected NF and NF service.

The NRF profile also contains authorisation information and the NRF only provides the profiles to a consumer that can discover specific the network function or service.

13.1.5 UDM—Unified data management function

The UDM is a front-end for the user subscription data stored in the UDR.

The UDM uses subscription data that may be stored in UDR to execute application logic like access authorization, registration management and reachability for terminating event e.g., SMS.

When a UE attaches to the system the UDM authorizes the access and performs several checks of supported features, barring and restrictions due to e.g., roaming.

The UDM generates the authentication credentials that the AUSF use to authenticate UEs. It also manages permanent identity privacy and can be used by other entities to resolve the concealed permanent identity (SUCI) to the real permanent identity (SUPI).

Different instances of the UDM can be used for same user in different transactions.

The UDM also UDM keeps track of which the AMF instance that is serving a specific UE and also the SMF(s) that is serving it's PDU sessions.

13.1.6 UDR—Unified data repository

The UDR is the database where various types of data is stored. Important data is of course the subscription data and data defining various types of network or user policies. UDR storage and access to data is offered as services to other network functions, specifically UDM, PCF, and NEF.

13.1.7 UDSF—Unstructured data storage function

The UDSF is an optional function that allows other NFs to store dynamic context data outside the NF itself. This is sometimes referred to as a "stateless" implementation.

Unstructured data refers to data for which the structure is not defined in 3GPP specifications. Each vendor using a UDSF may impose it's own specific structure of the data stored in the UDSF and it is not expected that an NF from a different vendors can read and understand the stored data.

13.1.8 AUSF—AUthentication server function

The AUSF provides three services and is in the subscriber's home network. It is responsible for handling the authentication in the home network, based on information received from the UE and information retrieved from the UDM. It provides security parameters to protect Steering of Roaming information and it also provides security parameters to protect information in the UE Update procedure.

13.1.9 5G-EIR—5G equipment identity registry

The 5G-EIR is a network function that can check if the Permanent Equipment ID (an ID of the actual device hardware) has been blacklisted or not. This can for example be used by operators to block access to the network if the device has been stolen and blacklisted.

13.1.10 PCF—Policy control function

The PCF provides policy control for Sessions management related functionality, for access and mobility related functionality, for UE access selection and PDU Session selection related functionality and supports Negotiation of future background data transfers.

For session management related the PCF interacts with application functions and the SMF to provides authorized QoS and charging control for service data flows, PDU Session related policy control and event reporting for PDU sessions.

The PCF interacts with the AMF to for the access and mobility policy control that include management of service area restrictions and the management of the RFSP (Radio Frequency Selection Priority, a parameter used by NG-RAN to differentiate the treatment of different UEs).

The PCF also provides policy information to the UE (via the AMF). These polices include discovery and selection policies for non 3GPP networks, Session continuity mode selection policy, network slice selection policy, data network name selection policy and more.

13.1.11 NSSF—Network slice selection function

The NSSF selects the (set of) network slice instances for the UE and the set of AMFs that should serve the UE. The AMF may be dedicated to one or a set of network slices and the NSSF that knows of all slices I the network assist the AMF with cross slice selection functionality:

13.1.12 NEF—Network exposure function

The NEF has a similar role as the SCEF in EPS and supports exposure of event and capabilities from the 5G system towards applications and network functions inside and outside the operator's network.

The NEF can support monitoring of specific events in the 5G System and making these events available to authorized applications and network functions. Examples of event that can be made available in 3GPP Release 15 are location of UE, reachability, roaming status, and loss of connectivity.

The NEF can also support provisioning of foreseen UE behavioral information, this information can be further used in e.g., the AMF to tune the system and UE behaviour.

The NEF can in addition support external applications to manage for specific QoS and/or charging. It can be used by authorized applications to request specific QoS/priority handling for a session, and for setting applicable charging party or charging rate.

A single NEF may support a subset of the functionalities and there may be NEFs with different capabilities in one network.

13.1.13 NWDAF—Network data analytics function

The NWDAF is a function that can collect data, perform analytics and provide the results to other network functions. The network functions may adapt their behaviour based on the reported results from the NDWAF. In 3GPP Release 15 the NWDAF is somewhat limited and only provides network slice data analytics (network slice load level information). The PCF and NSSF can consume network analytics from the NWDAF and e.g., NSSF may use the network slice load level information for slice selection.

13.1.14 SEPP—Security edge protection proxy

The SEPP is a non-transparent proxy that is used to protect the signaling between operators in roaming scenarios. The SEPP acts as a relay between the Service Producer and the Service Consumer and it hides the network topology from other operators and it supports Message filtering and policing.

13.1.15 N3IWF—Non-3GPP inter working function

The non-3GPP interworking function (N3IWF) is used for integrating non-3GPP accesses with the 5G core. It is used for non-3GPP access types such as WiFi and fixed. The N3IWF terminates the IKEv2 and IPsec protocols that is used towards the UE over NWu and relays the information needed to authenticate the user equipment and authorize its access to the 5G core over the N2 interface. It is connected to the 5G Core via the N2 and N3 interfaces for the control and user planes, respectively.

13.1.16 AF—Application function

The AF is a 3GPP representation of applications either inside outside the operator's network that interacts with the 3GPP Core Network. Applications may interact and influence some aspects of the 5G core. Applications may influence traffic routing (e.g., an edge computing application), they may access the exposure function interact with the PCF to influence QoS and charging policies.

Applications considered trusted by an operator may be allowed to interact directly with relevant Network Functions. Other Applications may use the external exposure framework via the NEF to interact with relevant Network Functions.

13.1.17 SMSF—Short message service function

The SMSF supports delivery of SMS between the UE and the 5G Core (via the AMF). The SMSF is responsible for checking subscription checking and terminates the SMS protocols (SM-RP/SM-CP) used for communication with the UE. The AMF provides access to the NAS connection to the UE as a service to the SMSF on top of which the SM-RP/SM-CP messages are transferred.

13.1.18 LMF—Location management function

The LMF provides functionality to determine the location of a UE. The LMF can retrieve location estimated from the UE and it can retrieve location measurements and other data from NG-RAN. Based on retrieved data, subscription and privacy checks the LMF can determine and provide the location of UEs to other NFs. To reach the UE and the NG-RAN the LMF uses services from the AMF to access to the NAS connection to the UE and the N2 connection NG-RAN respectively.

13.2 Services and service operations

The 5G Core Network Functions offers their capabilities to other 5G Core network functions as NF services, which are accessed through service based interfaces i.e., Restful APIs. The focus of the specifications is to define the behaviour of service producers and leave flexibility for the service consumers in order to allow reuse of services when relevant and possible. A simplified picture of the service in 5GC is shown in Fig. 13.1.

More details of the services described in the following chapters can be found in 3GPP TS 23.502, 3GPP TS 23.503, and 3GPP TS 33.501.

13.2.1 AMF services

The AMF acts as a service producer for four services Namf_Communication, Namf_EventExposure Namf_MT, and Namf_Location as shown in Fig. 13.2.

The Namf_Communication service is the main service of the AMF and it has numerous Service Operations. For example, the Namf_Communication service enables other NFs like the SMF and PCF to communicate with the UE and/or the NG-RAN through the AMF. It allows new AMFs to retrieve the UE context at mobility. It also allows subscription to status changes and has an operation that allows the SMF to request Bearer Identities.

The Namf_MT service allows other NFs make sure the UE is reachable. The Namf_Location allows other NFs to request location information for a UE. Namf_EventExposure allows other NFs subscribe to notifications of mobility related events and statistics in the AMF.

13.2.1.1 Namf_Communication service
The Namf_Communication service enables an NF to communicate with the UE through N1 NAS messages or to communicate with NG-RAN or other access networks.

Namf_Communication_UEContextTransfer service operation
The Namf_Communication_UEContextTransfer request is used by an AMF to retrieve the UE context from another AMF, it's e.g., used when a UE registers in a different AMF. The new AMF can use the service operation to pass an integrity protected message that it receives from the UE to the old AMF. The old AMF use the integrity protection to verify that the new AMF has received the message from a UE using the credentials from the old AMF. If successfully verified, the AMF provides UE context to the new AMF in the Namf_Communication_UEContextTransfer response.

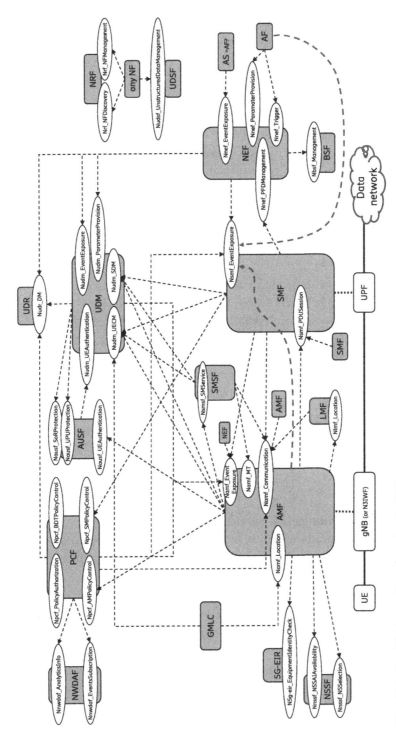

Fig. 13.1 Simplified overview of services in 5GC.

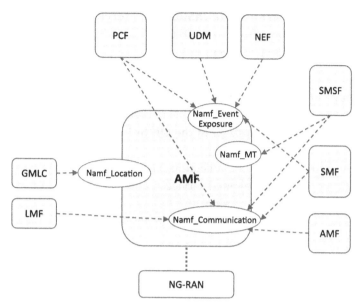

Fig. 13.2 AMF services.

Namf_Communication_RegistrationCompleteNotify service operation

The Namf_Communication_RegistrationCompleteNotify service operation is used by a new AMF to inform the old AMF that a UE context transfer succeeded and that the UE is now successfully registering with the new AMF. The old AMF marks the UE context as inactive.

The new AMF sends a Namf_Communication_RegistrationCompleteNotify acknowledgment to the consumer NF. The AMF is notified whether the AM Policy Association Information in the UE context will be used or not (i.e., new AMF may select a different PCF and then create a new AM Policy Association).

Namf_Communication_N1MessageSubscribe service operation

The Namf_ Communication_N1MessageSubscribe enables an NF e.g., SMSF to subscribe to the AMF to get notified of a particular N1 message type from the UE.

When an NF subscribes with the Namf_Communication_N1MessageSubscribe service operation the AMF checks. If the NF is allowed to subscribe to the requested N1 message type and stores a binding for the NF to deliver uplink N1 NAS messages, of the requested type, with the Namf_Communication_N1MessageNotify service operation.

Namf_Communication_N1MessageNotify service operation

The Namf_Communication_ N1MessageNotify service operation is used to pass uplink N1 NAS messages from the UE to an NF. The receiving NF has either explicitly subscribed to receiving the N1 NAS message type or the NF type is known to known to consume the received message type.

Namf_Communication_N1MessageUnSubscribe service operation

To stop receiving notifications from an AMF an NF can use the Namf_Communication_N1MessageUnSubscribe service operation.

Namf_Communication_N1N2MessageTransfer service operation

The Namf_Communication_N1N2MessageTransfer service operation can be used by NFs to transfer downlink N1 messages to the UE or N2 message to NG-RAN through the AMF.

If there is no active N1 or N2 connection (i.e., UE is in CM-IDLE state) the AMF need to invoke the network triggered service request procedure to page the UE and re-establish the N1 and N2 connection.

If there are N1 and N2 connections available the AMF sends the message to the UE and/or NG-RAN according to the request and responds to the requesting NF, with a Namf_Communication_N1N2MessageTransfer response, providing an indication of whether the transfer the N1 and/or the N2 message towards the UE and/or the NG-RAN was successful. Note that this means that the message has been sent from the AMF and it is not guaranteed that it is successfully received by UE or NG-RAN.

By including the N1N2TransferFailure Notification Target Address in the Namf_Communication_N1N2MessageTransfer service operation the requesting NF is implicitly subscribed to be notified of any transfer failures. When AMF detects that the UE fails to response to paging, the AMF invokes the Namf_Communication_N1N2TransferFailureNotification.

Namf_Communication_N1N2TransferFailureNotification service operation

The AMF uses the Namf_Communication_N1N2TransferFailureNotification service operation to notify an NF, that earlier initiated an Namf_Communication_N1N2MessageTransfer service operation, that the AMF failed to deliver the N1 message to the UE as the UE failed to respond to paging.

Namf_Communication_N2InfoSubscribe service operation

An NF can use the Namf_Communication_N2InfoSubscribe service operation to subscribe to N2 messages from NG-RAN of a specific type. The AMF creates a binding for the requesting NF and the requested N2 message type.

Namf_Communication_N2InfoUnSubscribe service operation

An NF can use the Namf_Communication_N2InfoUnSubscribe service operation to unsubscribe to N2 messages from NG-RAN. The AMF deletes the binding for the requesting NF and the requested N2 message type.

Namf_Communication_N2InfoNotify service operation

The AMF uses the Namf_Communication_N2InfoNotify service operation to send N2 message information to an NF that has subscribed to the particular N2 messages. This service operation is also used to redirect N2 messages to a new AMF that is currently serving the UE.

Namf_Communication_CreateUEContext service operation

The Namf_Communication_CreateUEContext service operation is used by a source AMF to create a UE context in a target AMF during handover procedures. The UE context information transferred from the source AMF to the target AMF includes key parameters needed by the new AMF like the 5G-GUTI, SUPI, DRX parameters, AM policy information, PCF ID, UE network capability, N1 security context information, event subscriptions by other consumer NF, and the list of SM PDU Session IDs along with the SMF handling the PDU Session.

Namf_Communication_ReleaseUEContext service operation

The Namf_ Communication_ReleaseUEContext service operation is used by a source AMF to release the UE context in a target AMF in case the handover fails and the handover cancel procedures are invoked.

Namf_Communication_EBIAssignment service operation

An SMF can use the Namf_Communication_EBIAssignment service operation to request and release EPS Bearer IDs (EBIs). EBIs are not needed in 5GS but they are needed to retain the bearers when performing inter system HO from 5GS to EPS. Since there is a limited number of bearers supported in EPS the AMF coordinates the allocation of EBIs since there may be several PDU sessions with in different SMFs for each UE.

The SMF invokes the Namf_Communication_EBIAssignment service operation when it determines that one or more EPS Bearer IDs are required for EPS QoS mapping for a PDU Session. The AMF uses the QoS parameters and the S-NSSAI to prioritize the EBI request.

If the SMF determines that some EBIs are not needed, the consumer NF indicates the EBI(s) that can be released in the Released EBI list.

Namf_Communication_AMFStatusChangeSubscribe service operation

The Namf_Communication_AMFStatusChangeSubscribe service operation is used by peer NFs to subscribe to updates of the AMF status e.g., if the AMF becomes unavailable or no longer serve the indicated GUAMI(s). The peer NFs (SMF, UDM, PCF) can use this service to detect if a GUAMI is served by a different AMF in case a group.

Namf Communication_AMFStatusChangeUnSubscribe service operation

The Namf_Communication_AMFStatusChangeUnSubscribe service operation is used by an NF to unsubscribe for AMF status change notification.

Namf_Communication_AMFStatusChangeNotify service operation

The Namf_Communication_AMFStatusChangeNotify service operation report AMF Status change (e.g., AMF unavailable) notification to NFs that has previously registered with the AMF using the Namf_Communication_AMFStatusChangeSubscribe service operation.

The notify message includes the GUAMI(s) impacted by the status change and it may also include an alternative AMF that can serve the GUAMI instead.

13.2.1.2 Namf_EventExposure service

The Namf_EventExposure service allows NFs to subscribe and get notified about AMF events. The Namf_EventExposure service has three service operations Namf_EventExposure_Subscribe, Namf_EventExposure_UnSubscribe and Namf_EventExposure_Notify.

The AMF can expose information about UE related events like:
- Location changes
- Time zone changes
- Access Type changes
- Registration state changes
- Connectivity state changes
- UE loss of communication
- UE reachability status

Event filter can be used by the requesting NF to narrow down the specific event of interest. For example, if the requesting NF is interested in knowing when a UE moves in and out of a specific tracking area it can subscribe to the location event and specify an event filter for the parameter tracking area and a specific tracking area ID value.

Namf_EventExposure_Subscribe service operation

The requesting NF can use the Namf_EventExposure_Subscribe service operation to subscribe to, or modify, event reporting for one UE, a group of UEs or all UEs I the AMF.

The requesting NF provides the target UEs, the Event IDs and associated Event filters and in addition, a Notification Correlation ID. The target UE(s) may be identified by SUPI(s), Internal Group ID or an indication that the AMF should report for all UEs.

When the AMF accepts the subscription, it responds with a Subscription Correlation ID that is used for managing or removing the subscription and possibly an expiry time for the subscription indicating when the AMF will stop further reporting. The AMF may also include a first event report, if available.

In the case that the requesting NF subscribes on behalf of another NF, the NF includes a Notification Target Address and correlation information for each of the Event IDs that should be notified directly to another NF.

If the requesting NF needs to modify a subscription, that it has previously created, it invokes Namf_EventExposure_Subscribe service operation with the Subscription Correlation ID and provides updated Event Filters with Event ID(s) to the AMF.

Namf_EventExposure_UnSubscribe service operation

An NF can use the Namf_EventExposure_UnSubscribe service operation to stop further event reporting for previously subscribed events. The Subscription Correlation ID that was received when subscribing to the event reporting is used as an input to the AMF to identify the specific events to stop reporting for.

Namf_EventExposure_Notify service operation

When the AMF detects an event corresponding to a subscription, it invokes the Namf_EventExposure_Notify service operation, for each subscribed NF that matches the event and event filter. The AMF includes information like AMF ID, Notification Correlation Information, Event ID, corresponding UE (SUPI and if available GPSI) and time stamp. The Notification Target Address and Notification Correlation ID helps the Receiving NF identify the event notification subscription. In addition, the AMF also included event specific parameter that indicates the type of event that has occurred and related information, e.g., Registration Area Update in new Registration Area.

13.2.1.3 Namf_MT service

The Namf_MT service allows an NF the service to make sure the UE is reachable to send e.g., MT SMS to a UE. The Namf_MT service also has a service operation that allows an NF to retrieve information to assist in terminating domain selection for IMS voice services.

Namf_MT_EnableUEReachability service operation

An NF can use the Namf_MT_EnableUEReachability service operation to request UE reachability from the AMF. This service operation is typically used by the SMSF to make sure the UE is ready to receive and SMS via the N1 NAS connection to the AMF.

If UE is in CM-CONNECTED state, the AMF responds to the requesting NF immediately. If the UE is in CM-IDLE state, the AMF may page the UE and respond to the consumer NF after the UE enters CM-CONNECTED state.

If paging fails and the UE is not reachable the AMF informs the requesting NF about the failure. If the AMF no longer serves the UE but the AMF knows which AMF that is serving the UE, the AMF provides redirection information which can be used by the NF consumer to retry vie the new AMF.

Namf_MT_ProvideDomainSelectionInfo service operation

The Namf_MT_ProvideDomainSelectionInfo service operation can be used by UDM to retrieve information that it can use to enhances the probability of reaching the UE with voice sessions.

When invoking the service operation UDM provides the SUPI of the UE and the AMF responds with an Indication of support for IMS voice over PS Session or not, Time stamp of the last radio contact with the UE and Current RAT type.

13.2.1.4 Namf_Location service

This Namf_Location service enables an NF to request location information for a target UE. The following are the key functionalities of this NF service.
- Allow NFs to request the current geodetic and optionally civic location of a target UE.
- Allow NFs to be notified of event information related to emergency sessions.
- Allow NFs to request Network Provided Location Information (NPLI) and/or local time zone corresponding to the location of a target UE.

Namf_Location_ProvidePositioningInfo service operation

The Namf_Location_ProvidePositioningInfo is typically triggered by the GMLC to request the position of a UE. The SUPI or PEI of the UE is provided as input to the Service operation and the response from the AMF includes positioning information for the UE e.g., Geodetic Location, Civic Location, Position Methods Used, Failure Cause.

To provide the positioning information the AMF may in turn use the Nlmf_Location_DetermineLocation service operation from the LMF that can trigger UE positioning procedures and provide positioning information.

Namf_Location_EventNotify service operation

The Namf_Location_EventNotify Service Operation is currently used to inform the GMLC of an emergency session and provides any UE location available to the AMF. It is also used to inform the GMLC when the emergency session has been released.

Namf_Location_ProvideLocationInfo service operation

The Namf_Location_ProvideLocationInfo service operation is typically used by the UDM to retrieve Network Provided Location Information of a target UE. The UE is identified by a SUPI and the AMF provides the available location information e.g., Cell Identity, Tracking Area Identity, Geographical/Geodetic Information, Current RAT Type, Local Time Zone.

13.2.2 SMF services

The SMF provides two services the Nsmf_PDUSession service and the Namf_EventExposure service as shown in Fig. 13.3. The Nsmf_PDUSession service provides

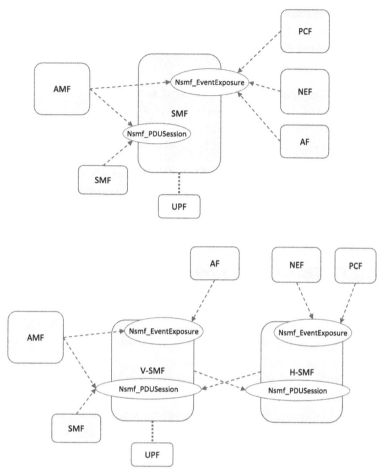

Fig. 13.3 SMF services.

the ability to manage PDU sessions and the Namf_EventExposure service provides the possibility to expose events from the SMF.

13.2.2.1 Nsmf_PDUSession_CreateSMContext service operation
The Nsmf_PDUSession_CreateSMContext service operation is used by the AMF to creates an AMF-SMF association to support a PDU Session. The AMF provides the SUPI, DNN, AMF ID and other parameters needed by the SMF to create the PDU session including the N1 SM message from the UE.

The SMF responds with a SM Context ID, PDU Session ID and any N1 SM messages for transfer to the UE and/or N2 messages to be transferred to the NG-RAN.

13.2.2.2 Nsmf_PDUSession_UpdateSMContext service operation
The Nsmf_PDUSession_UpdateSMContext service operation allows the AMF to update the AMF-SMF association to support a PDU Session and/or to provide SMF with

N1 or N2 SM information received from the UE or from the NG-RAN. The AMF includes the SM Context ID to identify the context in the SMF and N1 SM message, N2 message information or other parameters depending on the reason for update request.

13.2.2.3 Nsmf_PDUSession_ReleaseSMContext service operation

The Nsmf_PDUSession_ReleaseSMContext service operation is used by the AMF to release the AMF-SMF association when the PDU Session is being released.

13.2.2.4 Nsmf_PDUSession_SMContextStatusNotify service operation

The Nsmf_PDUSession_SMContextStatusNotify service operation is used by the SMF to notify the AMF when a PDU session is released (by e.g., the SMF or PCF) or when a PDU Session is handed over to a different system or access type.

13.2.2.5 Nsmf_PDUSession_Create service operation

The Nsmf_PDUSession_Create service operation is used in roaming scenarios between the V-SMF in the serving PLMN and H-SMF in the home PLMN. The Nsmf_PDUSession_Create Service Operation is triggered by the V-SMF because of the AMF invoking the Nsmf_PDUSession_CreateSMContext service operation to the V-SMF. The Nsmf_PDUSession_Create service operation is used to create a new PDU Session in the H-SMF or to create an association with an existing PDN connection in the home PGW-C + SMF.

The V-SMF SM Context ID in the input provides addressing information allocated by the V-SMF (to be used for service operations towards the V-SMF for this PDU Session).

13.2.2.6 Nsmf_PDUSession_Update service operation

The Nsmf_PDUSession_Update service operation is used between the V-SMF and H-SMF in roaming scenarios to Update the established PDU Session.

This service operation is invoked by the V-SMF towards the H-SMF in case the AMF has invoked the Nsmf_PDUSession_UpdateSMContext towards the V-SMF due to a UE or serving network requested PDU Session Modification. The Nsmf_PDUSession_Update Service Operation can also be used by the V-SMF to inform the H-SMF the access type changes are allowed for the PDU session.

The H-SMF invokes the Nsmf_PDUSession_Update Service Operation towards the V-SMF for UE and HPLMN initiated PDU Session Modifications and PDU Session Release to transfer SM PDU Session Modification request or SM PDU Session Release request messages to the UE. The service operation can also be used by the H-SMF towards the V-SMF to release resources e.g., during handover to EPS and.

13.2.2.7 Nsmf_PDUSession_Release service operation

The Nsmf_PDUSession_Release service operation is used in roaming scenarios by V-SMF to request the H-SMF to release the PDU Session and related resources during serving network initiated PDU Session release cases (e.g., implicit De-registration of UE in the serving network).

13.2.2.8 Nsmf_PDUSession_StatusNotify service operation

The Nsmf_PDUSession_StatusNotify Service Operation service operation is used in roaming scenarios by the H-SMF to notify V-SMF of changes to the status of a PDU Session e.g., PDU Session is released or handed over to EPS or a different access type.

13.2.2.9 Nsmf_PDUSession_ContextRequest service operation

The Nsmf_PDUSession_ContextRequest service operation is used by the AMF to fetch the SM Context during handover to EPS.

13.2.2.10 Nsmf_EventExposure service

The Nsmf_EventExposure service allows NFs to subscribe and get notified about events related to PDU sessions. The Nsmf_EventExposure service has three service operations Nsmf_EventExposure_Subscribe, Nsmf_EventExposure_UnSubscribe and Nsmf_EventExposure_Notify.

The SMF can expose information about PDU session related events like:
- UE IP address or Prefix change.
- PDU Session Release.
- UP path change.
- Change of Access Type.
- PLMN change.

Event filter can be used by the requesting NF to narrow down the specific event of interest. Event Filters specify the conditions to meet for triggering notification and can include one or more Parameters and the values that each parameter should match to trigger a notification.

The target of SMF event reporting may correspond to a single PDU Session ID, a UE ID an Internal Group Identifier or all UE on a specific DNN.

Nsmf_EventExposure_Subscribe service operation

The requesting NF can use the Nsmf_EventExposure_Subscribe service operation to subscribe to, or modify, event reporting for one UE, a group of UEs or all UEs I the SMF.

The requesting NF provides the target UEs, the Event IDs and associated Event filters and in addition, a Notification Correlation ID. The target UE(s) may be identified by SUPI(s), Internal Group ID or an indication that the SMF should report for all UEs.

When the SMF accepts the subscription, it responds with a Subscription Correlation ID that is used for managing or removing the subscription and possibly an expiry time for the subscription indicating when the SMF will stop further reporting. The AMF may also include a first event report, if available.

In the case that the requesting NF subscribes on behalf of another NF, the NF includes a Notification Target Address and correlation information for each of the Event IDs that should be notified directly to another NF.

If the requesting NF needs to modify a subscription, that it has previously created, it invokes Nsmf_EventExposure_Subscribe service operation with the Subscription Correlation ID and provides updated Event Filters with Event ID(s) to the SMF.

Nsmf_EventExposure_UnSubscribe service operation

An NF can use the Nsmf_EventExposure_UnSubscribe service operation to stop further event reporting for previously subscribed events. The Subscription Correlation ID that was received when subscribing to the event reporting is used as an input to the SMF to identify the specific events to stop reporting for.

Nsmf_EventExposure_Notify service operation

When the SMF detects an event corresponding to a subscription, it invokes the Nsmf_EventExposure_Notify service operation, for each subscribed NF that matches the event and event filter. The SMF includes information like SMF ID, Notification Correlation Information, Event ID, corresponding UE (SUPI and if available GPSI) and time stamp. The Notification Target Address and Notification Correlation ID helps the Receiving NF identify the event notification subscription. In addition, the SMF also included event specific parameter that indicates the type of event that has occurred and related information, e.g., Registration Area Update in new Registration Area.

13.2.3 PCF services

The PCF acts as a service producer for six services Npcf_AMPolicyControl, Npcf_Policy Authorization, Npcf_SMPolicyControl, Npcf_BDTPolicyControl, Npcf_UEPolicyControl, and Npcf_EventExposure, see Fig. 13.4.

The Npcf_AMPolicyControl service provides Access Control, network selection and Mobility Management related policies and UE Route Selection Policies to the AMF.

The Npcf_Policy Authorization authorizes and creates policies on request from an AF related to the PDU Session to which the AF session is bound to.

The Npcf_SMPolicyControl provides PDU session related policies to the SMF.

The Npcf_BDTPolicyControl, PCF service provides background data transfer policy to the NEF.

The Npcf_UEPolicyControl This PCF service provides the management of UE Policy Association to the NF consumers.

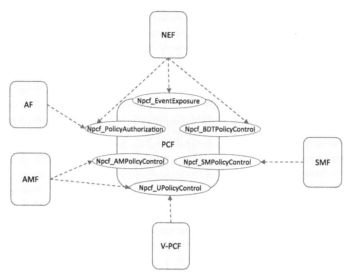

Fig. 13.4 PCF services.

The Npcf_EventExposure allows other NFs subscribe to notifications of PCF related events.

13.2.3.1 Npcf_AMPolicyControl service

The Npcf_AMPolicyControl service allows the AMF to create modify and delete per UE AM Policy Associations with the PCF. The PCF may provide the AMF with policy information per UE can may contain access and mobility related policy information as well as Policy Control Request Trigger conditions.

When the AMF detects that a Policy Control Request Trigger condition is met is will contact the PCF that may provide updated access and mobility related policy information and Policy Control Request Trigger conditions.

The Npcf_AMPolicyControl service also allows the PCF to send new AM policies for an established AM Policy Association.

Npcf_AMPolicyControl_Create service operation

The Npcf_AMPolicyControl_Create service operation allows the AMF to request the creation of an AM Policy Association with the PCF for a UE. When the PCF has created the AM Policy Association, the PCF may provide access and mobility related policy information and Policy Control Request Trigger conditions.

Npcf_AMPolicyControl_Update service operation

The AMF can use the Npcf_AMPolicyControl_Update service operation to request updated Policy information for a UE when a policy control request trigger is met, or the AMF is changed (due to mobility) but the same PCF is used. The AMF will include policy control trigger condition(s) that have been met in the request.

Npcf_AMPolicyControl_UpdateNotify service operation

The Npcf_AMPolicyControl_UpdateNotify service operation allows the PCF, at any time, to provide the AMF with updated access and mobility related Policy information for an AM Policy Association. This notification can be sent by the PCF to the AMF without prior explicit subscription. The creation of the AM policy association is used as an implicit subscription that allows the PCF to send the notifications to the AMF.

Npcf_AMPolicyControl_Delete service operation

The Npcf_AMPolicyControl_Delete service operation enables the AMF to remove the AM Policy Association. When the AMF initiates the AM Policy Association Termination procedure the PCF deletes the AM Policy Association for this SUPI. This service operation is for example used at UE deregistration.

13.2.3.2 Npcf_PolicyAuthorization service

Th Npcf_PolicyAuthorization service is used to authorize an AF request and to create policies as requested by the authorized AF for the PDU Session to which the AF session is bound. This service also allows the NF consumer to subscribe/unsubscribe the notification of events.

Npcf_PolicyAuthorization_Create service operation

The Npcf_PolicyAuthorization_Create service operation is invoked by an AF towards a PCF and allows the PCF to authorize the request, create an application session and optionally determine and install policies according to the information provided by the AF. To identify the application session the AF provides the IP address of the UE and additional information e.g., UE identity, Media type, Media format, bandwidth requirements, flow description and/or Application identifier or traffic filtering information. The PCF responds to the AF with the application context and an application session ID that is used to identify the application context subsequent service operations for the same application session.

If the PCF determines that an SM policy should be installed or updated due the AF request it can use the Npcf_SMPolicyControl service towards the SMF to install or modify the policies according to the AF request.

Npcf_PolicyAuthorization_Update service operation

The Npcf_PolicyAuthorization_Update service operation allows the AF to update an established application session. The AF provides the application session ID and new service information to the PCF. The PCF Provides updates an application context in with the accepted service information and use the Npcf_SMPolicyControl service to install any new the policies in the SMF. The PCF responds to the AF with an updated application context.

Npcf_PolicyAuthorization_Delete service operation

The Npcf_PolicyAuthorization_Delete Service Operation allows the AF to remove the application session and remove the application context in the PCF.

Npcf_PolicyAuthorization_Notify service operation

The Npcf_PolicyAuthorization_Notify service operation enables the PCF to notify an AF about events related to the application session. The PCF can inform the AF about events as outlined in the Npcf_PolicyAuthorization_Subscribe service operation below.

The PCF notifies the AF with an Event ID and Notification Correlation Information (information that enables the AF to identify the application session).

Npcf_PolicyAuthorization_Subscribe service operation

The Npcf_PolicyAuthorization_Subscribe service operation allows the AF to subscribe the notification of events related to applications sessions. The PCF supports event reporting to events like:
- Notification about application session context events.
- Notification about application session context termination.
- Notification about Service Data Flow QoS notification control.
- Notification about service data flow deactivation.
- Reporting usage for sponsored data connectivity.
- Notification of resources allocation outcome.

The AF includes one or more Event ID(s) as for events as listed above, information identifying the AF session e.g., UE IP address or SUPI, Notification Target Address and Notification Correlation ID.

The PCF responds with a Subscription Correlation ID when accepting the subscription. The AF can use the Subscription Correlation ID to refer to the subscription if it later wishes to modify or delete the subscription.

Npcf_PolicyAuthorization_Unsubscribe service operation

The Npcf_PolicyAuthorization_Unsubscribe service operation allows the NF to unsubscribe to notification of PCF events related to Npcf_PolicyAuthorization_Subscribe operation. The AF provides the Subscription Correlation information to the PCF so it can identify and remove the event subscription.

13.2.3.3 Npcf_SMPolicyControl service

The Npcf_SMPolicyControl service allows the SMF to create modify and delete per UE SM Policy Associations with the PCF. The PCF may provide the SMF with policy information per UE that may include PDU session related policy information as well as Policy Control Request Trigger conditions.

When the SMF detects that a Policy Control Request Trigger condition is met is will contact the PCF that may provide updated PDU session related policy information and Policy Control Request Trigger conditions.

The Npcf_SMPolicyControl service also allows the PCF to send new PDU Session policies for an established SM Session Policy Association.

Npcf_SMPolicyControl_Create service operation

The Npcf_SMPolicyControl_Create service operation allows the SMF to request the creation of an SM Policy Association with the PCF for a UE. When the PCF has created the SM Policy Association, the PCF may provide PDU Session related policy information and Policy Control Request Trigger conditions.

Npcf_SMPolicyControl_UpdateNotify service operation

The Npcf_SMPolicyControl_UpdateNotify service operation allows the PCF, at any time, to provide the SMF with updated PDU session related Policy information for an SM Policy Association. This notification can be sent by the PCF to the SMF without prior explicit subscription. The creation of the SM policy association is used as an implicit subscription that allows the PCF to send the notifications to the SMF.

Npcf_SMPolicyControl_Delete service operation

The Npcf_SMPolicyControl_Delete service operation enables the SMF to remove the SM Policy Association. When the SMF initiates the SM Policy Association Termination procedure the PCF deletes the SM Policy Association for this SUPI. This service operation is for example used at UE deregistration.

Npcf_SMPolicyControl_Update service operation

The SMF can use the Npcf_SMPolicyControl_Update service operation to request updated Policy information for a UE when a policy control request trigger is met. The SMF will include policy control trigger condition(s) that have been met in the request. The PCF may respond with provide updated policy information to the SMF.

13.2.3.4 Npcf_BDTPolicyControl service

The Npcf_BDTPolicyControl service provides background data transfer policy, which includes the following functionalities:
- Get background data transfer policies based on the request via NEF from AF; and
- Update background data transfer based on the selection provided by AF.

Npcf_BDTPolicyControl_Create service operation

The Npcf_BDTPolicyControl_Create service operation is used by the NEF (on request from an Application Service Provider) to request a background data transfer policy. The NEF provides an ID of the application service provider, the expected data volume per UE, The number of UE and desired time window. It may also indicate a network area where the UEs are located. The PCF responds with a one or more background data transfer policies and a Background Data Transfer Reference ID. The Reference ID can e.g., be used to request updates of the Background data transfer policy.

Npcf_BDTPolicyControl_Update service operation

The Npcf_BDTPolicyControl_Update service operation is used by the NEF to request updates to the background data transfer policy from the PCF. The NEF provides an identifier for the application service provider, background data transfer policy and the Background Data Transfer Reference ID. The PCF may respond with a new background data transfer policy to the NEF.

13.2.3.5 Npcf_UEPolicyControl service

The Npcf_UEPolicyControl service is used by the AMF to create and manage a UE Policy Association with the PCF through which the AMF Service Consumer receives Policy Control Request Trigger of UE Policy Association. The association allows PCF to provide UE access selection and PDU Session selection related policy information to the UE transparently through the AMF that provides using NAS messages to carry the UE policy container. In case of roaming the AMF use the Npcf_UEPolicyControl Service provided by the PCF in the visited network (V-PCF), the V-PCF will in turn use the Npcf_UEPolicyControl Service provided by the Home PCF (H-PCF).

As part of this service, the PCF may provide the NF Service Consumer, e.g., AMF, with policy information about the UE that may contain:

- UE access selection and PDU Session selection related policy information as defined in clause 6.6 of 3GPP TS 23.503. In the case of roaming, the URSP information is provided by H-PCF and the ANDSP information can be provided by V-PCF or H-PCF or both;
- Policy Control Request Trigger of UE Policy Association. When such a Policy Control Request Trigger condition is met, the NF Service Consumer, e.g., AMF, shall contact PCF and provide information on the Policy Request Trigger condition that has been met. In the case of roaming, the V-PCF may subscribe to AMF or the H-PCF may subscribe to AMF via V-PCF.

At Npcf_UEPolicyControl_Create, the NF Service Consumer, e.g., AMF, requests the creation of a corresponding "UE Policy Association" with the PCF (Npcf_UEPolicyControl_Create) and provides relevant parameters about the UE context to the PCF. When the PCF has created the UE Policy Association, the PCF may provide policy information as defined above.

When a Policy Control Request Trigger condition is met, the NF Service Consumer, e.g., AMF, requests the update (Npcf_UEPolicyControl_Update) of the UE Policy Association by providing information on the condition(s) that have been met. The PCF may provide updated policy information to the NF Service Consumer.

During the AMF relocation, if the target AMF receives the PCF ID from source AMF and the target AMF decides to contact with the PCF identified by the PCF ID based on the local policies, the target AMF requests the update (Npcf_UEPolicyControl_Update) of the UE Policy Association. If a Policy Control Request Trigger condition is met, the information matching the trigger condition may also be provided by the target AMF. The PCF may provide updated policy information to the target AMF.

The PCF may at any time provide updated policy information (Npcf_UEPolicyControl_UpdateNotify).

At UE deregistration the NF Service Consumer, e.g., AMF, requests the deletion of the corresponding UE Policy Association.

Npcf_UEPolicyControl_Create service operation

The Npcf_UEPolicyControl_Create service operation is used by the AMF to request the creation of a UE Policy Association. The AMF provides the SUPI of the UE and may include other parameters e.g., Access Type, Permanent Equipment Identifier, GPSI, User Location Information, UE Time Zone, Serving Network, RAT type, UE access selection and PDU session selection policy information including the list of PSIs, OS id and Internal Group. The PCF provides the Home PCF ID to the AMF and may also provide Policy Control Request Trigger for the UE Policy Association.

The UE Policy Association allows the PCF to send policy information to the UE via the AMF. The Policy Control Request Trigger may be provided to the AMF so that the AMF can inform the AMF when certain events occur, and the PCF may want to provide new policies to the UE. In case of roaming the AMF uses the Npcf_UEPolicyControl_Create Service Operation provided by the V-PCF and the V-PCF will in turn use the Npcf_UEPolicyControl_Create Service Operation provided by the H-PCF.

Npcf_UEPolicyControl_UpdateNotify service operation

The PCF may at any time use the Npcf_UEPolicyControl_UpdateNotify service operation to provide updated Policy information for the UE for an established UE Policy Association. In case of roaming he the H-PCF may invoke the Npcf_UEPolicyControl_UpdateNotify Service Operation towards the V-PCF and the V-PCF can, on its own initiative or because of a notify Service Operation from the H-PCF, invoke the Npcf_UEPolicyControl_UpdateNotify Service Operation towards the AMF.

Npcf_UEPolicyControl_Delete service operation

The Npcf_UEPolicyControl_Delete service operation allows the AMF to delete the UE policy control association in the PCF. In case of roaming AMF uses the Npcf_UEPolicyControl_Delete service operation provided by the V-PCF and the V-PCF will in turn use the Npcf_UEPolicyControl_Delete service operation provided by the H-PCF.

Npcf_UEPolicyControl_Update service operation

The Npcf_UEPolicyControl_Update service operation allows the AMF to request an update of the UE Policy Association to receive updated Policy information for the UE context. When the AMF detects that a Policy Control Request Trigger condition is met, the AMF requests an update by invoking the Npcf_UEPolicyControl_Update service operation of the UE Policy Association and provides information on the condition(s) that have been met. The PCF may provide updated policy information to the AMF.

The Npcf_UEPolicyControl_Update service operation can also be used during an AMF relocation, if the new AMF receives the PCF ID from source AMF and the target AMF decides to contact the PCF identified by the PCF ID, the new AMF invokes the Npcf_UEPolicyControl_Update service operation for the UE Policy Association. The PCF may provide updated policy information to the target AMF.

In case of roaming AMF uses the Npcf_UEPolicyControl_Update service operation provided by the V-PCF and the V-PCF will in turn use the Npcf_UEPolicyControl_Update service operation provided by the H-PCF.

13.2.3.6 Npcf_EventExposure service

The Npcf_EventExposure service allows an NF e.g., the NEF to subscribe, modify and get notified about PCF events for a group of UE(s) or all UEs sharing the same a DNN and S-NSSAI. The Npcf_EventExposure service has three service operations Npcf_EventExposure_Subscribe, Npcf_EventExposure_UnSubscribe, and Npcf_EventExposure_Notify.

The PCF can expose information about events like:
- PLMN identifier notification
- Change of Access Type

Event filter can be used by the requesting NF to narrow down the specific event of interest. Event Filters specify the conditions to meet for triggering notification and can include one or more Parameters and the values that each parameter should match to trigger a notification.

Npcf_EventExposure_Subscribe service operation

The requesting NF e.g., NEF can use the Npcf_EventExposure_Subscribe service operation to subscribe to or modify event reporting for a group of UE(s) or any UE accessing a combination of (DNN, S-NSSAI). The requesting NF also provides the Event IDs and associated Event filters and in addition, a Notification Correlation ID.

When the PCF accepts the subscription, it responds with a Subscription Correlation ID that is used for managing or removing the subscription and possibly an expiry time for the subscription indicating when the PCF will stop further reporting. The PCF may also include a first event report, if available.

If the requesting NF needs to modify a subscription, that it has previously created, it invokes Npcf_EventExposure_Subscribe service operation with the Subscription Correlation ID and provides updated Event Filters with Event ID(s) to the PCF.

Npcf_EventExposure_UnSubscribe service operation

An NF can use the Nsmf_EventExposure_UnSubscribe Service Operation to stop further event reporting for previously subscribed events. The Subscription Correlation ID that was received when subscribing to the event reporting is used as an input to the PCF to identify the specific events to stop reporting for.

Npcf_EventExposure_Notify service operation

When the PCF detects an event corresponding to a subscription, it invokes the Npcf_EventExposure_Notify service operation, for each subscribed NF that matches the event and event filter. The PCF includes information like PCF ID, Notification Correlation Information, Event ID, corresponding UE (SUPI and if available GPSI) and time stamp. The Notification Target Address and Notification Correlation ID helps the Receiving NF identify the event notification subscription. In addition, the PCF also included event specific parameter that indicates the type of event that has occurred and related information.

13.2.4 UDM services

The UDM acts as a service producer for five services Nudm_UEContextManagement, Nudm_SubscriberDataManagement, Nudm_UEAuthentication, Nudm_EventExposure, and Nudm_ParameterProvision as shown in Fig. 13.5. The UDM services are used by the AMF, SMF, SMSF, NEF, GMLC, and AUSF via the Nudm service based interface.

If the UDM is stateless and store information externally in a UDR it makes use of the Nudr services as described in Section 13.2.8.

The Nudm_UEContextManagement service is used for UE context management and allow NFs like the AMF, SMF and SMSF to register and deregister with UDM and can provide the NFs with information related to UE's e.g., a UE's serving NF

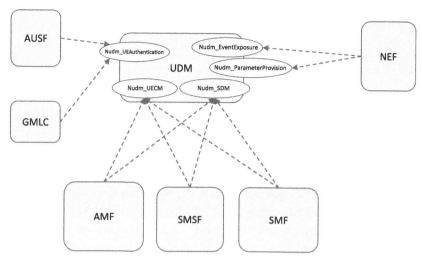

Fig. 13.5 UDM services.

identifier, UE status, etc. The Nudm_UEContextManagement Service is also known by a shorter version of the name, Nudm_UECM.

The Nudm_SubscriberDataManagement service is used to manage subscription data and enables NFs like AMF and SMF to retrieve user subscription data and allows the UDM to provided updates of subscriber data. The Nudm_SubscriberDataManagement Service is also known as the Nudm_SDM service.

The Nudm_UEAuthentication service provides authentication subscriber data to the e.g., AMF. For AKA based authentication, this operation can be also used to recover from security context synchronization failure situations. Used for being informed about the result of an authentication procedure with a UE.

The Namf_EventExposure service allows NFs to subscribe to events and can provides monitoring indication of the events to the subscribed NF consumer.

The Nudm_ParameterProvision service is used to provision information which can be used for the UE in 5GS.

13.2.4.1 Nudm_UECM (Nudm_UEContextManagement) service

The Nudm_UEContextManagement service is used by NFs (AMF, SMF, SMSF) to manage the registration of the serving NFs with UDM and to retrieve registration information e.g., to route terminating requests to the right serving NF.

The following service operations are defined for the Nudm_UEContextManagement service:
- Registration
- DeregistrationNotification

- Deregistration
- Get
- Update
- P-CSCF-RestorationNotification

Nudm_UECM_Registration service operation

The Nudm_UECM_Registration service operation is used by NFs (AMF, SMF, SMSF) to register at the UDM as the NF serving the UE (AMF and SMSF) or the NF serving a PDU session (SMF).

The AMF uses the Nudm_UECM_Registration to register as the serving NF for a UEs access and mobility management service during registration and in the same way the SMSF registers as the NF serving the UE for SMS. The SMF uses the Nudm_UECM_Registration register as the serving NF for session management services during PDU Session establishment.

The NF invoking the Nudm_UECM_Registration Service Operation provides it's NF ID, NF Type the SUPI of the UE. The UDM authorizes the request and if accepted the NF is set as the UE serving the UE.

When it's an AMF that uses the Nudm_UECM_Registration Service Operation, it is implicitly subscribed to be notified of deregistration's in UDM e.g., in case a UE moves to a different AMF. The notification are sent by UDM to a previously registered AMF with the Nudm_UECM_DeregistrationNotification service operation.

Nudm_UECM_DeregistrationNotification service operation

The Nudm_UECM_DeregistrationNotification service operation is used by UDM to notify an AMF that it has been deregistered as the serving NF for a UE e.g., due to mobility to a different AMF. The UDM provides the UEs SUPI and a deregistration reason e.g.:
- UE Initial Registration.
- UE Registration area change.
- Subscription Withdrawn.
- 5GS to EPS Mobility.

Nudm_UECM_Deregistration service operation

The Nudm_UECM_Deregistration service operation is used by the a previously registered NF (AMF, SMF or SMSF) to deregister from the UDM. The UDM receiving the request deletes the information related to the NF in the UE context and responds to the NF with an indication in the deregistration was successful or not.

When it's an AMF invoking the Nudm_UECM_Deregistration Service Operation it also means that the subscriptions to be notified when the NF is deregistered in UDM (i.e., Nudm_UECM_DeregistrationNotification) is also removed.

Nudm_UECM_Get service operation

The Nudm_UECM_Get Service Operation is used by NFs (e.g., NEF, GMLC, SMSF) to retrieve registration information from the UDM e.g., the NF ID where the UEs access and mobility management context or the PDU Sessions context can be reached. The requesting NF provides a UE ID the NF Type it's interested in. The Nudm_UECM_Get Service Operation uses the UE ID and NF type to search for the registered NF which is returned to the requestor.

SUPI, NF ID or SMS address of the NF corresponding to the NF type requested by NF consumer.

Nudm_UECM_Update service operation

The Nudm_UECM_Update service operation can be used by a registered NF (AMF or SMF) to update the stored registration information (e.g., UE capabilities, PGW-C + SMF FQDN for S5/S8 interface, etc.).

The consumer NF provides it's NF ID and type, the SUPI of the UE, a reference to the UE context information and the modification instruction.

Nudm_UECM_PCscfRestoration service operation

The Nudm_UECM_PCscfRestoration service operation is used to notify registered NFs (AMF, SMF) when UDM detects the need for P-CSCF restoration.

UDM notifies the AMF and/or SMF(s) that, during registration in UDM, indicated the need to be notified of P-CSCF Restoration.

13.2.4.2 Nudm_SubscriberDataManagement (SDM) service

The Nudm_SubscriberDataManagement service, also called Nudm_SDM Service, is used by NFs to retrieve subscription data from the UDM. The subscription data is structured into different data types and NFs retrieve the set of data types that they need for their operation. A key is used to identify the corresponding Subscription Data Type data.

Table 13.1 Lists the subscription data types, the main key and a non exhaustive list of the actual subscription data.

Nudm_SDM_Get service operation

The Nudm_SDM_Get Service operation is used by consumer NF to retrieve subscriber data. The Consumer NF indicates the subscription data types and corresponding keys and the UDM checks that the requesting NF is authorized to retrieve the requested subscription data. If the authorizations is successful the UDM responds with the requested data types.

Table 13.1 Subscription data types

Subscription data types	Data key	Example data
Access and Mobility Subscription data	SUPI	GPSI List, Group ID-list, Default S-NSSAIs, UE Usage Type, RAT restriction, Forbidden area, Service Area Restriction, Core Network type restriction, RFSP Index, UE behavioral information/Communication patterns, Subscribed DNN list, etc.
SMF Selection Subscription data	SUPI	S-NSSAI, Subscribed DNN list, Default DNN, LBO Roaming Information, Interworking with EPS indication list
UE context in SMF data	SUPI	PDU Session Id(s), DNN, SMF ID and address, PGW-C + SMF FQDN
SMS Management Subscription data	SUPI	SMS subscription, SMS barring list etc.
SMS Subscription data	SUPI	Indicates subscription to any SMS delivery service over NAS
UE Context in SMSF data	SUPI	AMF, Access Type, etc.
Session Management Subscription data	SUPI	GPSI List, Internal Group ID-list, S-NSSAI, Subscribed DNN list, DNN, UE Address, Allowed PDU Session Types, Default PDU Session Type, Allowed SSC modes, Default SSC mode, 5GS Subscribed QoS profile etc.
Identifier translation	GPSI	SUPI and optionally MSISDN
Slice Selection Subscription data	SUPI	Subscribed S-NSSAIs
Intersystem continuity Context	SUPI	List of (DNN + PGW FQDN)

Nudm_SDM_Notification service operation

The Nudm_SDM_Notification service operation is used by the UDM to notify NFs consumer of updates of previously retrieved subscriber data. The UDM includes the updated Subscription Data Type corresponding keys.

The UDM invokes the Nudm_SDM_Notification service operation when the subscriber data is updated at the UDM or when the UDM needs to deliver e.g., Steering of Roaming information, a new Routing Indicator or a new Default Configured NSSAI to a UE.

Nudm_SDM_Subscribe service operation

The Nudm_SDM_Subscribe service operation is used by NFs (AMF, SMF, and SMSF) to subscribe to updates of UE subscriber data. The AMF and SMSF subscribes to be notified of updates after successfully retrieving subscription data, with the Nudm_SDM_Get service operation, during the registration procedure. Similarly, the SMF subscribes to be notified of updates after successfully retrieving subscription data, with the Nudm_SDM_Get service operation, during a PDU session establishment. The subscribing NF includes subscription data type(s) and corresponding key(s).

Nudm_SDM_Unsubscribe service operation

The Nudm_SDM_Unsubscribe service operation is used by NFs e.g., AMF and SMF to unsubscribe to further notifications of updates to UE's subscriber data.

Nudm_SDM_Info service operation

The Nudm_SDM_Info service operation is used by the AMF to provide UDM with status information regarding subscription data management procedures towards the UE. It is e.g., used to provide UE acknowledgment of Steering of Roaming information from and UE acknowledgement of successful Network Slicing Configuration after delivery of the Network Slicing Subscription Change Indication via the AMF.

13.2.4.3 Nudm_UEAuthentication service

This service is used by the AUSF to get authentication data and provide UDM with the result of the authentication procedure success. If the concealed identity, SUCI, is used as an input the UDM will also provide the corresponding SUPI to the AUSF.

Nudm_UEAuthentication_Get service operation

The Nudm_UEAuthentication_Get service operation is used by the AUSF to retrieve authentication data from UDM. The UDM indicates the authentication method to use and the corresponding authentication data for a certain UE as identified by SUPI or SUCI. If SUCI is used, the UDM also returns the SUPI.

UEAuthentication_ResultConfirmation service operation

The UEAuthentication_ResultConfirmation service operation is used by the AUSF to inform UDM about of the result of an authentication procedure with a UE.

The AUSF provides the SUPI, a timestamp of the authentication, the authentication method and the serving network name to the UDM.

13.2.4.4 Nudm_EventExposure service

The Nudm_EventExposure service allows the NEF to subscribe, unsubscribe and get notified about events from the UDM. UDM can support events like e.g., UE reachability for SMS, change of PEI (i.e., the UE HW or SW has been changed) and roaming status.

Nudm_EventExposure_Subscribe service operation

The Nudm_EventExposure_Subscribe service operation allows the NEF to subscribe to or update event subscriptions. The NEF provides the target of the subscribe service operation: UE ID(s) (SUPI, GPSI, Internal Group Identifier or External Group Identifier, or indication that all UEs are targeted). The NEF also includes Event filters with Event Id(s) and Event Reporting Information.

When the subscription is accepted by the UDM it provides a Subscription Correlation ID and possibly an Expiry time. The UDM may also include an event report, if information is available.

Nudm_EventExposure_Unsubscribe service operation

The Nudm_EventExposure_Unsubscribe service operation allows the NEF to delete the subscription to event in UDM that it previously subscribed to. The NEF includes the Subscription Correlation ID, received in the Nudm_EventExposure_Subscribe Service Operation, to enable the UDM to identify the subscription to delete.

Nudm_EventExposure_Notify service operation

The Nudm_EventExposure_Notify service operation is used by UDM to report on events that was previously subscribed by the NEF. The UDM provides Event ID, Notification Correlation Information, time stamp and any event specific parameters.

13.2.4.5 Nudm_ParameterProvision service

The Nudm_ParameterProvision service allows the NEF (or rather AF via the NEF) to provision of information which can be used for the UE in 5GS.

Nudm_ParameterProvision_Update service operation

The Nudm_ParameterProvision_Update service operation allows provisioning of some parameters in the UDM e.g., Expected UE Behaviour, Network Configuration parameters. The NEF provides the GPSI, AF ID, Transaction Reference ID(s) and the parameters to provision e.g., Expected UE Behaviour parameters or at least one of the Network Configuration parameters and a Validity Time.

When accepted the UDM updates the respective subscription data types and updates any NF that has subscribed to updates of those subscription data types.

13.2.5 NRF services

The NRF and its services are key enablers in a service based architecture. The NRF centralizes and automates the configuration required for NF/NF services to discover, select and connect to peer NF/NF services with the correct capabilities. To do this the NRF provides three services Nnrf_NFManagement, Nnrf_NFDiscovery, and Nnrf_AccessToken. The Nnrf_NFManagement enables NFs to register and manage their NF

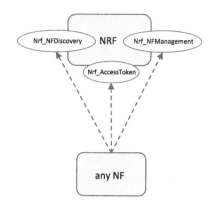

Fig. 13.6 NRF services.

services and capabilities in the NRF as shown in Fig. 13.6. The Nnrf_NFDiscovery allows NFs/NF Services to discover NFs/NF Services that match provided criteria. The Nnrf_AccessToken allows the NFs to request Auth2.0 access tokens that can be used to access services from other NFs.

13.2.5.1 Nnrf_NFManagement service
The Nnrf_NFManagement service has a set of service operations that allows NFs to register, update, deregister their NF profile including all NF Services in the NRF. It also has service operations that allows other NFs to subscribe to notifications of new, updated and removed NFs in the network. Note that as an alternative to the NF using the service Nnrf_NFManagement itself, another function e.g., OAM may use the service on behalf of the NF.

Nnrf_NFManagement_NFRegister service operation
The Nnrf_NFManagement_NFRegister service operation registers the NF and its NF services in the NRF by providing the NF profile of the consumer NF to NRF. All NF profiles contains information like: NF type, NF instance ID, NF service Names, PLMN ID and addressing information. In addition the NF profile contains information that is useful for discovery and selection and this information varies somewhat from NF to NF.

As result of a successful registration of the profile with the Nnrf_NFManagement_NFRegister service operation the NRF also marks the NF as available and will notify any NFs that has subscribed to this information.

Nnrf_NFManagement_NFUpdate service operation
The Nnrf_NFManagement_NFUpdate service operation allows NFs to update the NF profile in the NRF. The NF may replace the full NF profile, or it can update parts of the

NF profile by providing only the NF profile elements that needs to be updated and their new values.

As result of a successful update of the profile with Nnrf_NFManagement_NFUpdate the service operation the NRF also notifies any NFs that has subscribed to this information.

Nnrf_NFManagement_NFDeregister service operation

The Nnrf_NFManagement_NFDeregister service operation allows NFs to inform the NRF that they will no longer be available. The NRF marks the NF as unavailable, it may remove the profile information and inform any NF that has subscribed to information about the status of the NF.

Nnrf_NFManagement_NFStatusSubscribe service operation

The Nnrf_NFManagement_NFStatusSubscribe service operation allows NFs to subscribe to be notified if a new registered NF registers, updates its profile or deregisters. The NF can subscribe to information regarding:
- NF type, if NF status of a specific NF type is to be monitored
- NF instance ID, if NF status of a specific NF instance is to be monitored
- NF service, if NF status for NF which exposes a given NF service is to be monitored
In addition, the NF may further narrow down the subscription by providing additional parameters to match e.g., S-NSSAI(s) and the associated NSI ID(s), GUAMI(s) for AMF etc.

When the NRF accepts the subscription, it responds with a Subscription Correlation ID that is used for management of the subscription.

Nnrf_NFManagement_NFStatusNotify service operation

The Nnrf_NFManagement_NFStatusNotify service operation enables the NRF to notify subscribed NFs of newly registered NF along with its NF services, Updated NF profiles and Deregistered NF.

The NRF provides NF instance ID, NF Status and:
- The NF services (if the notification is for newly registered NF)
- The new NF profile (if the notification is for updated NF profile)
- Indication that an NF has deregistered.
Depending on the NF there may be additional parameters provided e.g., S-NSSAI(s) and the associated NSI ID(s), location of the NF, For AMF, list of GUAMI(s), TAI(s).

Nnrf_NFManagement_NFStatusUnsubscribe service operation

The Nnrf_NFManagement_NFStatusUnsubscribe service operation allows NF Consumers to unsubscribe from further notifications. The NF provides the Subscription Correlation ID to the NRF. The NRF uses the Subscription Correlation ID to identify the subscription deletes the associated resources.

13.2.5.2 Nnrf_NFDiscovery service

The Nnrf_NFDiscovery service is used to discovery of candidate NF instances with specific NF service or a target NF type. It also enables one NF service to discover a specific NF service. Based on the discovery result the NF can select a target NF/NF Service and initiate communication.

Nnrf_NFDiscovery_Request service operation

The Nnrf_NFDiscovery_Request service operation provides the requesting NF/NF Service a set of NF instances and its NF Services and additional information from the NF profile.

The NF service consumer provides one or more target NF service Name(s), NF type of the target NF, NF type of the NF requestor. If the NF service consumer intends to discover an NF service producer providing all the standardized services, it provides a wildcard NF service name. Depending on the NF and NF service the consumer wish to discover it may provide additional information like:
- S-NSSAI and the associated NSI ID
- DNN
- Target NF/NF service PLMN ID
- Serving PLMN ID
- NF service consumer ID
- NF location
- TAI
- The UE's Routing Indicator
- AMF region, AMF Set, GUAMI (for AMF)
- Group ID of the NF to discover

The NRF will search its internal database and match the input parameters and respond with a set of suitable NF instances, containing per NF Instance:
- NF Type
- NF Instance ID
- FQDN or IP address(es) of the NF instance
- List of NF Services Instances, each with:
  ○ Service Name
  ○ NF service instance ID
  ○ Optionally Endpoint Address(es) (list of IP addresses or an FQDN)

In addition, the NRF may, depending on the NF instance type, provide additional information from the NF profile, e.g.:
- If the target NF is BSF: Range(s) of (UE) IPv4 addresses or Range(s) of (UE) IPv6 prefixes.
- If the target NF stores Data Set(s) (e.g., UDR): Range(s) of SUPIs, range(s) of GPSIs, range(s) of external group identifiers, Data Set Identifier(s).

- If the target NF is UDM, UDR or AUSF, they can include UDM Group ID, UDR Group ID, AUSF Group ID.
- For UDM and AUSF, Routing Indicator.
- If the target NF is AMF, it includes list of GUAMI(s).
- If the target NF is CHF, it includes primary CHF instance and the secondary CHF instance pair(s).
- S-NSSAI(s) and the associated NSI ID(s).
- Location of the target NF.
- TAI(s).
- PLMN ID.

The NF Consumer uses the received set of suitable NFs to select a specific NF instance and NF service instance to initiate communication with. The NF consumer may also cash the discovered set of suitable NF instances. The cashed information may be reused for subsequent discovery requests that match the same criteria.

13.2.5.3 Nnrf_AccessToken service
The Nnrf_AccessToken service provides OAuth2 Access Tokens for NF to NF authorization for more information on OAuth2 authorization, see Chapter 8.

Nnrf_AccessToken_Get service operation
The Nnrf_AccessToken_Get Service Operation allows an NF consumer to request an NRF to authorize the consumer and provide an Access Token. The NF consumer can subsequently use the access token to show an NF Service Producer that it is authorized to use the service.

In the request the NF consumer provides NF Instance Id of the NF service consumer, NF producer service name(s), NF types of the NF producer instance and NF consumer. For roaming cases the home and visited PLMN IDs are also provided.

If authorization is successful the NRF provides an Access Token with appropriate claims, where the claims shall include NF Instance Id of NRF (issuer), NF Instance Id of the NF Service consumer (subject), NF type of the producers (audience), expected service name (scope) and expiration time (expiration).

13.2.6 AUSF services
The AUSF provides NF Service Consumers the following services:
- Nausf_UEAuthentication, authenticate the UE and provide keying material (AMF)
- Nausf_SoRProtection, protects the Steering Information List for the requester NF (UDM)
- Nausf_UPUProtection (UDM)

As shown in Fig. 13.7 the AMF and UDM are the NFs using the AUSF.

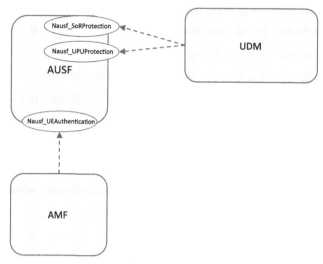

Fig. 13.7 AUSF services.

13.2.6.1 Nausf_UEAuthentication service

The Nausf_UEAuthentication service authenticates the UE and provides related keying material. It is used by the AMF and it has single service operation: Nausf_UEAuthentication.

Nausf_UEAuthentication service operation

The Nausf_UEAuthentication service operation is used by the AMF to initiate Authentication of the UE by providing the following information:
- UE id (SUPI or SUCI)
- Serving Network Name

Depending on the selected authentication method the AUSF executes either 5G-AKA or EAP-based authentication procedure. The initial requests create a resource. The content of the resource will depend on the procedure and will be returned to the AMF. The resources is used for subsequent requests carrying information as per the procedure executed for more information, see Chapter 8.

After executing the selected authentication procedure (several request/responses carrying the procedure messages) the AUSF will respond to the AMF with the authentication result and if success the master key which are used by AMF to derive NAS security keys and other security key(s). The AUSF will also provide the SUPI if the authentication was initiated with SUCI.

13.2.6.2 Nausf_SoRProtection service

The Nausf_SoRProtection service allows the UDM to request AUSF to provide protection parameters for the Steering of Roaming information (SoR). This prevents the

VPLMN to tamper with or remove the SoR information. This service also allows the AUSF to provide the UDM with information to verify that the UE received the Steering Information List.

Nausf_SoRProtection service operation

The Nausf_SoRProtection Service Operation is used by the UDM to protect SoR information. The UDM provides it's ID, SUPI of the UE and a SoR Header and optionally an ACK indication in the Request. The AUSF derives the SoR protection and returns it in the response to the UDM. If the ACK indication is included in request the AUSF also derives and returns information to validate the UE response in the SoR procedure. The UDM uses the received protection key to protect the SoR information in SoR procedures.

13.2.6.3 Nausf_UPUProtection service

The Nausf_UPUProtection service allows the UDM to request AUSF to provide protection parameters for the UE Parameters Update procedure.

Nausf_UPUProtection service operation

The Nausf_UPUProtection Service Operation provides security parameters to UDM for UE parameter updates. This operation provides the UDM with security parameters to protect the UE Parameters Update Data from being tampered with or removed by the VPLMN.

13.2.7 SMSF services

The SMSF provides the Nsmsf_SMService, as shown in Fig. 13.8 that allows activating and deactivating SM Service and to send uplink SMS messages.

13.2.7.1 Nsmsf_SMService service

The Nsmsf_SMService service allows the AMF to request the SMSF to activate and deactivate SM Service and to send uplink SMS messages.

Nsmsf_SMService_Activate service operation

The Nsmsf_SMService_Activate service operation is used by the AMF in the registration procedure to activate and authorize SM services. The AMF provides it's NF ID, the

Fig. 13.8 SMSF services.

SUPI and additional information in the request. The SMSF will register in the UDM and download subscription data and it will provide a successful response to the AMF if the UE is authorized to use SMS.

Nsmsf_SMService_Deactivate service operation

The Nsmsf_SMService_Deactivate service operation removes SMS service authorization from SMSF for a given SUPI. The Service operation is used by the AMF that includes a SUPI in the request. The SMSF deregisters with UDM and may delete data and resources related to the SUPI. The SMSF also responds with a deactivation result to the AMF.

Nsmsf_SMService_UplinkSMS service operation

The Nsmsf_SMService_UplinkSMS service operation is used by the AMF to pass uplink SMS messages from the UE to the SMSF. As a prerequisite the AMF and SMSF must have activated and authorized the SM Service. The AMF includes the SUPI and the SMS payload received from the UE in the request towards the SMSF. The SMSF sends the SM towards the SMS Service Centre and responds to the AMF with a transmission result.

13.2.8 UDR services

The Nudr_DataManagement (also called Nudr_DM) service, shown in Fig. 13.9, allows NF consumers to query, create, update, subscribe for change notifications, unsubscribe for change notifications and delete data stored in the UDR, based on the set of data applicable to the consumer.

Initially Data Sets and Data Set Identifiers have been specified for: Subscription Data, Policy Data, Application data and Data for Exposure. The Data Sets and Data Set Identifiers are intended to be extensible to cater for additional new identifiers as well as for operator specific identifiers and related data.

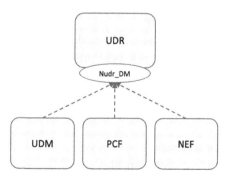

Fig. 13.9 UDR services.

13.2.8.1 Nudr_DataManagement (DM) service

The Nudr_DM service has the following service operations:

- Query
- Create
- Delete
- Update
- Subscribe
- Unsubscribe
- Notify

Common for all operation are that they can use the following parameters to specify the data they want to operate on:

- Data Set Identifier: Uniquely identifies the requested set of data within the UDR.
- Data Subset Identifier: It uniquely identifies the data subset within each Data Set Identifier.
- Data Keys (e.g., SUPI, GPSI, etc.).

For Nudr_DM_Subscribe and Nudr_DM_Notify operations:

- The Target of event reporting is made up of a Data Key and possibly a Data Sub Key
- The Data Set Identifier plus (if present) the (set of) Data Subset Identifier(s) corresponds to a (set of) Event ID(s).

An NF Service Consumer may include an indicator when it invokes Nudr_DM Query/Create/Update service operation to subscribe the changes of the data, to avoid a separate Nudr_DM_Subscribe service operation.

Nudr_DM_Query service operation

The Nudr_DM_Query service operation allows the NF service consumer (e.g., UDM) to requests a set of data from UDR. The NF Service Consumer provides Data Set Identifier and optionally Data Key(s), Data Subset Identifier(s), Data Sub Key(s). SUPI may also be included to identify which UE the latest list of stored PSIs belongs to. The UDM responds with the requested data.

Nudr_DM_Create service operation

The Nudr_DM_Create service Operation is used by NF service consumer to insert a new data record into the UDR, e.g., the NEF inserting a new application data record into the UDR. The requesting NF provides the Data Set Identifier, Data Key(s), optionally Data Subset Identifier(s), Data Sub Key(s) and the data. The UDR stores the data and responds with a result code.

Nudr_DM_Delete service operation

The Nudr_DM_Delete Service operation allows the NF service consumer to delete data stored in the UDR, e.g., a NEF service consumer want to delete an application data record. The requesting NF provides the Data Set Identifier, Data Key(s), optionally Data Subset Identifier(s), Data Sub Key(s) and the data. The UDR deletes the the data and responds with a result code.

Nudr_DM_Update service operation

The Nudr_DM_Update service operation allows the consumer to update that is stored data in the UDR. The requesting NF (e.g., UDM) provides the Data Set Identifier, Data Key(s), optionally Data Subset Identifier(s), Data Sub Key(s) and the data. The UDR updates the specified data and responds with a result code.

Nudr_DM_Subscribe service operation

The Nudr_DM_Subscribe Service operation enables the NF service consumer (e.g., UDM) to subscription to notification of data modified in the UDR. The events can be changes on existing data or addition of data. The requesting NF provides a Data Set Identifier, Notification Target Address, Notification Correlation ID and Event Reporting Information. In case of modification of an existing subscription the requesting NF also includes the previously received Subscription Correlation ID. The UDR accepts the subscription and responds with a Subscription Correlation ID.

Nudr_DM_Unsubscribe service operation

The Nudr_DM_Unsubscribe service operation allow the NF service consumer to remove previous subscriptions. The requesting NF provides the Subscription Correlation ID that allows the UDR to identify and remove the subscription information.

Nudr_DM_Notify service operation

The Nudr_DM_Notify service operation allows the UDR to notify previously subscribed notification targets about modification of data, when data in the UDR is added, modified or deleted. The UDR uses the Notification Target Address in the subscription and includes Notification Correlation Information, Data Set Identifier and the Updated Data.

13.2.9 5G-EIR services

The 5G-EIR provides a N5g-eir_EquipmentIdentityCheck service, as shown in Fig. 13.10, that is used by the AMF to check whether the Permanent Equipment ID (PEI) is on the black list or not. The N5g-eir_EquipmentIdentityCheck has a single service operation.

13.2.9.1 N5g-eir_EquipmentIdentityCheck_Get service operation

The N5g-eir_EquipmentIdentityCheck_Get service operation allows the AMF to check the PEI and determine whether the subscriber is allowed to use the equipment or not.

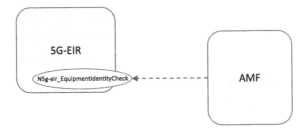

Fig. 13.10 5G-EIR services.

The AMF receives the PEI during the registration procedure and may use the N5g-eir_
EquipmentIdentityCheck_Get Service operation offered by the 5G-EIR. The AMF
provides the PEI and SUPI in the request. The 5G-EIR checks the PEI and responds
to the AMF that indicates if the PEI is white-, gray- or black-listed. Based on the result
the AMF determines if it can continue the registration procedure or reject the UE.

13.2.10 NWDAF services

The following NWDAF provides two services Nnwdaf_EventsSubscription and
Nnwdaf_AnalyticsInfo, as shown in Fig. 13.11. The Nnwdaf_EventsSubscription service
enables the NF service consumers to subscribe/unsubscribe for different type of information from NWDAF. The Nnwdaf_AnalyticsInfo service enables the NF service consumers to request different types information from NWDAF.

   In 3GPP Release 15, the NDWAF is limited to load level event in one or more Network Slice Instances and possibly Load Level Thresholds. In later releases it is expected
that the NWDAF will support additional events and event filters.

13.2.10.1 Nnwdaf_EventsSubscription service

The Nnwdaf_EventsSubscription service enables the consumer to subscribe/unsubscribe
to notification of load-based events in Network Slice instances. Periodic notification and
notification upon threshold exceeded can be subscribed.

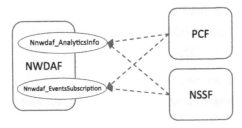

Fig. 13.11 NWDAF services.

Nnwdaf_EventsSubscription_Subscribe service operation

The Nnwdaf_EventsSubscription_Subscribe service operation allows e.g., the NSSF to subscribe to NWDAF events. The requesting NF provides S-NSSAI, Event ID(s), Notification Target Address, Notification Correlation ID and Event Reporting Information. Event Filter(s) e.g., Load Level Threshold value may also be included. In case of modification of an existing subscription the requesting NF also includes the previously received Subscription Correlation ID. The NWDAF accepts the subscription and responds with a Subscription Correlation ID.

Nnwdaf_EventsSubscription_Unsubscribe service operation

The Nnwdaf_EventsSubscription_Unsubscribe service operation allow the NF service consumer to remove previous subscriptions. The requesting NF provides the Subscription Correlation ID that allows the NWDAF to identify and remove the subscription information.

Nnwdaf_EventsSubscription_Notify service operation

The NWDAF notifies the notification target that an subscribed event has occurred. Depending upon type of subscription this notification is either on a periodic basis or triggered whenever a threshold (as defined in the subscribe operation) is crossed. The NWDAF uses the Notification Target Address and includes the Event ID, Notification Correlation ID, S-NSSAI and the Load level information for the Network Slice instance.

13.2.10.2 Nnwdaf_Analytics_Info service

The Nnwdaf_Analytics_Info service enables the consumer to request and get from NWDAF load level information of Network Slice instance(s). This service has a single Service operation.

Nnwdaf_AnalyticsInfo_Request service operation

The Nnwdaf_AnalyticsInfo_Request service operation allows NF consumers to request load information for one of more network slices. The requesting NF specifies Event ID: load level information and can include one or more network slice instance(s) in the event filter. The NWDAF responds with the requested load information for the specified network slice instances.

13.2.11 UDSF services

The UDSF services was only defined on a stage 2 level in 3GPP Release 15. There was no stage 3 protocol solution defined but it is expected that later releases will study suitable protocol solutions that can support the performance requirements on dynamic data access required by NFs using the UDSF.

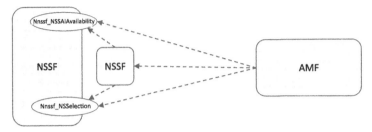

Fig. 13.12 NSSF services.

13.2.12 NSSF services

The NSSF produces two services Nnssf_NSSelection and Nnssf_NSSAIAvailability, as shown in Fig. 13.12. The Nnssf_NSSelection service provides the Network Slice information to the Requester and the Nnssf_NSSAIAvailability service provides the availability of S-NSSAIs on a per TA basis.

13.2.12.1 Nnssf_NSSelection service

The Nnssf_NSSelection service has a single service operation, Nnssf_NSSelection_Get.

Nnssf_NSSelection_Get service operation

The Nnssf_NSSelection_Get Service Operation allows the AMF to request allowed NSSAI and the Configured NSSAI for the Serving PLMN. The AMF may invoke the Nnssf_NSSelection_Get Service Operation during Registration procedure, during PDU Session Establishment procedure or during UE Configuration Update procedure. When invoked during Registration procedure it may possibly trigger AMF re-allocation. The request may in roaming cases the "relayed" from the NSSF in one PLMN to an NSSF in a different PLMN.

If this service operation is invoked during registration AMF provides the Subscribed S-NSSAI(s), default S-NSSAI, Home PLMN ID, TAI, NF type of the NF service consumer, Requester ID. If available requested NSSAI, Mapping of Requested NSSAI, Default Configured NSSAI Indication, Allowed NSSAI for current Access Type, Allowed NSSAI for the other Access Type, and the corresponding Mapping Of Allowed NSSAIs for current Access Type and other Access Type may be provided.

In other cases, the available and relevant information is provided in the request.

The NSSF uses the input information to determine suitable network slice information and provides a response that in the registration case may contains one or more of Allowed NSSAI, Configured NSSAI; Target AMF Set or, based on configuration, the list of candidate AMF(s).

The AMF stores the received information and will use different pieces information in internal functionality, in communication with the UE, RAN and at selection of SMF.

During Registration the AMF will e.g., determine if the it will trigger re-location, during PDU Session establishment.

13.2.12.2 Nnssf_NSSAIAvailability service

The Nnssf_NSSAIAvailability service has two service operations and it enables AMFs to update the NSSF and other AMFs on the availability of S-NSSAIs on a per TA basis.

Nnssf_NSSAIAvailability_Update service operation

The Nnssf_NSSAIAvailability_Update service operation enables the AMF to update the NSSF with the S-NSSAIs the AMF supports per TA. The AMF provides the supported S-NSSAIs per TAI, in the request towards the NSSF. The NSSF responds to the requesting AMF with a list of S-NSSAIs restricted per TAI (if any).

Nnssf_NSSAIAvailability_Notify service operation

The Nnssf_NSSAIAvailability_Notify service operation allows the NSSF to update the AMFs with any S-NSSAIs restricted per TAI and, subsequently remove any restriction per TAI. The NSSF provides a list of TAIs and the S-NSSAIs for which the status is changed (restricted/unrestricted) for each TAI to AMF. The AMF stores the updated information.

13.2.13 LMF services

The following LMF support one service, Nlmf_Location, as shown in Fig. 13.13. The Nlmf_Location service enables the AMF to request location determination for a target UE. It allows the AMF to request the current geodetic and optionally civic location of a target UE. The Nlmf_Location service has one service operation Nlmf_Location_DetermineLocation service operation.

13.2.13.1 Nlmf_Location_DetermineLocation service operation

The Nlmf_Location_DetermineLocation service operation Provides UE location information to the consumer NF. The AMF provides External Client Type and a LCS Correlation Identifier is may also include serving cell identifier, Location QoS, Supported GAD shapes and AMF identity.

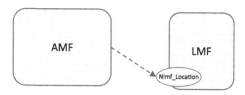

Fig. 13.13 LMF services.

The LMF may execute location procedures (e.g., by invoking other services). The result is provided by the LMF to the AMF and can include Geodetic Location, Civic Location, Position Methods Used.

13.2.14 NEF services

The NEF supports eight services as illustrated in Fig. 13.14 and listed below:
- Nnef_EventExposure, provides support for event exposure
- Nnef_PFDManagement, provides support for PFDs management
- Nnef_ParameterProvision, provides support to provision information which can be used for the UE in 5GS
- Nnef_Trigger, provides support for device triggering
- Nnef_BDTPNegotiation, provides support for negotiation about the transfer policies for the future background data transfer
- Nnef_TrafficInfluence, provide the ability to influence traffic routing
- Nnef_ChargeableParty, requests to become the chargeable party for a data session for a UE
- Nnef_AFsessionWithQoS, requests the network to provide a specific QoS for an AS session

13.2.14.1 Nnef_EventExposure service
Nnef_EventExposure_Subscribe service operation
The Nnef_EventExposure_Subscribe service operation allows internal or external AFs to subscribe to events. The service operation can also be used to update a previous subscription. The requesting AF provides a (Set of) Event ID(s), target of event reporting (GPSI or External Group Identifier), Event Reporting Information Notification Target Address and Notification Correlation ID. The requesting AF may also provide Event Filters to narrow down the event reporting. If the AF want to update a subscription it also includes

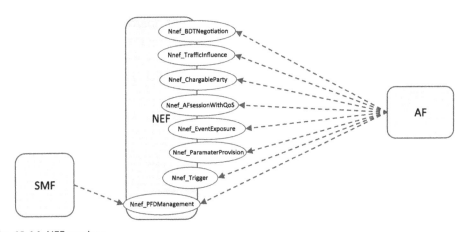

Fig. 13.14 NEF services.

the Subscription Correlation ID that it received when previously subscribing. The NEF does not produce event by itself, but it can in turn subscribe to relevant events from other NFs. The NEF accepts the subscription and responds with a Subscription Correlation ID and possibly an expiry time. The response may also include a first event report if data is available.

Nnef_EventExposure_Unsubscribe service operation
The Nnef_EventExposure_Unsubscribe service operation allows an AF to remove a previously subscription to Event Exposure. The AF invoking the Nnef_EventExposure_ Unsubscribe Service operation provides the Subscription Correlation ID. The NEF uses the Subscription Correlation ID to identify and remove the subscription. It acknowledges the removal back to the AF and may delete any related resources.

Nnef_EventExposure_Notify service operation
The Nnef_EventExposure_Notify Service operation allows the NEF to report the event to the consumer that has previously subscribed. The NEF used the Notification Target Address and includes Event ID, Notification Correlation Information, time stamp. The NEF may also include event information (depends on the specific Event).

13.2.14.2 Nnef_PFDManagement service
The Nnef_PFDManagement service provides AF the capability to create, update or remove PFDs via the NEF and for the SMF to fetch or subscribe to PFDs.

Nnef_PFDManagement_Fetch service operation
The Nnef_PFDManagement_Fetch service operation allows the SMF to fetch PFDs for one or more application identifier. The SMF provide the Application Identifier(s). The NEF responds with the Application Identifier(s) and the corresponding PFDs.

Nnef_PFDManagement_Subscribe service operation
The Nnef_PFDManagement_Subscribe service operation allows the SMF consumers to explicitly subscribe the notification of changes of PFDs for Application Identifier(s). The SMF provides Application Identifier(s). The NEF creates a subscription resource which is included in the response to the SMF.

Nnef_PFDManagement_Notify service operation
The Nnef_PFDManagement_Notify service operation is used by the NEF to inform subscribed SMFs of changes to PFDs related to the subscribed Application Identifiers. The NEF included Provides Update PFDs for Application Identifier in a notification to the SMF.

Nnef_PFDManagement_Unsubscribe service operation

The Nnef_PFDManagement_Unsubscribe service operation allows the SMF to remove a subscription. The SMF provides the Application Identifier(s) it wants to remove the subscription for. The NEF removes the subscription and may delete any related resources.

Nnef_PFDManagement_Create service operation

The Nnef_PFDManagement_Create service operation allows an AF to create PFDs. The AF provides an AF ID, Application Identifier(s) and corresponding PFDs. If accepted by the NEF it stores the Application Identifier(s) and corresponding PFDs and responds to the AF with a Transaction Reference ID.

Nnef_PFDManagement_Update service operation

The Nnef_PFDManagement_Update service operation allows an AF to update PFDs. The AF provides a Transaction Reference ID, Application Identifier(s) and corresponding PFDs. The NEF stores the new PFDs and may update any subscribed SMFs.

Nnef_PFDManagement_Delete service operation

The Nnef_PFDManagement_Delete service operation allows the AF to request deletion of PFDs. The AF provides Transaction Reference ID and the NEF deletes the corresponding information.

13.2.14.3 Nnef_ParameterProvision service

This service is for allowing external party to provision of information which can be used for the UE in 5GS.

13.2.14.4 Nnef_ParameterProvision_Update service operation

The Nnef_ParameterProvision_Update service operation allows an AF to update the UE related information i.e., Expected UE Behaviour. The AF provides the GPSI, AF ID, Transaction Reference ID and Expected UE Behaviour parameters.

13.2.14.5 Nnef_Trigger service
Nnef_Trigger_Delivery service operation

The Nnef_Trigger_Delivery service operation allows the consumer to request that a trigger be sent to an application on a UE and it also implicitly subscribes to be notified about result of the trigger delivery attempt. The AF provides GPSI, AF ID, Trigger Reference Number, Application Port ID and the NEF responds with a Transaction Reference ID.

Nnef_Trigger_DeliveryNotify service operation

Nnef_Trigger_DeliveryNotify service operation allows the NEF to report the status of the trigger delivery to the application on the UE. The NEF includes the Transaction Reference ID and a Delivery Report.

13.2.14.6 Nnef_BDTPNegotiation service
Nnef_BDTPNegotiation create service operation

The Nnef_BDTPNegotiation create service operation allows the AF to request a background data transfer policy. The AF provides ASP Identifier, Volume per UE, Number of UEs, Desired time window and possibly the expected Network Area. The NEF responds with a Transaction Reference ID and one or more background data transfer policies.

Nnef_BDTPNegotiation update service operation

The Nnef_BDTPNegotiation update service operation allows the NF requests the selected background data transfer policy to be used. This service is only used if the NEF responded with several possible background data transfer policies in the Nnef_BDTPNegotiation Create Service Operation. The AF provides the Transaction Reference ID, ASP Identifier, and the selected background data transfer policy.

13.2.14.7 Nnef_TrafficInfluence service

The Nnef_TrafficInfluence service allows the NEF to provide authorization of requests, parameter mapping and possibility to influence traffic routing decisions.

Nnef_TrafficInfluence_Create service operation

The Nnef_TrafficInfluence_Create service operation allows the NEF to authorize the request and forward the request for traffic influence the relevant NF (that can execute the influence in traffic). The AF includes an AF Transaction Id and parameters specifying the traffic and the subscribers that should be influenced and parameter describing how traffic should be influenced. If authorized by the NEF it identifies the impacted NFs and forwards the request.

Nnef_TrafficInfluence_Update service operation

The Nnef_TrafficInfluence_Update service operation allows the NEF to authorize and forward the updated the traffic influence request. The AF provides the AF Transaction Id and any parameters to update. The NEF identifies the NFs to be updated and provides the update.

Nnef_TrafficInfluence_Delete service operation

The Nnef_TrafficInfluence_Delete service operation allows the AF to the request deletion of previous request for traffic influence. The AF provides the AF Transaction Id. The NEF authorizes the requests and forwards it to the relevant NFs.

Nnef_TrafficInfluence_Notify service operation

The Nnef_TrafficInfluence_Notify service operation allow the NEF to forward the UP path management event reports to AF. The NEF includes the AF Transaction Id,

UP path management event. The AF Transaction Id identifies the AF request for traffic influence that the event report is related to.

13.2.14.8 Nnef_ChargeableParty service
The Nnef_ChargeableParty service allows an NF to become the chargeable party for a data session for a UE.

Nnef_ChargeableParty create service operation
The Nnef_ChargeableParty create service operation allows an AF to request to become the chargeable party for a data session for a UE. The AF provides AF Identifier, UE IP address, Description of the application flows, Sponsor Information, Sponsoring Status. When accepted the NEF responds with a Transaction Reference ID.

Nnef_ChargeableParty update service operation
The Nnef_ChargeableParty update service operation allows the AF to change the chargeable party of a data session for a UE that has been previously created. The AF provides an AF Identifier, Transaction Reference ID and Sponsoring Status.

Nnef_ChargeableParty notify service operation
The Nnef_ChargeableParty notify service operation allows the NEF to report bearer level event(s) to the AF.

13.2.14.9 Nnef_AFsessionWithQoS service
The Nnef_AFsessionWithQoS service allows an NF to request a specific QoS for a session.

Nnef_AFsessionWithQoS create service operation
The Nnef_AFsessionWithQoS create service operation allows an AF to requests the network to provide a specific QoS for a session. The AF provides an AF Identifier, UE IP address, Description of the application flows and a QoS Reference. The NEF responds with an Transaction Reference ID.

Nnef_AFsessionWithQoS notify service operation
The Nnef_AFsessionWithQoS notify service operation allows the NEF to report bearer level event(s) to the AF.

CHAPTER 14

Protocols

14.1 Introduction

This chapter outlines the main protocols used in the 5GS, with the aim of giving a high-level overview of these protocols and their basic properties.

14.2 5G non-access stratum (5G NAS)

14.2.1 Introduction

NAS denotes the main Control Plane protocols between the UE and the core network. The main functions of NAS are:

- Handling of UE registration and mobility, including generic functionality for access control such as connection management, authentication, NAS security handling, UE identification and UE configuration
- Support of Session Management procedures to establish and maintain PDU Session connectivity and QoS for the User Plane between the UE and the DN
- General NAS transport between UE and AMF to carry other types of messages that are not defined as part of the NAS protocol as such. This includes, e.g., transport of SMS, LPP protocol for location services, UDM data such as Steering of Roaming (SOR) messages, as well as UE policies (URSP).

NAS consists of two basic protocols to support the functionality above; the 5GS Mobility Management (5GMM) protocol and the 5GS Session Management (5GSM) protocol. The 5GMM protocol runs between the UE and the AMF and is the basic NAS protocol used for handling UE registrations, mobility, security and also transport of the 5GSM protocol as well as the general NAS transport of other types of messages. The 5GSM protocol runs between UE and SMF (via the AMF) and supports management of PDU Session connectivity. It is carried on top of the 5GMM protocol as shown in Fig. 14.1. The 5GMM protocol is also used to transport information between UE and PCF, UE and SMSF, etc. also shown in Fig. 14.1. The 5GMM and 5GSM protocols will be further described below.

With 5G, the NAS protocol is used over both 3GPP and non-3GPP access. This is a key difference compared to EPS/4G, where NAS was tailor-made for 3GPP access (E-UTRAN) only.

5G Core Networks
https://doi.org/10.1016/B978-0-08-103009-7.00014-4

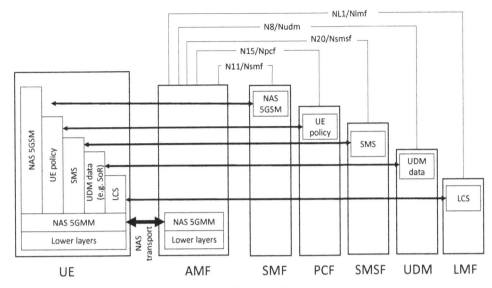

Fig. 14.1 NAS protocol stack with NAS-MM and NAS-MM protocols.

The NAS messages are transported by NGAP (used on N2 reference point) between AMF and the (R)AN and by access specific means between (R)AN and UE. NGAP is described in a Section 14.3 in this chapter.

The 5G NAS protocols are defined as new protocols in 5G but they have many similarities with the NAS protocols used for 4G/EPS and also the NAS protocols defined for 2G/3G/GPRS. The 5G NAS protocols are specified in 3GPP TS 24.501.

14.2.2 5G mobility management

5GMM procedures are used to keep track of the whereabouts of the UE, to authenticate the UE and control integrity protection and ciphering. The 5GMM procedures also allow the network to assign new temporary identities to the UE (5G-GUTI) and also request identity information (SUCI and PEI) from the UE. In addition, the 5GMM procedures provide the UE's capability information to the network and the network may also inform the UE about information regarding specific services in the network. The 5GMM protocol thus operates on a UE level (per Access Type) in contrast to the 5GSM protocol that is on a per PDU Session level. The 5GMM NAS signaling takes place between the UE and the AMF. The basic 5GMM procedures are:

- Registration
- Deregistration
- Authentication
- Security mode control
- Service request
- Notification

- Uplink NAS transport
- Downlink NAS transport
- UE configuration update (e.g., for 5G-GUTI re-allocation, TAI list update, etc.)
- UE identity request

The 5GS mobility management NAS message types used to support these procedures are listed in Table 14.1.

Table 14.1 NAS message types for mobility management.

Type of procedure	Message type	Direction
5GMM specific procedures	Registration request	UE → AMF
	Registration accept	AMF → UE
	Registration complete	UE → AMF
	Registration reject	AMF → UE
	Deregistration request (UE originating procedure)	UE → AMF
	Deregistration accept (UE originating procedure)	AMF → UE
	Deregistration request (UE terminated procedure)	AMF → UE
	Deregistration accept (UE terminated procedure)	UE → AMF
5GMM connection management procedures	Service request	UE → AMF
	Service reject	AMF → UE
	Service accept	AMF → UE
5GMM common procedures	Configuration update command	AMF → UE
	Configuration update complete	UE → AMF
	Authentication request	AMF → UE
	Authentication response	UE → AMF
	Authentication reject	AMF → UE
	Authentication failure	UE → AMF
	Authentication result	AMF → UE
	Identity request	AMF → UE
	Identity response	UE → AMF

Continued

Table 14.1 NAS message types for mobility management—cont'd

Type of procedure	Message type	Direction
	Security mode command	AMF → UE
	Security mode complete	UE → AMF
	Security mode reject	UE → AMF
	5GMM status	UE → AMF or AMF → UE
	Notification	AMF → UE
	Notification response	UE → AMF
	UL NAS transport	UE → AMF
	DL NAS transport	AMF → UE

The 5GMM procedures can only be performed if a NAS signaling connection has been established between the UE and the AMF. If there is no active signaling connection, the 5GMM layer has to initiate the establishment of a NAS signaling connection. The NAS signaling connection is established by a Registration or a Service Request procedure from the UE. For downlink NAS signaling, if there is no active signaling connection, the AMF first initiates a paging procedure that triggers the UE to execute the Service Request procedure. (See Chapter 15 for a description of these procedures.) The 5GMM procedures in turn rely on services from the underlying NGAP protocol between the (R)AN and AMF (i.e., N2) and access-specific signaling between UE and (R)AN such as RRC for 3GPP access to establish connectivity.

14.2.3 5G session management

5GSM procedures are used to manage the PDU Sessions and QoS for the User Plane. This includes procedures for establishing and releasing PDU Sessions as well as modification of PDU Sessions to add, remove or modify QoS rules. The 5GSM procedures are also used to carry out the secondary authentication for a PDU Session (see Chapter 6 for additional description of secondary authentication). The 5GSM protocol thus operates on a PDU Session level in contrast to the 5GMM protocol that works on a UE level.

The basic 5GSM procedures are:
- PDU Session establishment
- PDU Session release
- PDU Session modification
- PDU Session authentication and authorization
- 5GSM status (to exchange PDU Session status information)

The SM NAS message types supporting these procedures are listed in Table 14.2.

Table 14.2 NAS message types for session management.

Message type	Direction
PDU Session establishment request	UE → SMF
PDU Session establishment accept	SMF → UE
PDU Session establishment reject	SMF → UE
PDU Session authentication command	SMF → UE
PDU Session authentication complete	UE → SMF
PDU Session authentication result	SMF → UE
PDU Session modification request	UE → SMF
PDU Session modification reject	SMF → UE
PDU Session modification command	SMF → UE
PDU Session modification complete	UE → SMF
PDU Session modification command reject	UE → SMF
PDU Session release request	UE → SMF
PDU Session release reject	SMF → UE
PDU Session release command	SMF → UE
PDU Session release complete	UE → SMF
5GSM status	UE → SMF or SMF → UE

14.2.4 Message structure

The NAS protocols are implemented as standard 3GPP L3 messages in accordance with 3GPP TS 24.007. Standard 3GPP L3 according to 3GPP TS 24.007 and its predecessors have also been used for NAS signaling messages in previous generations (2G, 3G, 4G). The encoding rules have been developed to optimize the message size over the air interface and to allow extensibility and backwards compatibility without the need for version negotiation.

Each NAS message contains a Protocol Discriminator and a Message Type. The Protocol Discriminator is a value that indicates the protocol being used, i.e., for 5G NAS messages it is either 5GMM or 5GSM (to be precise, for 5G, an Extended Protocol Discriminator had to be defined as the available spare numbers of the original Protocol Discriminator was running out). The Message Type indicates the specific message that is sent, e.g., Registration request, Registration accept or PDU Session Modification request, as shown in Tables 14.1 and 14.2.

NAS 5GMM messages also contain a security header that indicates if the message is integrity protected and/or ciphered. 5GSM messages contain a PDU Session identity that identifies which PDU Session the 5GSM message refers to. The rest of the information elements in the 5GMM and 5GSM messages are tailored for each specific NAS message.

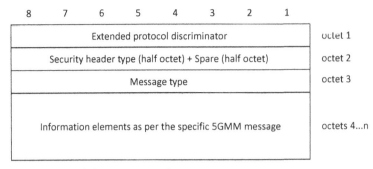

Fig. 14.2 Frame structure of plain 5GMM NAS message.

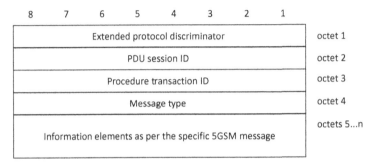

Fig. 14.3 Frame structure of plain 5GSM NAS message.

The organization of a plain 5GMM NAS message is shown in Fig. 14.2 and of a plain 5GSM message is shown in Fig. 14.3.

When an NAS message is security protected the plain NAS message is encapsulated as shown in Fig. 14.4. This format applies to all 5GSM messages since they are always security protected. It also applies to security-protected 5GMM messages. In these security protected NAS messages, the first Extended Protocol Discriminator indicates that it is

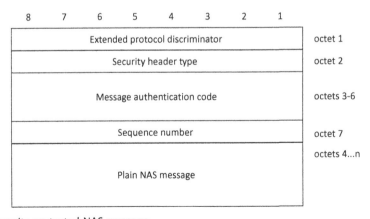

Fig. 14.4 Security protected NAS message.

a 5GMM message since NAS security is part of the 5GMM NAS protocol. The plain NAS message inside the security protected NAS message has additional Extended Protocol Discriminator(s) that indicate whether it is a 5GMM or a 5GSM message. Further encapsulation may be done in the plain NAS message inside the security protected NAS message. The plain NAS message could, e.g., be a UL NAS transport (5GMM) message that contains a PDU Session Establishment request (5GSM) message.

Further details on the EPS NAS messages and the information elements are available in 3GPP TS 24.501 and 3GPP TS 24.007.

14.2.5 Future extensions and backward compatibility

The UE and network are in principle specified to ignore information elements they do not understand. It is therefore possible for a later release of the system to add new information elements in the 5G NAS signaling without affecting the UEs and network that implement earlier releases of the specifications.

14.3 NG application protocol (NGAP)

14.3.1 Introduction

The NGAP protocol is designed for use on the N2 interface between the (R)AN and AMF. It can be noted that the 3GPP RAN groups have given the name NG to the RAN-AMF interface that in the overall system architecture is called N2. The protocol name NGAP is thus derived from the interface name NG with the addition of AP (Application Protocol), which is a term that has been used many times by 3GPP to denote a signaling protocol between two network functions.

14.3.2 Basic principles

NGAP supports all mechanisms necessary to handle the procedures between AMF and (R)AN, and it also supports transparent transport for procedures that are executed between the UE and the AMF or other core network functions. NGAP is applicable both to 3GPP access and non-3GPP accesses integrated with 5GC. This is a key difference to EPC where S1AP was designed for use only with 3GPP access (E-UTRAN) and not non-3GPP accesses. However, even though NGAP is applicable to any access, the design has been primarily targeted at 3GPP accesses (NG-RAN), which can also be noticed in the protocol specification defined in 3GPP TS 38.413. Support for specific parameters related to non-3GPP accesses have been added to the protocol when needed.

The NGAP interactions between AMF and (R)AN are divided into two groups:

- Non UE-associated services: These NGAP services are related to the whole NG interface instance between the (R)AN node and AMF. They are, e.g., used to establish the NGAP signaling connection between AMF and (R)AN, handle some overload situations and to exchange RAN and AMF configuration data.

- UE-associated services: These NGAP services are related to one UE. This NGAP signaling is thus related to procedures where a UE is involved, e.g., at Registration, PDU Session Establishment, etc.

The NGAP protocol supports the following functions:

- NG (i.e., N2) interface management functions, for example initial NG interface setup as well as Reset, Error Indication, Overload Indication and Load Balancing.
- Initial UE Context Setup functionality for establishment of an initial UE context in the (R)AN node.
- Provision of the UE capability information to the AMF (when received from the UE).
- Mobility functions for UEs in order to enable handover in NG-RAN, e.g., Path Switch request.
- Setup, modification, and release of PDU Session resources (User Plane resources)
- Paging, providing the functionality for 5GC to page the UE.
- NAS signaling transport functionality between the UE and the AMF
- Management of the binding between a NGAP UE association and a specific transport network layer association for a given UE
- Status transfer functionality (transfers PDCP Sequence Number status information from source NG-RAN node to target NG-RAN node (via AMF) in support of in-sequence delivery and duplication avoidance for handover).
- Trace of active UEs.
- UE location reporting and positioning protocol support.
- Warning message transmission.

14.3.3 NGAP elementary procedures

NGAP consists of Elementary Procedures. An Elementary Procedure is a unit of interaction between the (R)AN (e.g., NG-RAN node) and AMF. These Elementary Procedures are defined separately and are intended to be used to build up complete sequences in a flexible manner. The Elementary Procedures may be invoked independently of each other as standalone procedures, which can be active in parallel. Some elementary procedures are specifically related to only Non UE-associated services (e.g., the NG Setup procedure) while others are related to only UE-associated services (e.g., the PDU Session Resource Modify procedure). Some elementary procedures may be using either Non UE-associated or UE-associated signaling depending on the scope and the context, e.g., the Error Indication procedure that uses UE-associated signaling if the error was related to a reception of a UE-associated signaling message, while it uses Non UE-associated signaling otherwise. In some cases, the independence between some Elementary Procedures is restricted; in this case the particular restriction is specified in the NGAP protocol specification.

Tables 14.3 and 14.4 list the elementary procedures in NGAP. Some of the procedures are request-response type of procedures, where the initiator gets a response from the receiver of the request, indicating whether the request was successfully handled or

Table 14.3 NGAP elementary procedures with a response to indicate success or failure (table based on Table 8.1-1 in 3GPP TS 38.413).

Elementary procedure	Initiating NGAP message	Successful outcome NGAP response message	Unsuccessful outcome NGAP response message
AMF configuration update	AMF configuration update	AMF configuration update acknowledge	AMF configuration update failure
RAN configuration update	RAN configuration update	RAN configuration update acknowledge	RAN configuration update failure
Handover cancellation	Handover cancel	Handover cancel acknowledge	
Handover preparation	Handover required	Handover command	Handover preparation failure
Handover resource allocation	Handover request	Handover request acknowledge	Handover failure
Initial context setup	Initial context setup request	Initial context setup response	Initial context setup failure
NG reset	NG reset	NG reset acknowledge	
NG setup	NG setup request	NG setup response	NG setup failure
Path switch request	Path switch request	Path switch request acknowledge	Path switch request failure
PDU session resource modify	PDU session resource modify request	PDU session resource modify response	
PDU session resource modify indication	PDU session resource modify indication	PDU session resource modify confirm	
PDU session resource release	PDU session resource release command	PDU session resource release response	
PDU session resource setup	PDU session resource setup request	PDU session resource setup response	
UE context modification	UE context modification request	UE context modification response	UE context modification failure
UE context release	UE context release command	UE context release complete	
Write-replace warning	Write-replace warning request	Write-replace warning response	
PWS cancel	PWS cancel request	PWS cancel response	
UE radio capability check	UE radio capability check request	UE radio capability check response	

Table 14.4 Elementary NGAP procedures without response (table based on Table 8.1-2 in 3GPP TS 38.413).

Elementary procedure	NGAP message
Downlink RAN configuration transfer	Downlink RAN configuration transfer
Downlink RAN status transfer	Downlink RAN status transfer
Downlink NAS transport	Downlink NAS transport
Error indication	Error indication
Uplink RAN configuration transfer	Uplink RAN configuration transfer
Uplink RAN status transfer	Uplink RAN status transfer
Handover notification	Handover notify
Initial UE message	Initial UE message
NAS non delivery indication	NAS non delivery indication
Paging	Paging
PDU session resource notify	PDU session resource notify
Reroute NAS request	Reroute NAS request
UE context release request	UE context release request
Uplink NAS transport	Uplink NAS transport
AMF status indication	AMF status indication
PWS restart indication	PWS restart indication
PWS failure indication	PWS failure indication
Downlink UE associated NRPPa transport	Downlink UE associated NRPPA transport
Uplink UE associated NRPPa transport	Uplink UE associated NRPPA transport
Downlink Non UE associated NRPPa transport	Downlink non UE associated NRPPA transport
Uplink non UE associated NRPPa transport	Uplink non UE associated NRPPA transport
Trace start	Trace start
Trace failure indication	Trace failure indication
Deactivate trace	Deactivate trace
Cell traffic trace	Cell traffic trace
Location reporting control	Location reporting control
Location reporting failure indication	Location reporting failure indication

Table 14.4 Elementary NGAP procedures without response (table based on Table 81-2 in).—cont'd

Elementary procedure	NGAP message
Location report	Location report
UE TNLA binding release	UE TNLA binding release request
UE radio capability info indication	UE radio capability info indication
RRC inactive transition report	RRC inactive transition report
Overload start	Overload start
Overload stop	Overload stop

not. These are listed in Table 14.3. Other procedures are elementary procedures without a response. These messages are used, e.g., when AMF wants to only deliver a downlink NAS message. There is no need for RAN to provide a response in that case since error handling is handled on NAS level. The elementary procedures that do not have a response are listed in Table 14.4.

There is no version negotiation in NGAP. The forward and backwards compatibility of the protocol is instead ensured by a mechanism where all current and future messages, and IEs or groups of related IEs, include ID and criticality fields that are coded in a standard format that will not be changed in the future. These parts can always be decoded regardless of the standard version.

NGAP relies on a reliable transport mechanism and is designed to run on top of SCTP (SCTP is further described in Section 14.10).

14.4 Hypertext transfer protocol (HTTP)

14.4.1 Introduction

In today's world with the familiarity of the Internet, HTTP is maybe one of the most well-known protocols. It is the foundation of data communication on the Internet, where HTTP is used to "surf the web", giving access to web pages where hyperlinks lead to other resources that the user can easily access by clicking with the mouse or tapping the screen. At the same time, HTTP is a newcomer in the family of protocols in the 3GPP core networks, at least if we look only at the Control Plane interfaces. While the 2G/3G/4G Control Plane relied heavily on GTP-C, MAP and Diameter in the core network, the 5GC Control Plane relies almost exclusively on HTTP.

In this section we will give a high-level introduction to HTTP and mention some aspects for how it is applied in 5GC. There is however plenty of tutorial information available on the Internet that may be consulted by a reader interested in more detailed information.

HTTP was defined as an application protocol for distributing and getting access to hypermedia information. The original development of HTTP took place in the early days of the World Wide Web in the late 1980s and early 1990s and was done at CERN. The first documented version of HTTP was HTTP v0.9 from 1991 which was followed by HTTP 1.0 in 1996 and HTTP 1.1 in 1997. HTTP 1.1 was then improved, clarified and updated during several years. HTTP 1.1 was the first version that came into common use on the Internet and it is still in wide use today. A new revision, HTTP/2, was standardized on 2015 in IETF RFC 7540, and this is the version that 3GPP uses for the 5GC Control Plane. Most of the features described below are common to HTTP 1.1 and HTTP/2, but some features used by 3GPP were introduced with HTTP/2.

The main objective of HTTP/2 was to improve the performance in order to provide a better experience for web users. Key benefits of HTTP/2 compared to HTTP 1.1 is, e.-g., support for full multiplexing of requests and responses, support for compression of HTTP header fields to minimize protocol overhead, and support for request prioritization and server push. It can be noted that HTTP/2 does not modify the application semantics of HTTP. All the basic concepts that we will describe below, such as HTTP methods, status codes, URIs, and header fields, remain. What HTTP/2 does is changing (improving) the way information in the messages is formatted and transported between the client and server. That formatting is however not visible to the application layer.

14.4.2 Basic principles

HTTP is a request-response protocol that runs between a client and a server. The protocol is carried over TCP to ensure reliable transmission. HTTP 1.1 could also be carried over other transport protocols but HTTP/2 is specified to be carried only over TCP. 3GPP has discussed to use HTTP with QUIC transport (HTTP over QUIC is also referred to as HTTP/3) but it was left out since it was not mature enough for rel-15. Use of HTTP over QUIC is however being studied and will potentially be specified in a future 3GPP release.

The protocol stack for HTTP is shown in Fig. 14.5. Transport Layer Security (TLS) is optionally used to protect the HTTP traffic between client and server. HTTP over TLS is also referred to as HTTPS. TLS is further described in Section 14.5.

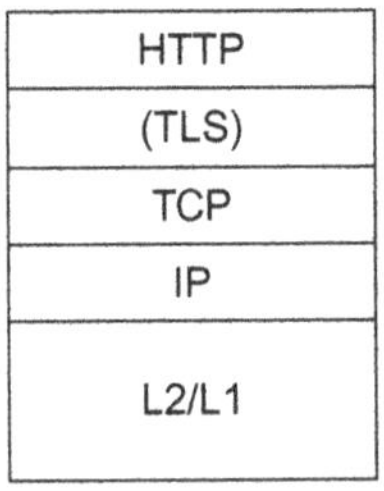

Fig. 14.5 HTTP protocol stack.

When a HTTP client wants to communicate with a server it first establishes a TCP connection towards a specific port on the server. The default port for HTTP is 80, while the default port for HTTPS is 443. With 5GC, a NF producing a service (i.e., acting as HTTP server) can register its FQDN and/or IP address and port for the service in NRF. The consumer NF (i.e., the entity acting as HTTP client) can then discover from NRF what server FQDN or IP address and optionally port number to connect to for a specific NF service. If a producer NF does not register any port number in NRF then the default port numbers for HTTP and HTTPS will be used.

Once the TCP connection is established the client can send a HTTP request to the server via that TCP connection. Multiple outstanding HTTP requests can be sent on a single TCP connection. This is one of the key benefits with HTTP/2 as it improves the possibility for multiplexing HTTP request/response pairs over a single TCP connection, allowing a fewer number of TCP connections to be used compared to HTTP 1.1. It can be mentioned that HTTP 1.1 includes a feature called pipelining that allows some level of multiplexing, but it has severe limitations and is not as capable as HTTP/2.

14.4.3 HTTP messages, methods, resources and URIs

As mentioned above, HTTP is a request-response protocol. The client sends requests to the server and the server sends responses. In a common use case on the Internet today, the client is a web browser and the server is a web server running on a computer in a data center. In a 5GC setting, the client and the server are NFs in the 5GC. Fig. 14.6 illustrates a simple HTTP exchange.

The target of an HTTP request is called a "resource". A resource represents something that is provided by the server. It can represent many things and it is not defined in further detail by the HTTP protocol. When it comes to web browsing, a resource can, e.-g., be a web page, a document or a photo. In 3GPP 5GC, a resource can, e.g., be a PDU Session (for services produced by SMF), a policy session (for services produced by PCF) or subscription data (for services produced by UDM). Each resource is identified by a Uniform Resource Identifier (URI). A very common form of the URI is the Uniform Resource Locator (URL), which is a special type of URI that both identifies a resource and also provides information on how to access that resources, e.g., using a protocol such as HTTP (e.g., https://www.3gpp.org).

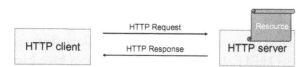

Fig. 14.6 Simple HTTP exchange.

In 5GC, the URI in the request message uniquely identifies the resource in the producer NF. An absolute URI for a Service Based interface in 5GC has the following structure:

{apiRoot}/{apiName}/{apiVersion}/{apiSpecificResourceUriPart}

We will briefly describe each part of the URI and give examples.

The "apiRoot" is a concatenation of several parts:

- scheme ("http" or "https")
- a fixed string "://"
- host and optional port (so called "authority")
- an optional deployment-specific string (API prefix) that starts with a "/" character.

The "apiName" defines the name of the API and the "apiVersion" indicates the version of the API. The "apiRoot", "apiName" and "apiVersion" together define the base URI of the API, and then each "apiSpecificResourceUriPart" defines a resource URI relative to this base URI. The structure and content of the "apiSpecificResourceUriPart" differs depending on the type service. For e.g. UDM the client NF sends requests for "provisioned" resources such as subscription data, with a URI that includes, e.g., SUPI. An example could look like "https://udm1.operatorX.com/nudm-sdm/v1/imsi-1234567890/sm-data", where the IMSI in this example is "1234567890". For, e.g., SMF and AMF the resources are instead dynamically created at run-time, e.g., when the UE registers or a PDU Session is established. The URI can thus not be as static as in the UDM example. Instead the producer NF (server) returns a resource reference to the consumer NF (client) in the first reply message. Later when the consumer wants to address that specific resource (e.g., PDU Session in SMF), a URI including that resource reference is used. An example could be "https://smf3.operatorY.com/nsmf-pdusession/v1/sm-contexts/347c3edf-129a-276e-e4c7-c48e7b515605", where the last part is the resource reference.

After receiving a HTTP request, the server, which has access to the resource, parses the request and may perform functions on behalf of the client, and then returns a response message to the client. The response includes status information about the transaction and typically contains the requested content (a representation of the requested resource).

In the HTTP request sent by the client, the client describes the resource it is targeting (via a URI) and also describes the specific action that the client is requesting to be performed on that resource. These actions are called "HTTP methods". A simple HTTP method, commonly used on the Internet for browsing the web, is GET. This method allows a client to request a specific document (resource) on a server and the server to send that document (resource) to the client in the reply message. The GET method does not modify the resource in any way. This method was one of the first methods to be defined in the early history of HTTP. Later several other methods have been added which are briefly described below.

- *GET*: This method requests a representation of the specified resource. Requests using GET should only retrieve data and should have no other effect.
- *HEAD*: This method is similar to GET in that it asks for a resource, but only the description of the resource and not the actual content. It can be useful if the client only wants to learn about the type of the resource without having the full content. (The HEAD method is so far not used in any of the 5GC APIs).
- *PUT*: This method requests that the content carried in the request from the client is stored under the provided URI. If the URI refers to a resource that already exist on the server, this resource is modified. If the URI however does not point to any existing resource, then the server can create the resource with that URI.
- *POST*: This method requests that the server takes the content carried in the request from the client and amends it under the resource identified by the URI. In an Internet setting, the content in the request may, e.g., be a comment to a news article, data for a web form or an item to add to an online database.
- *PATCH*: This method applies partial modifications to a resource.
- *DELETE*: This method deletes the specified resource.
- *TRACE*: This method echoes the received request so that a client can see what (if any) changes or additions have been made by intermediate servers or proxies. (The TRACE method is so far not used in any of the 5GC APIs).
- *OPTIONS*: This method returns information about communication capabilities. For example, supported HTTP methods, that the server supports for the specified URL. This can be used to learn about the functionality supported by a HTTP server.
- *CONNECT*: This method converts the request connection to a transparent TCP/IP tunnel, usually to facilitate encrypted communication (HTTPS) through an HTTP proxy.

14.4.4 RESTful design

3GPP has the goal to define all services in 5GC according to the REST "paradigm". "REST" (short for "Representational State Transfer") is a software architectural style introduced by Roy T. Fielding in his dissertation in 2000. It defines a set of software design rules for how to design the interoperability of computer systems across the internet. RESTful webservices use a uniform and predefined set of stateless operations to allow systems to access and manipulate textual representations of web-based resources.

Principles for RESTful design are summarized in the numbered list below, adapted from the 3GPP study made in 3GPP TS 29.891. In practice, existing APIs that claim to be RESTful follow these principles to varying degrees. The same applies in 3GPP. As will be commented on below, the different principles are more or less applicable to 5GC.

1. *Client/Server*: This refers to the split of responsibilities between client and server where a client sends a request to a server which processes the request and returns a response. Such a split allows for the separation of different tasks, e.g., the user interface generation from the data storage, which simplifies the single tasks and enhances portability and scalability.

In 5GC this is applied to a high degree since services are defined in such a way that provider NFs offer services to consumer NFs. However, services can include asynchronous notifications or requests from server to client that then may be modeled as a separate pair of client and server with reversed roles.

2. *Stateless*: Each request from client to server must contain all the information necessary to understand the request. Session state is therefore kept entirely on the client. The server does not keep history/memory of previous requests. Session state can be transferred by the server to another service such as a database to maintain a persistent state for a period. Different requests can be served by different servers, improving reliability and scalability.

Within 5GC this principle is only partially met. It is assumed that a server (NF) is selected to hold the state information associated with a resource. For example, an SMF will have state about PDU Sessions, AMF will have state about UE mobility context, PCF will have state about active policy sessions, etc. These are, however, application layer states and the underlying HTTP protocol can still be stateless.

3. *Cacheable*: If a response is cacheable, a client can cache a response and reuse that response data later, for equivalent requests. This makes it possible to eliminate some interactions as well as improving efficiency, scalability, and average latency.

In 5GC this principle is not really met, except for NRF services. Most interactions within 5GC relate to resources that may be changing frequently, e.g., due to UE mobility or end-user application usage. Caching related responses would thus offer little benefit. The exception is NRF, where NFs can cache NRF responses.

4. *Uniform interface*: A REST based interface is based on an identification of resources and allows for the manipulation of resources through representations of these resources. Individual resources are identified via requests, for example using URIs. The resources themselves are conceptually separate from the representations that are returned to the client. Messages are therefore self-descriptive (e.g., indicate format via MIME type).

By using a uniform interface, the overall system architecture is simplified, and the visibility of interactions improved. Implementations are also decoupled from the services they provide, which encourages independent evolvability. There is however a trade-off related to this principle, in that a uniform (general) interface typically impacts efficiency negatively compared to interfaces that have been optimized for each application's specific needs.

In 5GC this principle has been met to a fairly high degree as 3GPP agreed on common design rules for the NF services before starting the work on each specific NF service. The principles, common data types, etc. can be found in 3GPP TS 29.500, 3GPP TS 29.501 and 3GPP TS 29.571.

5. *Layered system*: The system is composed of hierarchical layers by constraining component behavior such that each component cannot "see" beyond the immediate layer with which they are interacting. Intermediary servers may improve system scalability by enabling load balancing and by providing shared caches. The client does not care about how the server provides the response.

   In 5GC this principle has been met to a high degree. A consumer NF is not aware of how a producer NF provides a response.

6. *Code on Demand (optional)*: REST allows client functionality to be extended by downloading and executing code in the form of applets or scripts. This simplifies clients by reducing the number of features required to be pre-implemented on the client.

   This principle is not met by 3GPP. The 5GC NF behaviors, both client side and server side, are specified in the 3GPP Technical Specifications and downloading executable code from an NF producer to a NF consumer has not been necessary.

14.4.5 HTTP protocol format

The protocol format of HTTP 1.1 and HTTP/2 messages differs significantly, but on a high level they follow similar semantics. They both have Information Elements carrying information about the HTTP method used, additional parameters relevant to Request or Response, as well as optional content related to the resource. The terminology and format differ however between HTTP 1.1 and HTTP/2. Below we will highlight some differences, but in general focus on HTTP/2 since this is the version used in 5GC.

HTTP 1.1 messages are encoded as plain-text human readable messages where even the division into separate Information Elements (the protocol framing) is done using plain text characters, e.g., white spaces, line feeds, carriage returns, etc. HTTP/2 on the other hand has a binary protocol framing, i.e., the division into Information Elements, etc. is using binary parameters. HTTP/2 also supports header compression to allow a more efficient transfer. Protocols with binary framing have certain benefits compared to pure textual protocols. They are e.g. more efficient to parse by a receiver and more compact when carried over the interfaces (require fewer IP packets). They are also less error-prone compared to textual protocols like HTTP 1.1, because they don't need to care about whitespace handling, capitalization, line endings, blank lines, etc. The Information Elements as such, e.g., HTTP headers, are however still encoded using text even with HTTP/2, unless header compression is applied.

A HTTP/2 message is a combination of one or two HEADER frames (carrying the HTTP headers), zero or more DATA frames (carrying actual content related to the resources), and one optional terminal HEADER frame (carrying the HTTP trailers). In HTTP 1.1. these parts are called header part and message body part.

Request and Response messages always include a HEADERS frame, carrying the HTTP/2 headers, but a DATA frame is included only if needed.

Request message:

- The HEADERS frame of a Request message includes HTTP header fields and so-called pseudo-header fields that carry information about the request. Pseudo-header fields contain information about the HTTP method as well as the targeted resource (in HTTP 1.1 this is included in a single "request line" instead). As an example, to request a web page "https://www.3gpp.org/specifications" with HTTP/2, the HEADERS frame contains:
  - a ":method" pseudo-header indicating GET,
  - a ":scheme" pseudo-header indicating "https",
  - a ":host" pseudo-header containing "www.3gpp.org" and
  - a ":path" pseudo-header containing "/specifications".

An example from 5GC to retrieve the SM subscription data from UDM for a SUPI with IMSI = 1234567890 could, e.g., be including the following pseudo-headers:
- ":method" indicating GET,
- ":scheme" indicating "https",
- ":host" containing "udm1.operatorX.com" and
- ":path" including "/nudm-sdm/v1/imsi-1234567890/sm-data".

- The HEADERS frame may also include HTTP header fields containing parameters that further describe the request. These header fields can indicate accepted languages for the response, the accepted encodings of the response, the content type and content length of the message body (if included), etc.
- The DATA frame of the Request can include the content to be provided to the server (e.g., for PUT and POST methods).

Response message:

- The Response message includes a HEADER frame with a ":status" pseudo-header field that contains the HTTP status code field. This pseudo-header field is included in all response messages. An example is "200 OK", which indicates that the client's request was successful, or "400 Bad Request" that indicates that the request could not be processed, e.g., due to a badly formatted request. Similar to the Request, the Response HEADER frame also includes other HTTP header fields that further describe the response. Here the most common header fields are the ones that describe the body, such as content type (MIME type), content encoding, content length, content language, etc.
- The DATA frame contains the content provided by the server, i.e., a representation of the requested resource such as a web page or the SM subscription data, as in the two examples above.

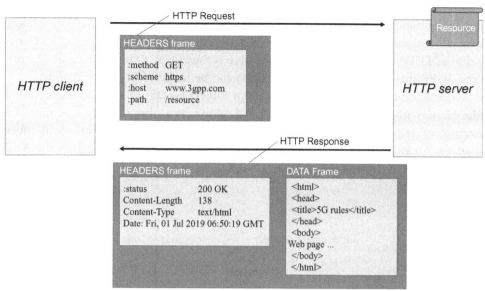

Fig. 14.7 Example of HTTP GET method.

An example of a HTTP/2 request-response exchange with a GET method is provided in Fig. 14.7.

14.4.6 Serialization protocol

The HTTP body may contain different types of information, including both plain-text parts as well as binary parts. HTTP is flexible enough to carry different types of content in the body. On the internet, it commonly carries text, pictures, audio, etc. The Content-Type HTTP header field is used to describe what content is included in the body. It is also possible to carry multiple different content types in a single HTTP body. One issue that arises in a 5GC context, is that there is a need to carry structured 3GPP-specific information elements between the NFs. This could, e.g., be a list of 3GPP parameters (e.g., SUPI, DNN, S-NSSAI, etc.) or more complex structures such as UE subscription data. To carry such information in a HTTP body, it needs to be encoded in a way that allows sender to generate the content in the HTTP body in a way that the receiver can parse it and extract the individual information elements. It is preferable that it should also be encoded in plain text format as it is easier for humans to read than binary.

It is also easier to transform than binary code. To do this there is a need to perform so called serialization of the data, i.e., to describe structured data such as Session management information, subscription data, policy rules, etc. in a text format that can be carried within a HTTP body. The "serialization" here is thus a way to transform a set of structured 3GPP data types into a text object that can be included in a HTTP body.

3GPP has agreed to use JSON (JavaScript Object Notation) as serialization format. JSON is described in IETF RFC 8259 and allows data objects to be transmitted as attribute-value pairs in text format using specified data types such as strings, numbers or arrays. An example of a HTTP Request from AMF to UDM to register the AMF as serving the UE is provided below.

```
HTTP HEADERS frame {
  :method          PUT
  :host            example-udm.com
  :path            /nudm—uecm/v1/imsi-<IMSI>/registrations/amf-3gpp-access
  :scheme          https
  Content-Length: xyz
  Content-Type:    application/json
}
HTTP DATA Frame
{
  "amfInstanceId": "777c3edf-129f-486e-a3f8-c48e7b515605",
  "deregCallbackUri": "/example-amf.com/<path>"
  "guami": {
    "plmnId": {
      "mcc": "46",
      "mnc": "000"
    },
    "amfId": "<AMF Id>"
  "ratType": "NR"
}
```

There may also be a need to carry binary elements in HTTP, such as SM NAS messages between AMF and SMF. Where multipart messages are used the binary content is included in the HTTP body using a 3GPP vendor specific content type. The JSON part in the HTTP body then contains the plain-text human readable information elements and a reference to the binary data part.

As an example, when a new PDU Session is established the AMF will forward the SM NAS message to the SMF, as well as additional information such as SUPI, requested DNN, requested S-NSSAI, UE location information, etc. This information is carried in the HTTP body part of a Request message with POST method where one part is provided via JSON encoding as attribute-value pairs and a separate part of the HTTP body includes the binary data. The HTTP Request from AMF to SMF could then look as in the example below:

```
HTTP HEADERS frame {
  :method        POST
  :host          example-smf.com
  :path          /nsmf-pdusession/v1/sm-contexts
  :scheme        https
  Content-Type:  multipart/related; boundary=--Boundary
  Content-Length: xyz
}
HTTP DATA Frame {
  ------Boundary
  Content-Type: application/json
  {
   "supi": "imsi-<IMSI>",
   "pduSessionId": 235,
   "dnn": "<DNN>",
   "sNssai": {
   "sst": 0
   },
   "servingNfId": "<AMF Id>",
   "n1SmMsg": {
   "contentId": "n1message"
   },
   "anType": "3GPP_ACCESS",
   "smContextStatusUri": "<URI>"
  }
  ------Boundary
  Content-Type: application/vnd.3gpp.5gnas
  Content-Id: n1message
  { N1 SM Message binary data }
  ------Boundary
}
```

14.4.7 Interface definition language

When HTTP APIs are defined it is common that the APIs are described using an Interface Definition Language (IDL). An IDL is specification language that allows an API to be described using formal rules in order to have a clear specification of resources, operations, information elements, data structures, data formats, etc. Using an IDL to formally describe an API helps avoid any ambiguity that may arise if the HTTP requests and responses would have been described using plain English with less formal rules. The benefit with an IDL is therefore that interoperability problems in the field that are caused by different vendors interpreting the standard text differently can be avoided.

The IDL is independent of the vendor and computer programming language that is used to implement the API in software in the actual products. This makes it suitable for defining an API and interactions between software written by different vendors and in different programming languages. Since it is a formal language with specific structures it can also be a tool in software development and actually used to generate parts of the code.

3GPP decided to use OpenAPI version 3 as the IDL for specification of HTTP-based services. OpenAPI Specification (OAS) is used to formally describe RESTful APIs, and the language itself is specified by the OpenAPI Initiative (OAI), an organization under the Linux Foundation. A description of an API using OpenAPI can, e.g., include:

- General information about the API
- The resources used, i.e., the paths in the URI
- Methods available on each resource (e.g., GET)
- The input and output parameters supported for each method and resource, their data types (e.g., integer, string, etc.), etc.

One can say that OpenAPI is a specification for how to specify an API. The OpenAPI specification for an API is written in a human readable text file. It can use either JSON or YAML to express the API specification. 3GPP selected YAML for its specifications of the service-based interfaces, mostly because YAML is easier for a human to read and write than JSON. OpenAPI specifications in YAML and JSON are however mostly equivalent.

The YAML description of a NF service is included as an Annex in each corresponding TS (e.g., 3GPP TS 29.518 for AMF services) as well as a separate YAML file distributed together with the TS. The YAML specification of a NF service tend to be rather lengthy, since they include a description of all NF service operations for a NF service, as well as all supported input and output parameters, parameter formats and possible parameter values, etc. Below we show a YAML file using probably the simplest example from 3GPP; the service produced by the Equipment Identity Register (EIR) taken from 3GPP TS 29.511. This NF produces only a single service with a single service operation.

```
openapi: 3.0.0
info:
  version: '1.0.1'
  title: '5G-EIR Equipment Identity Check'
  description: '5G-EIR Equipment Identity Check Service'
servers:
  - url: '{apiRoot}/n5g-eir-eic/v1'
    variables:
      apiRoot:
        default: https://example.com
        description: apiRoot as defined in 3GPP TS 29.501
security:
  - {}
  - oAuth2ClientCredentials:
      - n5g-eir-eic
paths:
  /equipment-status:
    get:
      summary: Retrieves the status of the UE
      operationId: GetEquipmentStatus
      tags:
        - Equipment Status (Document)
      parameters:
        - name: pei
          in: query
          description: PEI of the UE
          required: true
          schema:
            $ref: 'TS29571_CommonData.yaml#/components/schemas/Pei'
        - name: supi
          in: query
          description: SUPI of the UE
          required: false
          schema:
            $ref: 'TS29571_CommonData.yaml#/components/schemas/Supi'
        - name: gpsi
          in: query
          description: GPSI of the UE
          required: false
          schema:
            $ref: 'TS29571_CommonData.yaml#/components/schemas/Gpsi'
      responses:
        '200':
          description: Expected response to a valid request
          content:
            application/json:
              schema:
                $ref: '#/components/schemas/EirResponseData'
        '400':
          $ref: 'TS29571_CommonData.yaml#/components/responses/400'
        '401':
          $ref: 'TS29571_CommonData.yaml#/components/responses/401'
        '404':
          description: PEI Not Found
          content:
            application/problem+json:
              schema:
                $ref: 'TS29571_CommonData.yaml#/components/schemas/
                  ProblemDetails'
        '414':
          $ref: 'TS29571_CommonData.yaml#/components/responses/414'
        '429':
          $ref: 'TS29571_CommonData.yaml#/components/responses/429'
        '500':
          $ref: 'TS29571_CommonData.yaml#/components/responses/500'
        '503':
          $ref: 'TS29571_CommonData.yaml#/components/responses/503'
        default:
          description: Unexpected error
```

Introduction describing the base URI for the service

Key for authorization

Only a single service operation for this service, to get the equipment status

Input parameters (PEI, SUPI and GPSI) for the service operation

Reply codes available for the service operations

```
components:
  securitySchemes:
    oAuth2ClientCredentials:
      type: oauth2
      flows:
        clientCredentials:
          tokenUrl: '{nrfApiRoot}/oauth2/token'
          scopes:
            n5g-eir-eic: Access to the N5g-eir_EquipmentIdentityCheck API
  schemas:
    EirResponseData:
      type: object
      required:
        - status
      properties:
        status:
          $ref: '#/components/schemas/EquipmentStatus'
    EquipmentStatus:
      type: string
      enum:
        - WHITELISTED
        - BLACKLISTED
        - GREYLISTED

externalDocs:
  description: 3GPP TS 29.511 V15.3.0; 5G System;
Equipment Identity Register Services; Stage 3
  url: 'http://www.3gpp.org/ftp/Specs/archive/29_series/29.511/'
```

Parameters related to the Oauth2-based access authorization to the API

Data type definitions used from other parts of the API specification

Reference to stage 3 TS describing the API

14.5 Transport layer security (TLS)

14.5.1 Introduction

Transport Layer Security (TLS) is a cryptographic protocol that aims at providing secure communication over an IP network. It is a very common protocol today since it is used to secure HTTP communication between web browsers and web servers. When HTTP is protected using TLS, it is usually referred to as HTTPS (HTTP Secure).

The TLS protocol can provide ciphering and data integrity between two communicating entities and also mutual authentication of the two end-points. It is application protocol independent so that higher-level protocols can run on top of TLS transparently. TLS is specified by the IETF and several revisions of TLS have been defined. The first version TLS 1.0 was defined in 1999 and it built on previous work on Secure Socket Layer (SSL). TLS 1.1 was then released in 2006. A common TLS versions used on the Internet today is TLS 1.2 (defined in IETF RFC 5246 from 2008) but support for TLS 1.3 (defined in IETF RFC 8446 from 2018) is becoming more common. As described in Chapter 8, TLS is used in the 5GC to protect the HTTP-based interfaces. 3GPP allows TLS 1.1, TLS 1.2 and TLS 1.3 to be used, even though the use of TLS 1.1 is not recommended (see 3GPP TS 33.210 for further details).

TLS consists of two primary components:

- A TLS Handshake protocol that authenticates the two end-points. This component also negotiates cryptographic parameters and generates keying material. The handshake protocol is designed to be resistant to attacks, i.e., an attacker should not be able to influence the security negotiation between the two end-points. The TLS Handshake protocol is further described in Section 14.5.2.

- A TLS Record protocol that uses the parameters established by the handshake protocol to protect traffic between the end-points. The record protocol divides the data traffic into a series of records. Each of these records is then protected using the keys established during the Handshake phase. The TLS Record protocol is further described in Section 14.5.3.

14.5.2 TLS handshake protocol

Before a client and server can start communication protected by TLS, they must mutually authenticate and derive key material. They must also agree on the version of the TLS protocol to use as well as which cipher to use for encrypting data. As mentioned above, the TLS protocol defines the TLS Handshake protocol for this purpose, during which mutual authentication, key derivation, TLS version negotiation as well as negotiation of cipher suites takes place. In many cases, public key cryptography is used for the authentication. This allows the two end-points to authenticate each other via certificates. The use of certificates and the established chain of trust built into the certificate infrastructure with certificate authorities avoids the need for the two communicating parties to have a prior explicit knowledge of each other. This is one of the key properties that makes TLS/ HTTPS work in practice for web traffic on the Internet.

The TLS Handshake runs on top of TCP. Once the TLS Handshake has completed, the two sides have enough information to start communication that is protected by TLS. The simple call flow in Fig. 14.8 illustrates how TLS 1.2 Handshake can be performed.

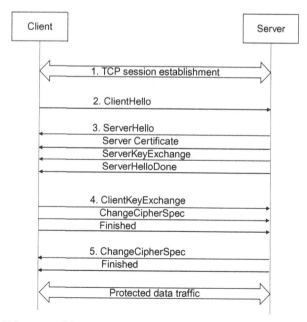

Fig. 14.8 Example TLS 1.2 Handshake procedure.

1. The TCP three-way handshake is initiated in order to setup a TCP connection.
2. The consumer (client) sends a TLS ClientHello message indicating the version of the TLS protocol it is running, the list of supported ciphersuites, and other TLS options it may want to use.
3. The producer (server) selects the TLS protocol version to use for the rest of the communication, selects a ciphersuite from the list provided by the consumer, includes additional TLS extensions if needed and sends the response (ServerHello) to the consumer. Then the server sends its certificate in Server Certificate message. The server may also include a request for the consumer's certificate. Depending on the negotiated key exchange algorithm, the server may send ServerKeyExchange message which includes, e.g., ephemeral Diffie-Hellman parameters. The server then sends ServerHelloDone.
4. If the client and server could agree on the TLS version and ciphersuite, the key exchange continues by the client sending, e.g., ephemeral Diffie-Hellman parameters in a ClientKeyExchange message. The aim of the key exchange is to derive cryptographic keys on both client and server sides for the ciphering and integrity protection of data. Different TLS versions support different types of key exchange methods, but most methods rely on RSA and/or the Diffie-Hellman key exchanges. The client then sends ChangeCipherSpec and Finished messages to take the negotiated security parameters and key in use and to integrity protect the previous handshake messages.
5. The server verifies the Finished message from the client (a Message Authentication Code, MAC, is included) and processes the information received in the message. If everything works well, a ChangeCipherSpec and encrypted and integrity protected Finished message are sent to the client.
6. The client decrypts the Finished message and verifies the MAC. This completes the TLS Handshake. Communication protected by TLS can now start.

As mentioned above, TLS was designed to operate on top of a reliable transport protocol such as TCP. However, it has also been adapted to run over datagram protocols such as UDP. The Datagram Transport Layer Security (DTLS) version 1.2 protocol, defined in IETF RFC 6347, is based TLS and can provide similar security properties but using a datagram transport. As mentioned in Chapter 8, DTLS is supported by 3GPP in addition to IPSec for protecting the signaling on N2, i.e., NGAP running over SCTP.

14.5.3 TLS record protocol

Once the TLS Handshake is complete, the actual application data can be sent protected by the TLS layer. The TLS Record protocol takes care of this process, supporting a number of functionalities. For sending data, the TLS Record protocol:

- Splits outgoing messages into manageable blocks (records),
- Optionally compresses outgoing blocks,
- Applies a Message Authentication Code (MAC) to outgoing messages, and
- Encrypts the messages.

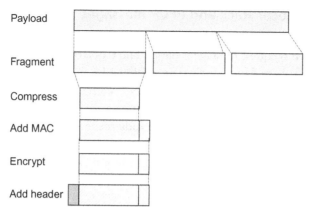

Fig. 14.9 TLS Record protocol for protecting data.

The procedure is illustrated in Fig. 14.9. The data including the TLS header is then passed to the TCP layer for transmission to the destination. At the receiving side, the same functionalities but in reverse order are performed. The TLS Record protocol will:

- Decrypt incoming messages.
- Verify incoming messages using the MAC, and
- Decompress incoming blocks,
- Reassemble incoming messages,

14.6 Packet forwarding control protocol (PFCP)

14.6.1 Introduction

The Packet Forwarding Control Protocol is used over the N4 reference point between SMF and UPF for controlling the UPF. PFCF was defined for EPC when CP-UP split of the SGW and PGW was introduced in 3GPP Release 14. Here the reference point between SGW-C/PGW-C and SGW-U/PGW-U is referred to as Sx. One of the design goals of that work was to define a protocol that was future-proof and re-usable and when the 5G work was done in 3GPP, it was decided to re-use the same protocol also for the N4. PFCP has however been evolved to fulfil the 5GS requirements, e.g., to support Ethernet PDU Session type which was at that time not available in EPS, and the 5GS QoS model.

Below we will describe PFCP as protocol over the N4 reference point between SMF and UPF. However, most aspects also apply to the Sx reference point between PGW-C/ SGW-C and PGW-U/SGW-U. For more general descriptions about CP-UP split in EPC and 5GS, refer to Chapter 4 and Chapter 6, Section 6.3 respectively.

14.6.2 PFCP protocol stack and PFCP messages

PFCP is specified in 3GPP TS 29.244. It runs over UDP and the UDP destination port number for a PFCP Request message is 8805. The Control Plane protocol stack is illustrated in Fig. 14.10. As will be described in Section 14.6.5, it is also possible to carry user plane packets between SMF and UPF via the N4 reference point.

There are two types of PFCP procedures (and related messages); Node related procedures and Session related procedures. The Node related procedures are used to establish a node-level association between CP function (SMF) and UP function (UPF) and to communicate node level information between the CP and UP functions. The Session related procedures are used to manage PFCP Sessions corresponding to individual PDU Sessions.

A node-level PFCP Association is set up between a CP function and a UP function prior to establishing any PFCP Sessions on that UP function. Either the CP function or the UP function can take the initiative to initiate a PFCP Association (it is mandatory in products to support CP-initiated and optional to support UP-initiated PFCP Association setup). In addition to PFCP Association setup, additional node related procedures are supported, e.g., for heartbeats, node level configuration and load/overload handling. The following PFCP Node related procedures are supported:

- PFCP Association Setup procedure
- PFCP Association Update procedure
- PFCP Association Release procedure
- PFCP Node Report procedure
- Heartbeat procedure
- Load Control procedure
- Overload Control procedure
- PFCP Packet Flow Description (PFD) management procedure

PFCP Sessions are used to manage the User-Plane for PDU Sessions. (It can actually also be used to manage the User Plane for traffic not related to PDU Sessions, e.g., in case

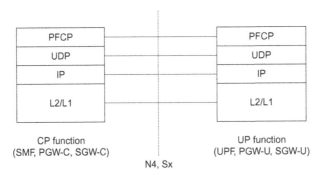

Fig. 14.10 PFCP Control Plane protocol stack.

SMF needs to communicate with DHCP or AAA servers in the DN via a UPF). When a new PDU Session is created, the SMF selects at least one UPF to serve the PDU Session. The SMF then initiates a PFCP Session establishment towards this UPF and provides the necessary information for instructing the UPF how to handle User Plane traffic for the PDU Session. If the SMF selects additional UPFs for a PDU Session, e.g., acting as I-UPF or ULCL, the SMF initiates a PFCP Session also towards those UFPs. There is thus one PFCP Session per PDU Session for each UPF on the path for the PDU Session.

When a PFCP Session is created, each end-point (SMF and UPF) generates a Session Endpoint Identifier (SEID). The SEID uniquely identifies a PFCP session at an IP address of a PFCP entity. In each session-related message, the destination SEID is included to allow the receiver to identify the PFCP Session for which the message applies.

The following PFCP Session related procedures are supported:

- PFCP Session Establishment procedure
- PFCP Session Modification procedure
- PFCP Session Deletion procedure
- PFCP Session Report procedure

Node-related and Session-related PFCP messages contain a PFCP message header and may contain additional Information Element(s) dependent on the type of message. The format of a PFCP message is depicted in Fig. 14.11. The format differs somewhat depending on whether the PFCP message is node-related (applicable to the SMF or UPF as such) or session-related (applicable to a specific PFCP session). Session related messages include the SEID, while node-related messages do not. The Version element

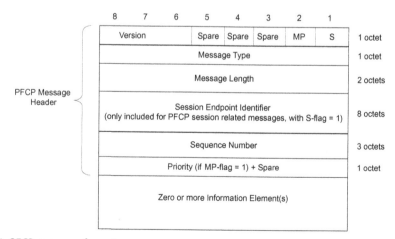

Fig. 14.11 PFCP message format.

is set to 1 in the current PFCP version. The MP-flag indicates whether a Message Priority is included on the header or not, and the S-flag indicates whether a SEID is included or not.

14.6.3 Packet forwarding model and PFCP rules
14.6.3.1 General
The SMF controls the packet processing in the UP function by establishing, modifying or deleting PFCP Session contexts and by provisioning (i.e., adding, modifying or deleting) specific rules that describes the User Plane actions that shall be performed by the UPF. In a sense it is like the SMF is programming the UPF via the N4 rules for how packet processing shall be performed. The N4 rules are included as Information Elements in PFCP Session-related messages.

The N4 rules supported on N4 (and Sx) are listed and briefly described below. We will later describe them in more detail.

- *Packet Detection Rule (PDR)*: This rule instructs the UPF how to detect incoming user data traffic (PDUs) and how to classify the traffic. The PDR contains Packet Detection Information (e.g., IP filters) used in the traffic detection and classification. There are separate PDRs for uplink and downlink.
- *QoS Enforcement Rule (QER)*: This rule contains information on how to enforce QoS, e.g., bit rate parameters.
- *Usage Reporting Rule (URR)*: This rule contains information on how the UPF shall measure (e.g., count) packets and bytes and report the usage to the SMF. The URR also contains information on events that shall be reported to SMF.
- *Forwarding Action Rule (FAR)*: This rule contains information for how a packet (PDU) shall be forwarded by the UPF, e.g., towards the Data Network in uplink or towards RAN in downlink.
- *Buffering Action Rule (BAR)*: This rule contains information for how buffering of packets shall be done, e.g., in case a UE is in CM-IDLE state.

The basic packet processing model in the UPF is that the PDRs are applied to incoming packets to find a match against the Packet Detection Information in a PDR. The Packet Detection Information can, e.g., be an IP 5-tuple describing a specific service. There are thus in general multiple PDRs associated to a PFCP Session. Each PDR contains pointers to associated QER(s), URR(s) and FAR. A match against a PDR thus leads to these other rules that describe how a packet shall be enforced, measured/counted and forwarded. Multiple PDRs can point to the same QER, URR and FAR, to support cases where multiple IP flows are enforced or forwarded as an aggregate, e.g., multiple IP flows should be forwarded on the same QoS flow, or multiple IP flows should be enforced using a common maximum bit-rate. The packet processing model is shown in Fig. 14.12 and the high-level structure of the PFCP rules are shown in Fig. 14.13. A more detailed description of each PFCP rule is provided below.

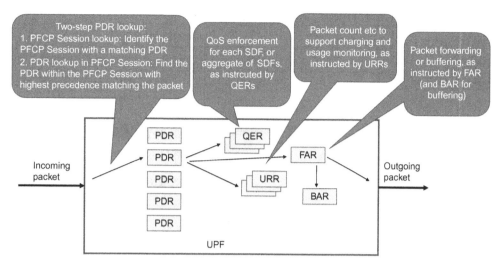

Fig. 14.12 UPF packet processing model.

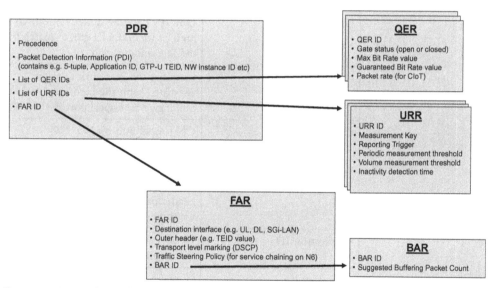

Fig. 14.13 PFCP rules and relations.

14.6.3.2 Packet Detection Rule (PDR)

The PDR contains information for how to classify packets and also includes links to other rules (QERs, URRs and FAR) describing how to treat the packet that match the Packet Detection Information in the PDR.

Table 14.5 shows the main elements of the PDR, with brief description of each parameter.

Table 14.5 Main elements of Packet Detection Rule.

Attribute		Description
PDR ID		Unique identifier to identify this rule
Precedence		Determines the order, in which the detection information of all rules is applied
Packet detection information (information against which incoming packets are matched)	Source interface	Indicates the logical source interface where packets are to be matched. Contains, e.g., the values "access side", "core side", "SMF", "N6-LAN" Identifies the interface for incoming packets from the access side (i.e., uplink), from the core side (i.e., downlink), from the SMF or from N6-LAN (i.e., the DN or the local DN).
	UE IP address	One IPv4 address and/or one IPv6 prefix with prefix length, to match for the incoming packet
	Network instance	Identifies the Network instance associated with the incoming packet, can, e.g., be a logical port in the UPF
	CN tunnel info	CN tunnel info on N3, N9 interfaces, i.e., GTP-U F-TEID
	Packet filter set	Filter information describing a packet flow, e.g., IP 5-tuple or Ethernet header parameters, to match for the incoming packet.
	Application ID	Identity of application to match for the incoming packet.
	QoS flow ID	Contains the value of 5QI or non-standardized QFI to match for the incoming packet.
	Ethernet PDU session information	Indication that UPF shall use MAC address learning to determine what MAC addresses are available on a PDU Session
	Framed route information	Used in case the PDU session has a framed route

Table 14.5 Main elements of Packet Detection Rule—cont'd

Attribute	Description
Outer header removal	Instructs the UP function to remove one or more outer header(s) (e.g., GTP-U tunnel header), from the incoming packet.
FAR ID	The Forwarding Action Rule ID identifies a forwarding action that has to be applied.
List of URR ID(s)	Every Usage Reporting Rule ID identifies a measurement action that has to be applied. Several URR IDs are included if a matched packet shall be counted against multiple counters.
List of QER ID(s)	Every QoS Enforcement Rule ID identifies a QoS enforcement action that has to be applied. Several QER IDs are included if a matched packet shall be enforced against multiple rules, e.g., for both per IP flow maximum bit-rate and Session AMBR.
Activate predefined rules	Indicates if Rule(s) predefined in UPF shall be activated for this PDR. Multiple Predefined Rules names may be provided.

14.6.3.3 Forwarding Action Rule (FAR)

The FAR contains information for how a PDU shall be forwarded by the UPF, e.g., sent towards RAN or to N6. Multiple PDRs may refer to the same FAR, e.g., in case multiple IP flows described by different PDRs will be forwarded over the same GTP-U tunnel towards RAN.

Table 14.6 shows the main elements of the FAR, with brief description of each parameter.

14.6.3.4 QoS Enforcement Rule (QER)

The QER contains information for how a PDU shall be treated in terms of bit rate limitation and packet marking for QoS purposes. Multiple PDRs may refer to the same

Table 14.6 Main elements of Forwarding Action Rule.

Attribute		Description
FAR ID		Unique identifier to identify this rule.
Action		Indicates whether the packet is to be forwarded, duplicated, dropped or buffered. Depending on the action, other attributes in the FAR need to be included to describe the behaviour of the UPF. For forwarding and duplication actions, Forwarding Parameters and possibly Duplicating parameters need to be included. For buffering action, a Buffer Action Rule is also included.
Forwarding parameters	Network instance	Identifies the Network instance associated with the outgoing packet.
	Destination interface	Contains the values "access side", "core side", "SMF", "N6-LAN. Identifies the interface for outgoing packets towards the access side (i.e., downlink), the core side (i.e., uplink), the SMF, or to the N6-LAN (i.e., the DN or the local DN).
	Outer header creation	Instructs the UP function to add an outer header (e.g., IP+UDP+GTP+QFI, VLAN tag), IP+possibly UDP to the outgoing packet. Contains the CN tunnel info, N6 tunnel info or AN tunnel info of peer entity (e.g., NG-RAN, another UPF, SMF, local access to a DN represented by a DNAI). Any extension header stored for this packet shall be added
	Send end marker packet(s)	Instructs the UPF to construct GTP-U end marker packet(s) and send them out.
	Transport level marking	Transport level packet marking in the uplink and downlink, e.g., setting the DiffServ Code Point.
	Forwarding policy	Reference to a preconfigured traffic steering policy or http redirection. Contains one of the following policies identified by a TSP ID: - an N6-LAN steering policy to steer the subscriber's traffic to the appropriate N6 service functions deployed by the operator, or - a local N6 steering policy to enable traffic steering in the local access to the DN. or a Redirect Destination.

Table 14.6 Main elements of Forwarding Action Rule—cont'd

Attribute		Description
	Request for Proxying in UPF	Indicates that the UPF shall perform ARP proxying and / or IPv6 Neighbour Solicitation Proxying. Applies to the Ethernet PDU Session type.
	Container for header enrichment	Contains information to be used by the UPF for header enrichment. Only relevant for the uplink direction.
Duplicating parameters	Destination interface	When the Apply Action requests the packets are to be duplicated (e.g., for Lawful Intercept purposes), the Duplicating Parameters are included. They contain the forwarding instructions to be applied by the UP function for the traffic to be duplicated.
	Outer header creation	
	Transport level marking	
	Forwarding policy	
Buffering action rule ID		When the Apply Action requests the packets are to be buffered, this information element contains a reference to a Buffering Action Rule defining the buffering instructions to be applied by the UPF.

QER, e.g., to have an aggregate rate limit of multiple IP flows. This means that all PDRs that refer to the same QER will share the same QoS resources, e.g., MFBR.

Table 14.7 shows the main elements of the QER, with brief description of each parameter.

14.6.3.5 Usage Reporting Rule (URR)

The URR contains instructions to request the UPF to:

- Measure the network resources usage in terms of traffic data volume, duration (i.e., time) and/or events, according to the provisioned Measurement Method; and
- Send a usage report to the CP function when a Reporting Trigger is armed, e.g., when the measurement reaches a certain threshold, periodically or when detecting a certain event.

The SMF provides Reporting Trigger(s) in the URR. In addition, the SMF may also request an immediate report, e.g., at RAT type change to generate CDR container.

The UPF will provide a Session Report message to UPF including the measurements, event reports, etc. as requested by SMF.

Table 14.7 Main elements of QoS Enforcement Rule.

Attribute	Description
QER ID	Unique identifier to identify this rule.
Gate status UL/DL	Instructs the UP function to let the flow pass or to block the flow. Values are: open, close, close after measurement report.
Maximum bitrate	The uplink/downlink maximum bitrate to be enforced for the packets. This field may, e.g., contain any one of: • Session-AMBR (for a QER that is referenced by all relevant Packet Detection Rules of the PDU Session) • QoS Flow MBR (for a QER that is referenced by all Packet Detection Rules of a QoS Flow) • SDF MBR (for a QER that is referenced by the uplink/downlink Packet Detection Rule of a SDF) • Bearer MBR (for a QER that is referenced by all relevant Packet Detection Rules of a bearer) (only applicable for EPC interworking).
Guaranteed bitrate	The uplink/downlink guaranteed bitrate authorized for the packets. This field may, e.g., contain: • QoS Flow GBR (for a QER that is referenced by all Packet Detection Rules of a QoS Flow) • Bearer GBR (for a QER that is referenced by all relevant Packet Detection Rules of a bearer) (only applicable for EPC interworking).
QoS flow identifier	Includes the QFI that shall be inserted by the UPF
Reflective QoS	Included if the UPF is required to insert a Reflective QoS Identifier to request reflective QoS for uplink traffic.
Paging Policy Indicator (PPI)	Included if the UPF is required to set the Paging Policy Indicator (PPI) in outgoing packets
Averaging window	The time duration over which the Maximum and Guaranteed bitrate shall be calculated. This is for counting the packets received during the time duration.
Down-link flow level marking	Flow level packet marking in the downlink. For UPF, this is for controlling the setting of the RQI in the encapsulation header as described in clause 5.7.5.3.
Packet rate	Number of packets per time interval to be enforced (only applicable to EPC interworking) This field contains any one of: • downlink packet rate for Serving PLMN Rate Control • uplink/downlink packet rate for APN Rate Control

Table 14.8 shows the main elements of the URR, with brief description of each parameter. The URR however contains more information specifically included to support requirements coming from online charging and how to handle the interface towards CHF. It would go beyond the scope of this book to go into details of these charging aspects, but an interested reader may have a look at the complete URR content in clause 7.5.2.4 in 3GPP TS 29.244.

Table 14.8 Main elements of Usage Reporting Rule.

Attribute	Description
URR ID	Unique identifier to identify this information.
Measurement method	Indicates the method for measuring the network resources usage, i.e., whether the data volume, duration (i.e., time), combined volume/duration, or event shall be measured.
Reporting triggers	Indicates the trigger(s) for reporting network resources usage to the CP function, e.g., periodic reporting or reporting upon reaching a threshold, or envelope closure. One or multiple of the events can be activated for the generation and reporting of the usage report. Applicable events include, e.g.: • Start/stop of traffic detection • Deletion of last PDR for a URR • Periodic measurement threshold reached • Volume/Time/Event measurement threshold reached; • Immediate report requested; • Threshold of discarded traffic reached • MAC address reporting in the UL traffic
Measurement period	Included if periodic reporting is required. When present, it shall indicate the period for generating and reporting usage reports.
Volume threshold	Included if volume-based measurement is used and reporting is required upon reaching a volume threshold. When present, it shall indicate the traffic volume value after which the UPF shall report network resources usage to the SMF for this URR.
Volume quota	Included if volume-based measurement is used and the SMF needs to provision a Volume Quota in the UPF. When present, it shall indicate the Volume Quota value.
Event threshold	Included if event-based measurement is used and reporting is required upon reaching an event threshold. When present, it shall indicate the number of events after which the UPF shall report to the SMF for this URR.

Continued

Table 14.8 Main elements of Usage Reporting Rule—cont'd

Attribute	Description
Event quota	Included if event-based measurement is used and the SMF needs to provision an Event Quota in the UP function. When present, it shall indicate the Event Quota value.
Time threshold	Included if time-based measurement is used and reporting is required upon reaching a time threshold. When present, it shall indicate the time usage after which the UPF shall report network resources usage to the SMF for this URR.
Time quota	Included if time-based measurement is used and the SMF needs to provision a Time Quota in the UP function. When present, it shall indicate the Time Quota value.
Dropped DL traffic threshold	Included if reporting is required when the DL traffic being dropped exceeds a threshold. When present, it shall contain the threshold of the DL traffic being dropped.
Inactivity detection time	Defines the period of time after which the time measurement shall stop, if no packets are received. Timer corresponding to this duration is restarted at the end of each transmitted packet.
Event based reporting	Points to a locally configured policy which identifies event(s) trigger for generating usage report.
Linked URR ID(s)	Points to one or more other URR ID. This enables the generation of a combined Usage Report for this and other URRs by triggering their reporting.
Ethernet inactivity timer	Included if Ethernet traffic reporting is used and the SMF requests the UP function to also report inactive UE MAC addresses. When present, it shall contain the duration of the Ethernet inactivity period.

14.6.3.6 Buffering Action Rule (BAR)

The BAR contains information for how a PDU shall be buffered by UPF, e.g., in case the UE is in CM-IDLE mode and there is no User Plane path towards RAN for the PDU Session. After receiving a downlink packet with the BAR activated, the UPF will notify the SMF which in turn will notify AMF so that AMF can page the UE.

Table 14.9 shows the main elements of the BAR, with brief description of each parameter.

Table 14.9 Main elements of Buffering Action Rule.

Attribute	Description
BAR ID	Unique identifier to identify this rule.
Suggested buffering packets count	When present, it shall contain the number of packets that are suggested to be buffered in UPF when the Apply Action parameter requests to buffer the packets. The packets that exceed the limit shall be discarded.

14.6.4 Reporting from UPF to SMF

The UPF will report to SMF based on the triggers provided by SMF in the URR. The UPF will send a Session Report to SMF containing information about the reported event.

Table 14.10 shows the main elements of the Session Report, with brief description of each parameter. The Session Report may contain different types of reports; Downlink

Table 14.10 Main elements of Session Report.

Attribute	Description
Report type	Indicates the type of the report (Downlink Data Report, Usage Report, Error Indication Report, User Plane Inactivity Report)
Downlink Data Report	Included if the Report Type indicates a Downlink Data Report, i.e., report that UPF has received and buffered a downlink packet. See Table 14.11 for more details.
Usage Report	Included if Report Type indicates a Usage Report. Several Usage Reports may be included in the same message. See Table 14.12 for more details.
Error Indication Report	Included if the Report Type indicates an Error Indication Report, e.g., that an error indication has been received form a remote GTP-U endpoint. See Table 14.13 for more details.
Load control information	Can be included if the UPF supports the load control feature and the feature is activated in the network. Includes a load metric providing SMF with information about UPF load level.
Overload control information	If overload control feature is supported and activated, the UPF may include this information during an overload condition.
Additional usage reports information	Indicates to SMF that additional usage reports will follow. When present, this IE indicates the total number of usage reports that need to be sent in PFCP Session Report Request messages.

Table 14.11 Main content of Downlink Data Report.

Attribute	Description
PDR ID	Identifies the PDR for which downlink data packets have been received at the UPF.
Downlink data service information	Includes information about the buffered packet (the DSCP value).

Data Report, Usage Report, Error Indication Report and User Plane Inactivity Report. The Session Report is also used to report load and overload information. The content of Downlink Data Report, Usage Report and Error Report are further described in Tables 14.11–14.13, respectively. The User Plane Inactivity Report does not contain any specific information except the basic Session Report.

Table 14.12 Main elements of Usage Report.

Attribute	Description
URR ID	Identifies the URR for which usage is reported.
UR-SEQN	A unique identifier of this particular Usage Report for the URR.
Usage report trigger	Identifies the trigger for this report, e.g., Periodic Reporting, Volume Threshold, Immediate Report, Start or Stop of Application Traffic, etc.
Start time	When present, this IE includes the timestamp when the collection of the information in this report was started.
End time	When present, this IE includes the timestamp when the collection of the information in this report was generated.
Volume measurement	Included if a volume measurement needs to be reported. The parameters in this IE contain the total uplink and downlink number of octets.
Duration measurement	Included if a duration measurement needs to be reported. The IE contains the used time in seconds.
Application detection information	Included if application detection information needs to be reported. The parameters in this IE contain the Application ID, Application Instance ID and (if deducible) flow information of the detected application such as IP 5-tuple.
Time of first packet	Indicates the time stamp for the first IP packet transmitted for a given usage report.

Table 14.12 Main elements of Usage Report—cont'd

Attribute	Description
Time of last packet	Indicates the time stamp for the last IP packet transmitted for a given usage report.
Usage information	Included if the UPF reports Usage Reports before and after a Monitoring Time, or before and after QoS enforcement. When present, it shall indicate whether the usage is reported for the period before or after that time, or before or after QoS enforcement.
Event time stamp	Included if the report is related to an event. When present, it shall be set to the time when the event occurs. Several IEs with the same IE type may be present to report multiple occurrences for an event for this URR ID.
Ethernet traffic information	Included if Ethernet Traffic Information needs to be reported. The IE includes information about MAC Address(es) that have been newly detected and MAC address(es) that been removed, MAC address(es) that have been inactive for a duration exceeding the Ethernet inactivity Timer.

Table 14.13 Main elements of Error Indication Report.

Attribute	Description
Remote F-TEID	Identifies the remote F-TEID of the GTP-U tunnel for which an Error Indication has been received at the UPF.

14.6.5 Data forwarding between SMF and UPF

Even though N4 (and Sx) is primarily used as a control interface towards the UPF and for receiving reports from UPF, it is also possible to instruct the UPF to forward user data packets (PDUs) to SMF. The SMF will then provide PFCP rules to UPF for setting up a GTP-U tunnel over N4 between SMF and UPF.

Data forwarding between SMF and UPF is, e.g., used when the SMF needs to communicate with DHCP server or AAA server on the Data Network via N6. Such DHCP/AAA traffic is then carried as User Plane traffic from SMF to UPF (via GTP-U) and then forwarded by UPF towards the DN via N6. Data forwarding on N4 is also used for IPv6 and IPv4v6 PDU Session types where SMF generates Router Advertisements to the UE. The Router Advertisements are sent via GTP-U to the UPF and then forwarded via N3 towards the UE. Fig. 14.14 illustrates a few of the scenarios with data forwarding between SMF and UPF. See Section 14.7 for more information on GTP-U.

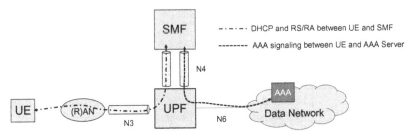

Fig. 14.14 Two scenarios for data forwarding between SMF and UPF.

14.7 GPRS tunneling protocol for the User Plane (GTP-U)

The two main components of GTP are the Control Plane part of GTP (GTP-C) and the User Plane part of GTP (GTP-U). GTP-C is the control protocol used in 3G/GPRS and 4G/EPS to control and manage PDN Connections and the User Plane tunnels that build up the User Plane path. The GTP-U uses a tunnel mechanism to carry the user data traffic and runs over UDP transport. In 5GS, GTP-U has been re-used to carry User Plane data over N3 and N9 (and N4) but the control protocol to manage the tunnel identities, etc. is instead using HTTP/2 and NGAP, which were described above. GTP-C is only used when 5GC is interworking with EPC. Here we will therefore only describe GTP-U. A reader interested in GTP-C may, e.g., consult a book on EPC such as Olsson et al. (2012).

GTP-U tunnels are used between two corresponding GTP-U nodes to separate traffic into different communication flows. A local Tunnel Endpoint Identifier (TEID), the IP address, and the UDP port uniquely identify a tunnel endpoint in each node, where the TEID assigned by the receiving entity must be used for the communication.

In 5GC, GTP-U tunnels are established by providing GTP-U TEIDs and IP addresses between (R)AN and SMF. This signaling is carried by HTTP/2 between SMF and AMF and by NGAP between AMF and (R)AN. There is thus no use of GTP-C in 5GC to manage GTP-U tunnels. The user plane protocol stack for a PDU Session is shown in Fig. 14.15.

A GTP path is identified in each node with an IP address and a UDP port number. A path may be used to multiplex GTP tunnels and there may be multiple paths between two entities supporting GTP.

The TEID that is present in the GTP-U header indicates which tunnel a particular payload belongs to. Thus, packets are multiplexed and demultiplexed by GTP-U between a given pair of Tunnel Endpoints. The GTP-U header is shown in Fig. 14.16. The GTP-U protocol is defined in 3GPP TS 29.281.

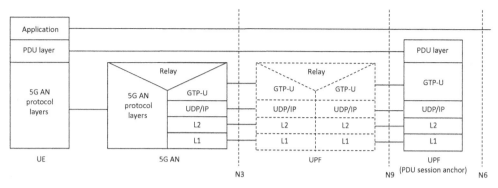

Fig. 14.15 User plane protocol stack for a PDU session.

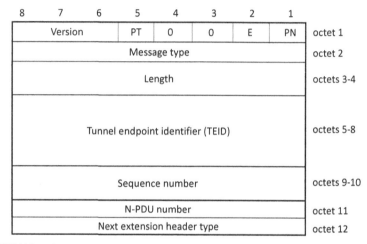

Fig. 14.16 GTP-U header.

14.8 Extensible Authentication Protocol (EAP)

14.8.1 General

The Extensible Authentication Protocol (EAP) is a protocol framework for performing authentication, typically between a UE and a network. It was first introduced in IETF for the Point-to-Point Protocol (PPP) in order to allow additional authentication methods to be used over PPP. Since then it has also been introduced in many other scenarios, for example as an authentication protocol for IKEv2, as well as for authentication in Wireless LANs using the IEEE 802.11i and 802.1x extensions.

EAP is extensible in the sense that it supports multiple authentication protocols and allows for new authentication protocols to be defined within the EAP framework. EAP is not an authentication method in itself, but rather a common authentication framework that can be used to implement specific authentication methods. These authentication methods are typically referred to as EAP methods.

The base EAP protocol is specified in IETF RFC 3748. It describes the EAP packet format as well as basic functions such as the negotiation of the desired authentication mechanism. It also specifies a few simple authentication methods, for example based on one-time passwords as well as a challenge-response authentication. It is possible to define additional EAP methods in addition to the EAP methods defined in IETF RFC 3748. Such EAP methods may implement other authentication mechanisms and/or utilize other credentials such as public key certificates or (U)SIM cards. A few of the EAP methods standardized by IETF are briefly described below:

- EAP-TLS is based on TLS and defines an EAP method for authentication and key derivation based on public key certificates. EAP-TLS is specified in IETF RFC 5216.
- EAP-AKA is defined for authentication and key derivation using the UMTS SIM card and is based on the UMTS AKA procedure. EAP-AKA is specified in IETF RFC 4187.
- EAP-AKA' is a small revision of EAP-AKA that provides for improved key separation between keys generated for different access networks. EAP-AKA' is defined in IETF RFC 5448.

In addition to the standardized methods, there are also proprietary EAP methods that have been deployed, e.g., in corporate WLAN networks.

As further described in Chapter 8, 5GS makes extensive use of EAP-AKA' for authentication over both 3GPP accesses and non-3GPP accesses.

14.8.2 EAP operation

The architecture for the EAP protocol distinguishes three different entities:

1. The EAP peer. This is the entity requesting access to the network, typically a UE. For EAP usage in WLAN, this entity is also known as the supplicant.
2. The authenticator. This is the entity performing access control, such as the SEAF or a WLAN access point or an ePDG.
3. The EAP server. This is the back-end authentication server providing authentication service to the authenticator. In 5GS it is the AUSF.

The EAP protocol architecture is illustrated in Fig. 14.17.

EAP is often used for network access control and thus the exchange takes place before the UE is allowed access and before the UE is provided with IP connectivity. Between the UE (EAP peer) and the authenticator, EAP messages are typically transported directly over data link layers, without requiring IP transport. Instead, the EAP messages are encapsulated directly in the underlying link-layer protocol. In 5GS, EAP is transported inside NAS during, e.g., the Registration procedure. In Wi-Fi, EAP is carried over Layer

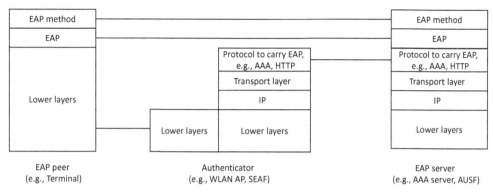

EAP method				EAP method
EAP				EAP
		Protocol to carry EAP, e.g., AAA, HTTP		Protocol to carry EAP, e.g., AAA, HTTP
		Transport layer		Transport layer
Lower layers		IP		IP
	Lower layers	Lower layers		Lower layers

EAP peer (e.g., Terminal)   Authenticator (e.g., WLAN AP, SEAF)   EAP server (e.g., AAA server, AUSF)

Fig. 14.17 EAP protocol structure.

2 in EAP-over-LAN (EAPoL) frames. EAP can also be used for authentication with IKEv2 as is the case with access to N3IWF, and in this case EAP is transported over IKEv2 and IP (the IKEv2 and IPsec layers are not shown in Fig. 14.17).

Between the authenticator and the EAP server, the EAP messages are carried in HTTP/2 using Nausf services in 5GS. In other systems the EAP messages may, e.g., be carried in a AAA protocol such as RADIUS or Diameter.

The EAP communication between the peer and the server is basically transparent to the authenticator (SEAF in 5GS). The authenticator therefore need not support the specific EAP method used but needs only to forward the EAP messages between the peer and the EAP server (AUSF in 5GS).

An EAP authentication typically begins by negotiating the EAP method to be used. After the EAP method has been chosen by the peers, there is an exchange of EAP messages between the UE and the EAP server where the actual authentication is performed. The number of round trips needed and the types of EAP messages exchanged depend on the particular EAP method used. When the authentication is complete, the EAP server sends an EAP message to the UE to indicate whether the authentication was successful or not. The authenticator is informed about the outcome of the authentication using the AAA protocol. Based on this information from the EAP server, the authenticator can provide the UE with access to the network or continue blocking access.

Depending on the EAP method, EAP authentication is also used to derive keying material in the EAP peer and the EAP server. This keying material can be transported from the EAP server to the authenticator. The keying material can then be used in the UE and in the authenticator (SEAF) for deriving the access-specific key needed to protect the access link. Further description on key derivation and EAP authentication in 5GS is available in Chapter 8. Chapter 8 also includes an example call flow describing authentication using EAP-AKA.

14.9 IP security (IPSec)

14.9.1 Introduction

IPsec is a very wide topic and many books have been written on this subject. It is not the intention or ambition of this chapter to provide a complete overview and tutorial on IPsec. Instead, we will give a high-level introduction to the basic concepts of IPsec focusing on the parts of IPsec that are used in 5GS.

IPsec provides security services for both IPv4 and IPv6. It operates at the IP layer, offers protection of traffic running above the IP layer, and it can also be used to protect the IP header information on the IP layer. 5GS uses IPsec to secure communication on several interfaces, in some cases between nodes in the core network and in other cases between the UE and the core network. For example, IPsec is used to protect traffic in the core network as part of the NDS/IP framework (see Chapter 8). IPsec is also used between the UE and the N3IWF to protect NAS signaling and User Plane traffic.

In the next section we give an overview of basic IPsec concepts. We then discuss the IPsec protocols for protecting user data: the ESP and the AH. After that we discuss the Internet Key Exchange (IKE) protocol used for authentication and establishing IPsec Security Associations (SAs). Finally, we briefly discuss the IKEv2 Mobility and Multi-homing Protocol (MOBIKE).

14.9.2 IPsec overview

The IPsec security architecture is defined in IETF RFC 4301. The set of security services provided by IPsec include:
- Access control
- Data origin authentication
- Connection-less integrity
- Detection and rejection of replays
- Confidentiality
- Limited traffic flow confidentiality.

By access control we mean the service to prevent unauthorized use of a resource such as a particular server or a particular network. The data origin authentication service allows the receiver of the data to verify the identity of the claimed sender of the data. Connection-less integrity is the service that ensures that a receiver can detect if the received data has been modified on the path from the sender. However, it does not detect if the packets have been duplicated (replayed) or reordered. Data origin authentication and connection-less integrity are typically used together. Detection and rejection of replays is a form of partial sequence integrity, where the receiver can detect if a packet has been duplicated. Confidentiality is the service that protects the traffic from being read by unauthorized parties. The mechanism to achieve confidentiality with IPsec is encryption, where the content of the IP packets is transformed using an encryption algorithm so that it becomes unintelligible. Limited traffic flow confidentiality is a service whereby IPsec can be used to protect some information about the characteristics of the traffic flow, e.g., source and destination addresses, message length, or frequency of packet lengths.

In order to use the IPsec services between two nodes, the nodes use certain security parameters that define the communication, such as keys, encryption algorithms, and so on. In order to manage these parameters, IPsec uses Security Associations (SAs). A SA is the relation between the two entities, defining how they are going to communicate using IPsec. A SA is unidirectional, so to provide IPsec protection of bidirectional traffic a pair of SAs is needed, one in each direction. Each IPsec SA is uniquely identified by a Security Parameter Index (SPI), together with the destination IP address and security protocol (AH or ESP; see below). The SPI can be seen as an index to a Security Associations database maintained by the IPsec nodes and containing all SAs. As will be seen below, the IKE protocol can be used to establish and maintain IPsec SAs.

IPsec also defines a nominal Security Policy Database (SPD), which contains the policy for what kind of IPsec service is provided to IP traffic entering and leaving the node. The SPD contains entries that define a subset of IP traffic, for example using packet filters, and points to a SA (if any) for that traffic.

14.9.3 Encapsulated Security Payload and Authentication Header

IPsec defines two protocols to protect data, the Encapsulated Security Payload (ESP) and the Authentication Header (AH). The ESP protocol is defined in IETF RFC 4303 and AH in IETF RFC 4302, both from 2005.

ESP can provide integrity and confidentiality while AH only provides integrity. Another difference is that ESP only protects the content of the IP packet (including the ESP header and part of the ESP trailer), while AH protects the complete IP packet, including the IP header and AH header. See Figs. 14.18 and 14.19 for illustrations of ESP- and AH-protected packets. The fields in the ESP and AH headers are briefly described below. ESP and AH are typically used separately but it is possible, although not common, to use them together. If used together, ESP is typically used for confidentiality and AH for integrity protection.

The SPI is present in both ESP and AH headers, and is a number that, together with the destination IP address and the security protocol type (ESP or AH), allows the receiver to identify the SA to which the incoming packet is bound. The Sequence number contains a counter that increases for each packet sent. It is used to assist in replay protection. The Integrity Check Value (ICV) in the AH header and ESP trailer contains the cryptographically computed integrity check value. The receiver computes the integrity check value for the received packet and compares it with the one received in the ESP or AH packet.

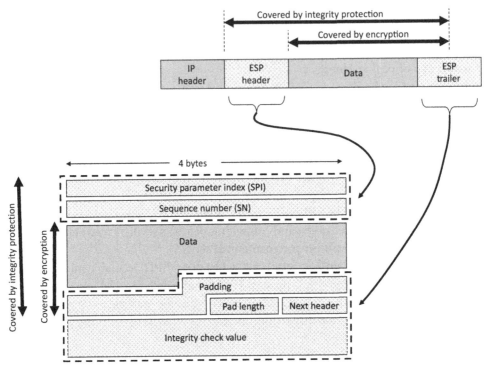

Fig. 14.18 IP packet (data) protected by ESP.

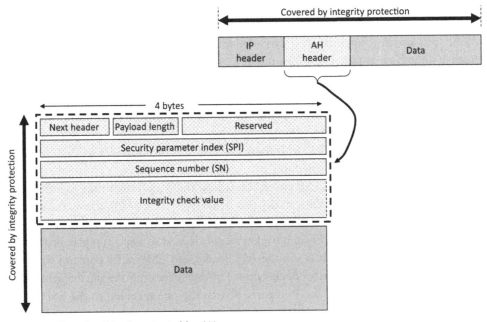

Fig. 14.19 IP packet (data) protected by AH.

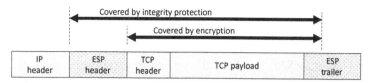

Fig. 14.20 Example of IP packet protected using ESP in transport mode.

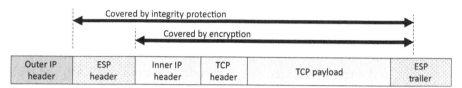

Fig. 14.21 Example of IP packet protected using ESP in tunnel mode.

ESP and AH can be used in two modes: transport mode and tunnel mode. In transport mode ESP is used to protect the payload of an IP packet. The Data field as depicted in Fig. 14.17 would then contain, for example, a UDP or TCP header as well as the application data carried by UDP or TCP. See Fig. 14.20 for an illustration of a UDP packet that is protected using ESP in transport mode. In tunnel mode, on the other hand, ESP and AH are used to protect a complete IP packet. The Data part of the ESP packet in Fig. 14.17 now corresponds to a complete IP packet, including the IP header. See Fig. 14.21 for an illustration of a UDP packet that is protected using ESP in tunnel mode.

Transport mode is often used between two endpoints to protect the traffic corresponding to a certain application. Tunnel mode is typically used to protect all IP traffic between security gateways or in VPN connections where a UE connects to a secure network via an unsecure access.

14.9.4 Internet key exchange

In order to communicate using IPsec, the two parties need to establish the required IPsec SAs. This can be done manually by simply configuring both parties with the required parameters. However, in many scenarios a dynamic mechanism for authentication, key generation, and IPsec SA generation is needed. This is where Internet Key Exchange (IKE) comes into the picture. IKE is used for authenticating the two parties and for dynamically negotiating, establishing, and maintaining SAs. (One could view IKE as the creator of SAs and IPsec as the user of SAs.) There are in fact two versions of IKE: IKE version 1 (IKEv1) and IKE version 2 (IKEv2).

IKEv1 is based on the Internet Security Association and Key Management Protocol (ISAKMP) framework. ISAKMP, IKEv1, and their use with IPsec are defined in IETF RFC 2407, RFC 2408, and RFC 2409. ISAKMP is a framework for negotiating, establishing, and maintaining SAs. It defines the procedures and packet formats for authentication and SA management. ISAKMP is, however, distinct from the actual key exchange protocols in order to cleanly separate the details of security association management (and key management) from the details of key exchange. ISAKMP typically uses IKEv1 for key exchange but could be used with other key exchange protocols. IKEv1 has subsequently been replaced by IKEv2, which is an evolution of IKEv1/ISAKMP. IKEv2 is defined in a single document, IETF RFC 7296. Improvements compared to IKEv1 have been made in areas such as reduced complexity of the protocol, reduced latency in common scenarios, and support for Extensible Authentication Protocol (EAP) and mobility extensions (MOBIKE).

The establishment of a SA using IKEv1 or IKEv2 occurs in two phases. (On this high level, the procedure is similar for IKEv1 and IKEv2.) In phase 1 an IKE SA is generated that is used to protect the key exchange traffic. Also, mutual authentication of the two parties takes place during phase 1. When IKEv1 is used, authentication can be based on either shared secrets or certificates by using a public key infrastructure (PKI). IKEv2 also supports the use of the EAP and therefore allows a wider range of credentials to be used, such as SIM cards (see Section 14.8 for more information on EAP). In phase 2, another SA is created that is called the IPsec SA in IKEv1 and child SA in IKEv2 (for simplicity we will use the term IPsec SA for both versions). This phase is protected by the IKE SA established in phase 1. The IPsec SAs are used for the IPsec protection of the data using ESP or AH. After phase 2 is completed, the two parties can start to exchange traffic using EPS or AH.

The original standards for NDS/IP (see Chapter 8) in 3GPP allowed both IKEv1 and IKEv2, but in later 3GPP releases the support for IKEv1 has been removed. It is also IKEv2 that is used on the interface between UE and N3IWF.

14.9.5 IKEv2 mobility and multi-homing

In the IKEv2 protocol, the IKE SAs and IPsec SAs are created between the IP addresses that are used when the IKE SA is established. In the base IKEv2 protocol, it is not possible to change these IP addresses after the IKE SA has been created. There are, however, scenarios where the IP addresses may change. One example is a multi-homing node with multiple interfaces and IP addresses. The node may want to use a different interface in case the currently used interface suddenly stops working. Another example is a scenario where a mobile UE changes its point of attachment to a network and is assigned a different IP address in the new access. In this case the UE would have to negotiate a new IKE SA and IPsec SA, which may take a long time and result in service interruption.

In 5GS this may occur if a user is using Wi-Fi to connect to an N3IWF. The NAS signaling and user traffic carried between the UE and the N3IWF is protected using ESP in tunnel mode. The IPsec SA for ESP has been set up using IKEv2. If the user now moves to a different network (e.g., to a different Wi-Fi hotspot) and receives a new IP address from the new Wi-Fi network, it would not be possible to continue using the old IPsec SA. A new IKEv2 authentication and IPsec SA establishment have to be performed.

The MOBIKE protocol extends IKEv2 with possibilities to dynamically update the IP address of the IKE SAs and IPsec SAs. MOBIKE is defined in IETF RFC 4555.

MOBIKE is used on the interface between UE and N3IWF to support scenarios where the UE moves between different untrusted non-3GPP accesses.

14.10 Stream Control Transmission Protocol (SCTP)

14.10.1 Introduction

The SCTP is a transport protocol, operating at an equivalent level in the stack as UDP (User Datagram Protocol) and TCP. Compared to TCP and UDP, SCTP is richer in functionality and also more tolerant against network failures. Even though both TCP and UDP are used as transport protocols in 5GS, we will not describe them in any detail in this book since we assume that most readers have a basic understanding of these protocols. The SCTP, on the other hand, used as transport protocol over the N2 interface, is a less known transport protocol and therefore is presented briefly in this section.

Compared to UDP from 1980 and TCP from 1981, SCTP is a rather new protocol. The original version was specified in IETF RFC 2960 in 2000 but it as has since then been obsoleted by a new version in IETF RFC 4960 from 2007. The motivation for designing SCTP was to overcome a number of limitations and issues with TCP that are of particular relevance in telecommunication environments. These limitations, as well as similarities and differences between UDP/TCP and SCTP, are discussed below.

14.10.2 Basic protocol features

SCTP shares many basic features with UDP or TCP. SCTP provides (similarly to TCP and in contrast to UDP) reliable transport ensuring that data reaches the destination without error. Also, similarly to TCP and in contrast to UDP, SCTP is a connection-oriented protocol, meaning that all data between two SCTP endpoints are transferred as part of a session (or association, as it is called by SCTP).

The SCTP association must be established between the endpoints before any data transfer can take place. With TCP, the session is set up using a three-way message exchange between the two endpoints. One issue with TCP session setup is that it is

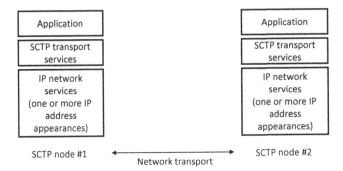

Fig. 14.22 SCTP association.

vulnerable to so-called SYN flooding attacks that may cause the TCP server to overload. SCTP has solved this problem by using a four-way message exchange for the association setup, including the use of a special "cookie" that identifies the association. This makes the SCTP association setup somewhat more complex but brings additional robustness against these types of attacks. An SCTP association, as well as the position of SCTP in the protocol stack, is illustrated in Fig. 14.22. As is also indicated in the figure, an SCTP association may utilize multiple IP addresses at each endpoint (this aspect is further elaborated below).

Like TCP, SCTP is rate adaptive. This means that it will decrease or increase the data transfer rate dynamically, for example depending on the congestion conditions in the network. The mechanisms for rate adaptation of a SCTP session are designed to behave cooperatively with TCP sessions attempting to use the same bandwidth.

Like UDP, SCTP is message-oriented, which means that SCTP maintains message boundaries and delivers complete messages (called chunks by SCTP). TCP, on the other hand, is byte-oriented in the sense that it provides the transport of a byte stream without any notion of separate messages within that byte stream. This is desirable to deliver, for example, a data file or a web page, but may not be optimal to transfer separate messages. If an application sends a message of X bytes and another message of Y bytes over a TCP session, the messages would be received as a single stream of X + Y bytes at the receiving end. Applications using TCP must therefore add their own record marking to separate their messages. Special handling is also needed to ensure that messages are "flushed out" from the send buffer to ensure that a complete message is transferred in a reasonable time. The reason is that TCP normally waits for the send buffer to exceed a certain size before sending any data. This can create considerable delays if the two sides are exchanging short messages and must wait for the response before continuing.

A comparison between SCTP, TCP, and UDP is provided in Table 14.14. More details on multi-streaming and multi-homing are provided below.

Table 14.14 Comparison between SCTP, TCP and UDP.

	SCTP	TCP	UDP
Connection oriented	Yes	Yes	No
Reliable transport	Yes	Yes	No
Preserves message boundary	Yes	No	Yes
In-order delivery	Yes	Yes	No
Un-ordered deliver	Yes	No	Yes
Data checksum	Yes (32-bit)	Yes (16-bit)	Yes (16-bit)
Flow and congestion control	Yes	Yes	No
Multiple streams within a session	Yes	No	No
Multi-homing support	Yes	No	No
Protection against SYN flooding attacks	Yes	No	N/A

14.10.3 Multi-streaming

TCP provides both reliable data transfer and strict order-of-transmission delivery of data, while UDP does not provide either reliable transport or strict order-of-transmission delivery. Some applications need reliable transfer but are satisfied with only partial ordering of the data and other applications would want reliable transfer but do not need any sequence maintenance. For example, in telephony signaling it is only necessary to maintain the ordering of messages that affect the same resource (e.g., the same call). Other messages are only loosely correlated and can be delivered without having to maintain a full sequence ordering for the whole session. In these cases, the so-called head-of-line blocking caused by TCP may result in unnecessary delay. Head-of-line blocking occurs, for example, when the first message or segment was lost for some reason. In this case the subsequent packets may have been successfully delivered at the destination but the TCP layer on the receiving side will not deliver the packets to the upper layers until the sequence order has been restored.

SCTP solves this by implementing a multi-streaming feature (the name Stream Control Transmission Protocol comes from this feature). This feature allows data to be divided into multiple streams that can be delivered with independent message sequence control. A message loss in one stream will then only affect the stream where the message loss occurred (at least initially), while all other streams can continue to flow. The streams are delivered within the same SCTP association and are thus subject to the same rate and congestion control. The overhead caused by SCTP control signaling is thus reduced.

Multi-streaming is implemented in SCTP by decoupling the reliable transfer of data from the strict order of transmission of the data (Fig. 14.23). This is different from TCP, where the two concepts are coupled. In SCTP, two types of sequence numbers are used. The Transport Sequence Number is used to detect packet loss and to control the retransmissions. Within each stream, SCTP then allocates an additional sequence number, the

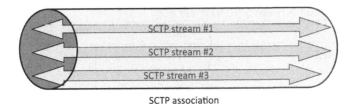

SCTP association

Fig. 14.23 Multi-streaming with SCTP.

Stream Sequence Number. Stream Sequence Numbers determine the sequence of data delivery within each independent stream and are used by the receiver to deliver the packets in sequence order for each stream.

SCTP also makes it possible to bypass the sequenced delivery service completely, so that messages are delivered to the user of SCTP in the same order they successfully arrive. This is useful for applications that require reliable transport but do not need sequential delivery or have their own means to handle sequencing of received packets.

14.10.4 Multi-homing

Another key aspect of SCTP that is an enhancement compared to TCP is the multihoming features. In a telecommunications network it is very important to maintain reliable communications paths to avoid service outage and other problems due to core network transmission problems. Even though the IP routing protocols would be able to find alternative paths in the case of a network failure, the time delays until the routing protocol converge and the connectivity is recovered are typically unacceptable in a telecommunications network. Also, if a network node is single homed, i.e., it has only a single network connection, the failure of that particular connection would make the node unreachable. Redundant network paths and network connections are thus two components in widely available telecommunications systems.

A TCP session involves a single IP address at each endpoint and if one of those IP addresses becomes unreachable, the session fails. It is therefore complicated to use TCP to provide widely available data transfer capability using multi-homed hosts, i.e., where the endpoints are reachable over multiple IP addresses. SCTP, on the other hand, is designed to handle multi-homed hosts and each endpoint of an SCTP association can be represented by multiple IP addresses. These IP addresses may also lead to different communication paths between the SCTP endpoints. For example, the IP addresses may belong to different local networks or to different backbone carrier networks. (It can be noted that TCP extensions have been developed during recent years to enable multi-path operation also for TCP.)

During the establishment of an SCTP association, the endpoints exchange lists of IP addresses. Each endpoint can be reached on any of the announced IP addresses. One of

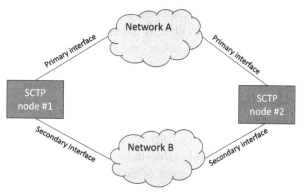

Fig. 14.24 Multi-homing with SCTP.

the IP addresses at each endpoint is established as the primary and the rest become secondary. If the primary should fail for whatever reason, the SCTP packets can be sent to the secondary IP address without the application knowing about it. When the primary IP address becomes available again, the communications can be transferred back. The primary and secondary interfaces are checked and monitored using a heartbeat process that tests the connectivity of the paths (Fig. 14.24).

14.10.5 Packet structure

The SCTP packet is composed of a Common Header and chunks. A chunk contains either user data or control information (Fig. 14.25).

The first 12 bytes make up the Common Header. This header contains the source and destination ports (SCTP uses the same port concept as for UDP and TCP). When an

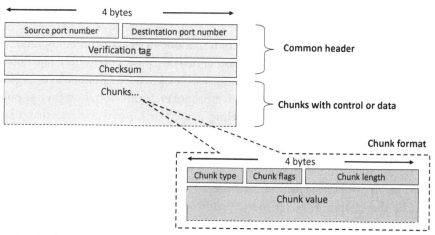

Fig. 14.25 SCTP header and chunk format.

SCTP association is established, each endpoint assigns a Verification Tag. The Verification Tag is then used in the packets to identify the association. The last field of the Common Header is a 32-bit checksum that allows the receiver to detect transmission errors. This checksum is more robust than the 16-bit checksum used in TCP and UDP.

The chunks containing control information or user data follow the Common Header. The chunk type field is used to distinguish between different types of chunks, i.e., whether it is a chunk containing user data or control information, and also what type of control information it is. The chunk flags are specific to each chunk type. The chunk value field contains the actual payload of the chunk. IETF RFC 4960 defines 13 different chunk type values and the detailed format of each chunk type.

14.11 Generic routing encapsulation (GRE)

14.11.1 Introduction

The GRE is a protocol designed for performing tunneling of a network layer protocol over another network layer protocol. It is generic in the sense that it provides encapsulation of one arbitrary network layer protocol (e.g., IP or MPLS) over another arbitrary network layer protocol. This is different from many other tunneling mechanisms, where one or both of the protocols are specific, such as IPv4-in-IPv4 (IETF RFC 2003) or Generic Packet Tunneling over IPv6 (IETF RFC 2473).

GRE is also used for many different applications and in many different network deployments outside the telecommunications area. It is not the intention of this book to discuss aspects for all those scenarios. Instead, we focus on the properties of GRE that are most relevant to 5GS.

14.11.2 Basic protocol aspects

The basic operation of a tunneling protocol is that one network protocol, which we call the payload protocol, is encapsulated in another delivery protocol. It should be noted that encapsulation is a key component of any protocol stack where an upper layer protocol is encapsulated in a lower layer protocol. This aspect of encapsulation, however, should not be considered as tunneling. When tunneling is used, it is often the case that a layer-3 protocol such as IP is encapsulated in a different layer-3 protocol or another instance of the same protocol. The resulting protocol stack may look like that shown in Fig. 14.26.

We use the following terminology:

- *Payload packet and payload protocol*: The packet and protocol that needs to be encapsulated (the three topmost boxes in the protocol stack in Fig. 14.26).
- *Encapsulation (or tunnel) protocol*: The protocol used to encapsulate the payload packet, i.e., GRE (the third box from the bottom in Fig. 14.26).

| Application layer |
| Transport layer (e.g., UDP) |
| Network layer (e.g., IP) |
| Tunneling layer (e.g., GRE) |
| Network layer (e.g., IP) |
| Layers 1 and 2 (e.g., Ethernet) |

Fig. 14.26 Example of protocol stack when GRE tunneling is used.

- *Delivery protocol*: The protocol used to deliver the encapsulated packet to the tunnel endpoint (the second box from the bottom in Fig. 14.26).

The basic operation of GRE is that a packet of protocol A (the payload protocol) that is to be tunneled to a destination is first encapsulated in a GRE packet (the tunneling protocol). The GRE packet is then encapsulated in another protocol B (the delivery protocol) and sent to the destination over a transport network of the delivery protocol. The receiver then decapsulates the packet and restores the original payload packet of protocol type

In 5GS, GRE is primarily used to carry the packets (PDUs) between UE and N3IWF. GRE here allows the QFI value and the RQI indicator for reflective QoS to be carried in the GRE header together with the encapsulated PDU. The QFI and RQI are included in the GRE key field (see below). Fig. 14.27 shows an example of an PDU carried in a GRE tunnel between UE and N3IWF over an IP delivery protocol.

GRE is specified in IETF RFC 2784. There are also additional RFCs that describe how GRE is used in particular environments or with specific payload and/or delivery protocols. One extension to the basic GRE specification that is of particular importance

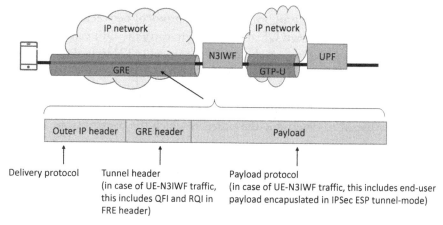

Fig. 14.27 Example of GRE tunnel between two network nodes with IPv4 delivery protocol.

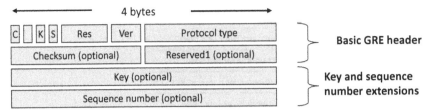

Fig. 14.28 GRE header format including the basic header as well as the key and sequence number extensions.

for EPS is the GRE Key field extension specified by IETF RFC 2890. The Key field extension is further described as part of the packet format below.

14.11.3 GRE packet format

The GRE header format is illustrated in Fig. 14.28.

The C flag indicates whether the Checksum and Reserved1 fields are present. If the C flag is set, the Checksum and Reserved1 fields are present. In this case the Checksum contains a checksum of the GRE header as well as the payload packet. The Reserved1 field, if present, is set to all zeros. If the C flag is not set, the Checksum and Reserved1 fields are not present in the header.

The K and S flags respectively indicate whether or not the Key and/or Sequence number is present.

The Protocol Type field contains the protocol type of the payload packet. This allows the receiving endpoint to identify the protocol type of the decapsulated packet.

The intention of the Key field is to identify an individual traffic flow within a GRE tunnel. GRE in itself does not specify how the two endpoints establish which Key field(s) to use. This is left to implementations or is specified by other standards making use of GRE. The Key field could, for example, be statically configured in the two endpoints, or be dynamically established using some signaling protocol between the endpoints. In 5GS the key field is used between UE and N3IWF to carry the QFI value and the RQI. The QFI takes 6 bits and RQI a single bit out of the available 32 bits in the key field. This is described in further detail in 3GPP TS 24.502.

The Sequence number field is used to maintain the sequence of packets within the GRE tunnel. The node that performs the encapsulation inserts the Sequence number and the receiver uses it to determine the order in which the packets were sent.

CHAPTER 15

Selected call flows

15.1 Introduction

Call flows (aka Procedures) are a very important tool for describing and understanding how a telecommunication system works. This applies also in a Service-Based system where interactions are described between Service Consumers and Service Producers. Even if in principle a service can be consumed by any NF, in order to have a system that functions end-to-end and provides the expected features, there is still a need to describe how the services are "stitched together" into complete procedures.

Some procedures have already been described in previous chapters of this book, together with the presentation of the 5GS key concepts. In this chapter, we provide further information and some additional procedures used in 5GS. It should be noted, however, that it is not feasible within this book to include a complete description of all procedures that exist in 5GS. Instead, we have chosen a few key procedures that should give a good overview of some of the most important use cases. We have also simplified the description of the procedures to present the main components of each procedure and avoid getting bogged down into details. The description provided below should thus only be used to get an overall understanding of how each procedure works. There are many aspects, options, and conditions that are not described. Readers interested in the full description, as well as procedures not described in this book, can consult the 3GPP technical specification 3GPP TS 23.502 for more information.

The following procedures are described below:
- Registration and Deregistration
- Service Request (UE-triggered and Network-triggered)
- UE Configuration Update
- PDU Session Establishment

- Handover procedures (Xn-based and N2-based handovers)
- Procedures for interworking with EPC
- Procedures for untrusted non–3GPP access

15.2 Registration and deregistration

15.2.1 Registration (initial-, periodic-, mobility-registration)

Registration is the first procedure the UE executes after being switched on. The procedure is performed to make it possible to receive services from the network. But the Registration procedure is also performed during the time the UE is connected to the network. There are several usages of the Registration procedure:

- *Initial Registration:* used by the UE to connect to the network after power-on.
- *Periodic Registration:* used by the UE that is in CM-IDLE state to show to the network that the UE is still there. The periodicity is based on a time value received from the AMF.
- *Mobility Registration:* used by the UE in case it moves out of the Registration Area, or when the UE needs to update its capabilities or other parameters that are negotiated in Registration procedure with or without changing to a new TA.
- *Emergency Registration:* used by the UE when it wants to register for emergency services only.

For more details on the different scenarios, see Chapter 7.

It can be noted that in EPS, the first case was supported using the Initial Attach procedure while the second and third cases were supported using Tracking Area Update (TAU) procedure. In 5G however, the three cases are supported using the Registration procedure. One benefit with that approach is that a mobility registration can be handled as an initial registration, with full authentication, etc., if there is some problem during the procedure. In EPS, a TAU that fails causes the MME to reject it and ask the UE to initiate an Attach procedure instead (Fig. 15.1).

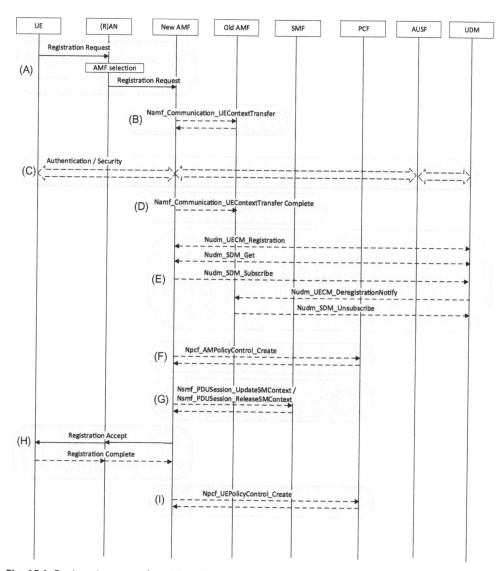

Fig. 15.1 Registration procedure (simplified).

The procedure is briefly described in the following steps:

A. The UE sends an NAS Registration Request message to AMF via the (R)AN. If temporary UE identities (5G-S-TMSI or GUAMI) are included and the (R)AN can map these to a valid AMF, the (R)AN forwards the NAS message to that AMF. Otherwise the (R)AN selects an AMF, based on Requested NSSAI (see Chapter 11) or a configured default AMF, and forwards the NAS message to that AMF.

B. In case a new AMF is selected (e.g., because the UE registers in an area not served by the old AMF), and the UE provided a GUAMI containing the identity of the old AMF, the new AMF retrieves the UE context from the old AMF.

C. Authentication is carried out, either using 5G AKA or EAP-AKA, as described in Chapter 8.

D. In case a new AMF has been selected, the new AMF indicates to the old AMF that it is now taking over as serving AMF for the UE.

E. The AMF registers as serving AMF for the UE in the specific access technology (3GPP access or non-3GPP access), using Nudm_UECM service. The AMF also requests subscription data and subscribes to subscription data updates using Nudm_SDM service. The UDM notifies the old AMF that it is deregistered in the UDM.

F. In case Access and Mobility management policies are deployed, the AMF initiates establishment of the AM policy association with PCF and retrieves the AM policies, as described in Chapter 10.

G. If the UE has indicated that it wants to activate User Plane connection for existing PDU Sessions, the AMF invokes the Nsmf_PDUSession_UpdateSMContext service operation for those PDU Sessions. If there is a mismatch in the UE and AMF PDU Session state, the AMF invokes the Nsmf_PDUSession_ReleaseSMContext service operation to notify the affected SMFs about those PDU Sessions.

H. If the Registration procedure is successful so far, the AMF provides a NAS Registration Accept to the UE. In some cases, the UE sends a NAS Registration Complete message to the AMF. This is done, e.g., to acknowledge the reception of a new 5G-GUTI or a new Configured NSSAI.

I. In case UE policies (ANDSP and/or URSP) are deployed, the AMF initiates establishment of the UE policy association with PCF. This allows the PCF to provide UE policies to the UE, as described in Chapter 10.

In addition, there are some additional steps that may be executed as part of the Registration procedure. For example, if the UE's temporary ID (GUTI) is unknown in both the old AMF and new AMF (after steps A and B), the new AMF will request the UE to send its SUCI, as shown in Fig. 15.2. The AMF may also use the Identity Request to request the UE to send the PEI.

The AMF may also check the ME identity with an Equipment Identity Register (EIR) as shown in Fig. 15.3, typically between steps D and E. The EIR can be used to blacklist, for example, stolen UEs. Depending on the response from the EIR, the AMF may continue the attachment procedure or reject the UE.

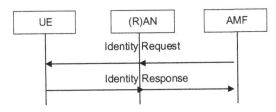

Fig. 15.2 Identity request procedure.

Fig. 15.3 Equipment identity check procedure.

15.2.2 Deregistration

The Deregistration procedure allows the UE to inform the network that it does not want to access the 5GS any longer, and the network to inform the UE that it does not have access to the 5GS any longer. Because of the Deregistration procedure the mobility and PDU Session contexts are removed. The procedure is, e.g., initiated by the UE when it is being turned off. The network may initiate the procedure, e.g., when the UE has not performed periodic registration because it is out of coverage or due to O&M triggers. The UE-initiated deregistration procedure is shown in Fig. 15.4. The network-initiated deregistration procedure is not explicitly shown, but an interested reader may refer to 3GPP TS 23.502, Clause 4.2.2.3.

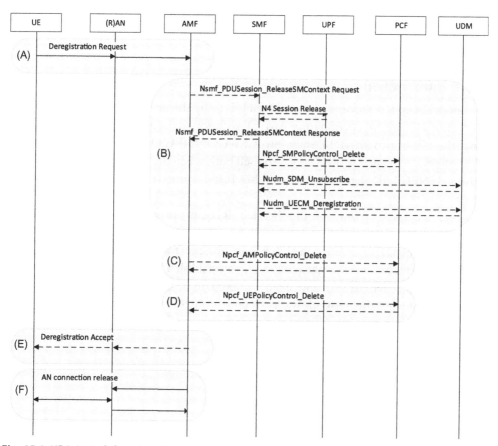

Fig. 15.4 UE-initiated deregistration.

The procedure is briefly described in the following steps:

A. The UE sends an NAS Deregistration Request message to AMF via the (R)AN.

B. The AMF notifies each SMF with active SM context that the corresponding SM context is released. The SMF then in turn notifies other NFs about the release of the PDU Session:

- The N4 session and associated User Plane resources are released.
- The SM policy association with PCF is released.
- The SMF deregisters from UDM for this PDU Session ID.
- If it is the last PDU Session for a specific DNN and S-NSSAI, the SMF also unsubscribes from subscription data updates.

C. The AMF releases the AM policy association with PCF (if any).

D. The AMF releases the UE policy association with PCF (if any).

E. The AMF sends a Deregistration Accept message to the UE, unless the UE indicated switch-off, i.e., the UE in that case does not wait for any acceptance from the network.

F. Finally, the AMF instructs (R)AN to release the N2 UE context. If there is still a (R)AN-level association between UE and (R)AN, the (R)AN may request the UE to release it.

15.3 Service Request

15.3.1 Introduction

The Service Request procedure is used by a UE or by the network to request the establishment of a secure connection to an AMF. Execution of the Service Request procedure brings the UE from CM-IDLE state into CM-CONNECTED state.

The Service Request procedure is also used both when the UE is in CM-IDLE and in CM-CONNECTED states to activate a User Plane connection for an established PDU Session.

There are two variants of the Service Request procedure: UE-triggered and Network-triggered Service Request. As the name indicates, the UE-triggered procedure is initiated by the UE, e.g., when the UE has up-link data to send. The network-triggered procedure is initiated by the network, e.g., to establish User Plane connection when down-link data have arrived and been buffered in a UPF, or when the UE is in CM-IDLE state and the network wants to send a NAS message to the UE.

15.3.2 UE triggered Service Request

In Fig. 15.5 the Service Request procedure is described. It can be noted that the Service Request may involve more steps than shown in the figure. For example, there are cases where an I–UPF has to be inserted, removed, or relocated. In those cases, there will be additional interactions in box C for selecting I–UPF and setting up User Plane tunnels between the different UPFs for forwarding of buffered packets. We have however chosen to show a simple example without I–UPF to describe the main principles. The full procedure is available in 3GPP TS 23.502, Clause 4.2.3.

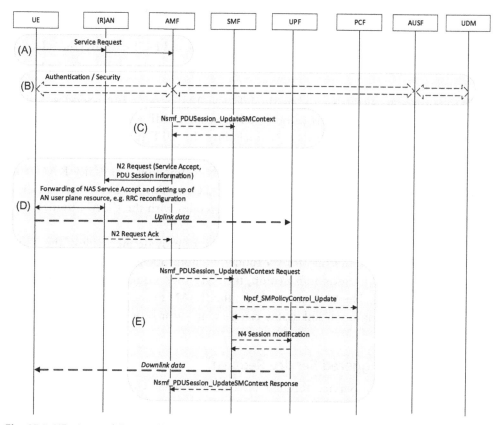

Fig. 15.5 UE-triggered Service Request.

The procedure is briefly described in the following steps:

A. When the UE wants to initiate the Service Request procedure, it sends a NAS Service Request message to AMF via the (R)AN. If the UE wants to establish User Plane connections for one or more existing PDU Sessions, the UE includes information to AMF about the PDU Session IDs for these PDU Sessions.

B. The network may optionally reauthenticate the UE during this procedure.

C. If the UE indicated in step A that it wants to establish User Plane connection for one or more PDU Sessions, the AMF notifies each SMF serving the corresponding PDU Session. In the simplest case the SMF can reply to this message with the UPF tunnel endpoint identifiers. In other cases, e.g., if the UE location is outside the serving area of the current UPF, the SMF may have to select a new I-UPF and then this step C becomes more involved, with signaling towards the old I-UPF (if any), new I-UPF and the anchor (PSA) UPF. For simplicity we have not shown this additional N4 signaling.

D. If the procedure was triggered by the UE, the AMF sends a NAS Service Accept to the UE. If User Plane is to be established, the AMF also forwards the PDU Session information to the (R)AN, including the UPF tunnel endpoint identifiers. The (R)AN configures the User Plane connection towards the UE. The exact details of how this is done depends on the AN technology. For 3GPP RAN, this may, e.g., be done via RRC reconfiguration. Once this has been done, up-link traffic can start to be sent. The (R)AN then replies with the (R)AN tunnel endpoint identifier(s) to the AMF.

E. The AMF now need to notify each SMF again, to provide the result of the User Plane establishment and the (R)AN tunnel endpoint identifier(s). If the PCF has subscribed to UE location information, the SMF then notifies the PCF about the new UE location. The SMF also provides the (R)AN tunnel endpoint identifier(s) to the UPF so that down-link traffic can be sent towards (R)AN.

15.3.3 Network triggered Service Request

The network triggered Service Request procedure is used when the network wants to trigger establishment of the signaling connection between UE and AMF (for UE in CM-IDLE), e.g., to send a down-link NAS message. The procedure is also used when the network wants to establish a User Plane connection for an existing PDU Session (for UE in CM-IDLE or CM-CONNECTED), e.g., when there is a down-link packet arriving.

In Fig. 15.6 the Network triggered Service Request procedure is described. Similar to the UE-triggered Service Request procedure, there may be additional steps in some cases, and other variants, e.g., depending on whether the UE is simultaneously connected in 3GPP and non-3GPP access with a common AMF or is only connected via one of the accesses. In Fig. 15.6 we show an example of single access.

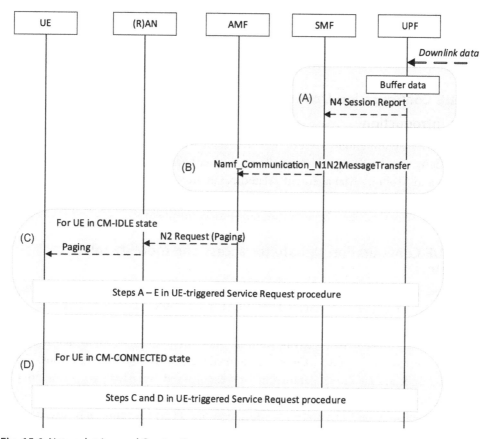

Fig. 15.6 Network-triggered Service Request.

The procedure is briefly described in the following steps:

A. If the UPF receives down-link data and has been instructed by SMF to buffer the packets, the UPF does the buffering and notifies SMF about the received data.

B. The SMF sends PDU Session information (UPF tunnel endpoint information and QoS information) to AMF for forwarding to (R)AN.

C. If the UE is in CM-IDLE state, the AMF needs to page the UE. The AMF stores the PDU Session information and sends a paging request to NG-RAN. The NG-RAN then pages the UE. When the UE receives the page, the UE sends a Service Request message to the network. That message and the rest of the procedure then follows the UE-triggered Service Request described in Section 15.3.2.

D. If the UE is in CM-CONNECTED state, there is no need to page the UE. Instead the AMF simply forwards the PDU Session information received from SMF to the (R)AN so that (R)AN can proceed with setting up the User Plane. This step and the rest of the procedure also follows the UE-triggered Service Request described in Section 15.3.2, but here only the steps (C) and (D) are needed in the UE-triggered Service Request.

15.4 UE Configuration Update

15.4.1 Introduction

The network may sometimes need to update specific parts of the UE configuration. The UE Configuration Update procedure allows the network to update:

- Access and Mobility Management related parameters
- UE Policy provided by the PCF.

The two cases are described by two different procedures below.

15.4.2 UE Configuration Update for access and mobility related parameters

The Access and Mobility Management related parameters that can be updated using this procedure include, e.g., 5G-GUTI, TAI List, Allowed NSSAI, Mapping Of Allowed NSSAI, Configured NSSAI for the Serving PLMN, Mapping Of Configured NSSAI, rejected S-NSSAIs, Network Identity and Time Zone, Mobility Restrictions, LADN Information, MICO, Operator-defined access category definitions, SMS Subscribed Indication. These parameters are determined by AMF, e.g., due to UE mobility change, Network policy, reception of Subscriber Data Update Notification from UDM, change of Network Slice configuration. The procedure may also be used by AMF to trigger the UE to initiate a *Re*-Registration procedure. The UE Configuration Update is shown in Fig. 15.7.

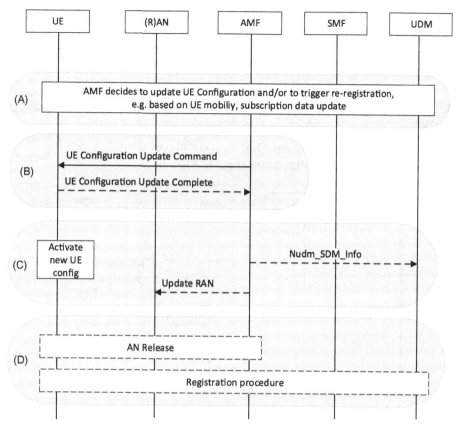

Fig. 15.7 UE Configuration Update of access and mobility-related parameters.

The procedure is briefly described in the following steps:

A. The AMF decides, e.g., based on updated subscription data, that UE configuration needs to be changed or a re-registration is needed.

B. The AMF sends a UE Configuration Update command that may include the updated configuration. It may also include indications whether or not UE shall send an acknowledgement (UE Configuration Update Complete) or whether or not Registration procedure shall be performed. Most UE Configuration Updates require an acknowledgement since AMF need to ensure that the UE received the update.

C. The UE then applies the new parameters. Depending on updated parameters, the AMF also may need to notify other entities. For example, if 5G-GUTI is changed, AMF needs to inform (R)AN about the new temporary identity. If the update was due to a Network Slicing Subscription Change Indication, then UDM is informed that UE has received the update.

D. Depending on the updated parameters, the AMF may release the AN association (e.g., Allowed NSSAI or Configured NSSAI are updated in a way that affects existing connectivity to Network Slices). The UE may also initiate a re-registration, e.g., to allow the UE to be connected to a new set of Network Slices.

15.4.3 UE Configuration Update Procedure for transparent UE policy delivery

This procedure is initiated when the PCF wants to update the UE policy (i.e., ANDSF and/or URSP) in the UE configuration (Fig. 15.8).

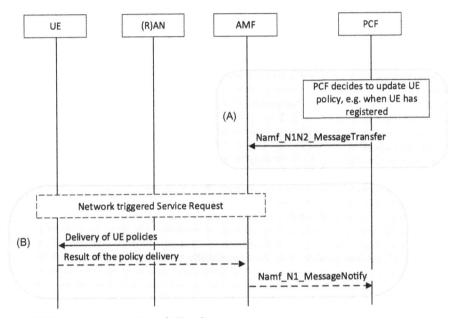

Fig. 15.8 UE Configuration Update of UE policy.

The procedure is briefly described in the following steps:

A. The PCF decides to update the UE policies, e.g., triggered by an AM Policy Session Establishment from AMF when UE has registered. The PCF sends the updated UE policies in a container to the AMF. If the UE policy is larger than can be sent in a single NAS message, the PCF splits the content and sends multiple requests to AMF.

B. If the UE is IDLE in 3GPP access, the AMF pages the UE (using the Network triggered Service Request procedure) in order to establish the NAS signaling connection. The AMF then sends DL NAS message including the UE policy container received from PCF. The UE implements the policy and sends the result back to the AMF. If the PCF has subscribed to be notified of the reception of the UE Policy container then the AMF forwards the response of the UE to the PCF using Namf_N1MessageNotify.

15.5 PDU Session Establishment

The PDU Session Establishment procedure (Fig. 15.9) is initiated by the UE when it wants to create a new PDU Session or handover a PDU Session between non-3GPP access to 3GPP access. The PDU Session Establishment procedure is also used for handing over a PDU Session from EPC to 5GC in case of EPC interworking without

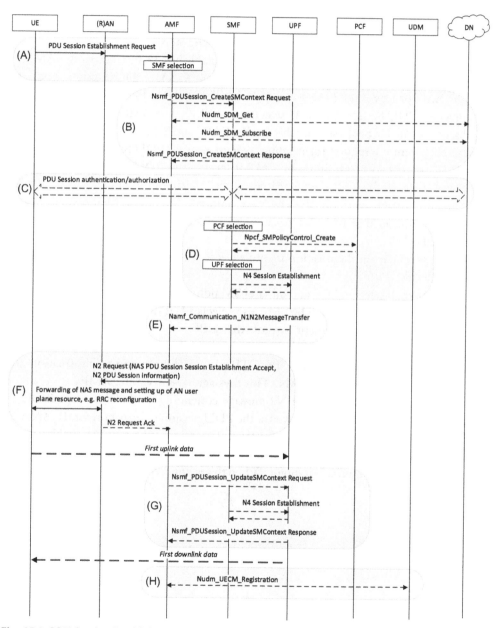

Fig. 15.9 PDU Session Establishment.

N26. The PDU Session Establishment procedure is always initiated by the UE, but it may be triggered by the network by sending a device trigger message to an application on the UE side. Based on that information contained in the device trigger message, the application on the UE side may decide to trigger the PDU Session Establishment procedure.

The procedure described below can be used over 3GPP access or non–3GPP access. Specific aspects related to PDU Session Establishment over untrusted non–3GPP access is available in Section 15.9.3.

The procedure is briefly described in the following steps:

A. The UE sends a 5GSM NAS PDU Session Establishment message to the AMF, including PDU Session Id, DNN-requested S–NSSAI, PDU Session type, etc. The AMF processes the NAS security. If the PDU Session Establishment is a request for a new PDU Session, the AMF selects a new SMF. The AMF may use the NRF to discover available SMFs serving the specific DNN and S–NSSAI. If the PDU Session Establishment is a request for handing over an existing PDU Session, the AMF uses its UE context to determine what SMF is serving the PDU Session Id.

B. The AMF then forwards the 5GSM container (containing the PDU Session Establishment message) to the SMF. The SMF retrieves the Session Management related UE subscription data from UDM and also subscribes to subscription data updates from UDM.

C. In case secondary authentication is applied, it is performed now. (see Chapter 8 for more details).

D. The SMF then selects a PCF and initiates SM policy session establishment to retrieve the initial set of PCC rules. The SMF also selects a UE IP address and UPF and initiates N4 session establishment towards that UPF.

E. The SMF then sends a 5GSM NAS PDU Session Establishment accept message towards the UE as well as the UPF GTP-U tunneling endpoint information and QoS information towards (R)AN. This message is sent via the AMF.

F. The AMF creates and sends the N2 message containing the NAS message (PDU Session Establishment accept) as well as the PDU Session information for (R)AN, i.e., GTP-U tunneling information and QoS information. The (R)AN establishes the required resources towards the UE and replies to the AMF with information about the (R)AN GTP-U tunnel endpoint. When this is done, the up-link data path is ready for use.

G. The AMF forwards the PDU Session information received from (R)AN to the SMF so that SMF can provide the (R)AN GTP-U tunnel endpoint to the UPF for down-link forwarding. Now also the down-link data path is ready for use.

H. If all works well, the SMF then registers itself in UDM, as serving the PDU Session Id in UDM.

15.6 Inter-NG-RAN handover

15.6.1 Introduction

Handover procedures in 5GS are used to hand over a UE from a source NG-RAN node to a target NG-RAN node. Like EPS, handover procedures are available both with and without Control Plane connection between source and target NG-RAN nodes. In the former case, the handover is "Xn-based" since the Xn interface between source and target NG-RAN nodes is used to manage the handover. The latter case is referred to as "N2-based" since the N2 interface between NG-RAN and AMF is used to manage the handover.

In the descriptions below, it may seem as the Xn-based handover is simpler (fewer messages) than N2-based handover. It should however be kept in mind that we do not show the interactions within the NG-RAN in case of Xn-based handover.

15.6.2 Xn-based inter-NG-RAN handover

This procedure is used to hand over a UE from a source NG-RAN to target NG-RAN using Xn. The Xn-based handover thus assumes that there is an Xn interface between source and target NG-RAN nodes. It only applies for intra-AMF mobility, i.e., Xn handover cannot be used if there is a need for AMF relocation.

During the procedure there may be a need to insert and/or remove an Inter-mediate UPF in the data path of PDU Sessions(s). This may happen, e.g., if the hand-over results in that the UE moves out of the service area of the UPF that currently has the N3 interface towards NG-RAN. To show the main principles of Xn hand-over and avoid complicating the call flow with details about N4/UPF signaling, we focus on the simplest case with no I-UPF insertion/removal/change below and only highlight the step where the UPF procedures would be inserted if needed (Fig. 15.10).

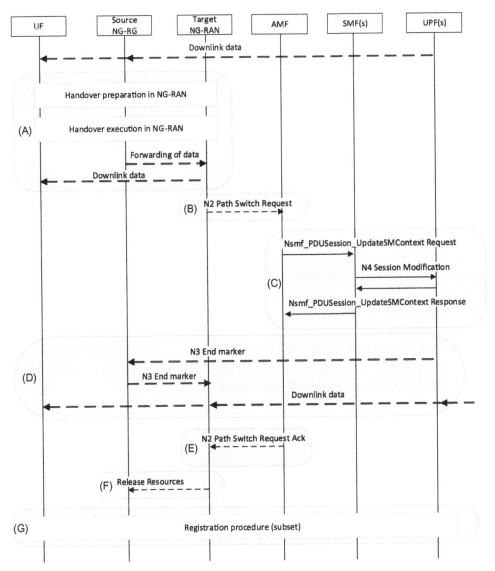

Fig. 15.10 Xn-based handover.

The procedure is briefly described in the following steps:

A. Before initiating the handover procedure towards the core network, the necessary handover preparation and handover execution signaling is performed in NG-RAN and between NG-RAN and the UE. We will not go into details for how this is performed, and an interested reader may refer to a book about 5G RAN, or consult, e.g., 3GPP TS 38.300.

B. When handover is confirmed in NG-RAN, the NG-RAN sends an N2 Path Switch Request to inform that the UE has moved to a new target cell. This message includes information about the PDU Sessions to be switched (including new NG-RAN N3 tunnel information) and also about PDU Sessions that have been rejected by target RAN (if any). The reason for target RAN to reject PDU Sessions may, e.g., be that the target RAN cannot provide the required QoS.

C. The AMF notifies each SMF that has a PDU Session that is affected by the handover. This includes both PDU Sessions that will be switched to target cell, as well as those PDU Sessions that failed to be handled by the target cell. For the PDU Sessions that are switched, the SMF(s) provides the NG-RAN N3 tunnel information to the UPF(s) so that down-link data can be sent to the new NG-RAN node. For the PDU Sessions that are deactivated, the SMF will later release those PDU Sessions using the PDU Session release procedure.

As mentioned above, the call flow shows the simple example of no change to the UPFs serving the PDU Session. However, if there is a need to insert a new I-UPF, release an old I-UPF, or both, these actions are performed at this point as well. In that case there will be additional N4 interactions, i.e., with new/old I-UPF and the anchor UPF, but on overall level the Xn handover procedure is the same.

The SMF then replies to AMF and includes the UPF N3 tunnel information.

D. The UPF has now switched the down-link path towards the target NG-RAN node and down-link packets will thus be sent to the target NG-RAN. However, to assist the reordering function in the Target NG-RAN, the UPF sends one or more "end marker" GTP-U packets for each N3 tunnel on the old path (i.e., to source NG-RAN) immediately after switching the path. This allows the NG-RAN to know when the last down-link packet is arriving on the old path so that the forwarding tunnel between source and target NG-RAN can be removed and the target NG-RAN can ensure that down-link packets are sent to the UE in order.

E. Once response is received from all the SMFs in step C, the AMF aggregates received UPF N3 tunnel information and sends this aggregated information to NG-RAN. This allows the NG-RAN to configure the up-link N3 data path(s).

F. The target NG-RAN then informs source NG-RAN that the handover is successfully completed so that source NG-RAN can release the associated resources.

G. In some cases, the UE needs to initiate a mobility Registration procedure after the handover is completed, e.g., if the handover resulted in that the UE moved outside its Registration Area.

15.6.3 N2-based inter-NG-RAN handover

15.6.3.1 General

N2-based handover is used in case there is no Xn signaling interface between Source NG-RAN and target NG-RAN, e.g., due to deployment or implementation aspects. In this case the handover signaling is carried via the core network, and also the communication between source NG-RAN and target NG-RAN need to be sent via the core network. It is the source NG-RAN that decides to initiate an N2-based handover to the target NG-RAN and this can be triggered, e.g., due to new radio conditions or due to load balancing.

N2 based handover has two phases: a preparation phase and an execution phase. The preparation phase is used to prepare the target NG-RAN and the SMF/UPFs about the handover to ensure that the execution phase can be performed fast and successfully.

Even though signaling is always carried via the core network when N2-based handover is done, data forwarding can be done either in RAN directly between source NG-RAN and target NG-RAN (so-called direct forwarding) or via a UPF in the core network (so-called indirect forwarding). A direct forwarding path can be used if there is IP connectivity between the source and target NG-RAN and security association(s) is in place, i.e., even if Xn-signaling between source and target NG-RAN is not supported.

In the procedure below, we show an example where the AMF is relocated, from a source AMF to a target AMF. Such relocation is needed if the target NG-RAN does not have N2 connectivity to the source AMF, i.e., the UE is moving outside the source AMF Service Area due to the handover. In case source AMF can continue to serve the UE after the handover, the call flow is simplified since the interactions between source and target AMF are not needed.

The procedure also shows an example where the I-UPF is relocated, from a source I-UPF to a target I-UPF. Such relocation is needed if the target NG-RAN does not have N3 connectivity to the source I-UPF, i.e., the UE is moving outside the source I-UPF Serving Area due to the handover. This will cause extra N4 signaling to manage the N3 and N9 tunneling. When there is no I-UPF, or the existing I-UPF can be maintain, the call flow is simplified somewhat since the amount of N4 signaling is reduced.

15.6.3.2 Preparation phase

The procedure is briefly described in the following steps (Fig. 15.11):

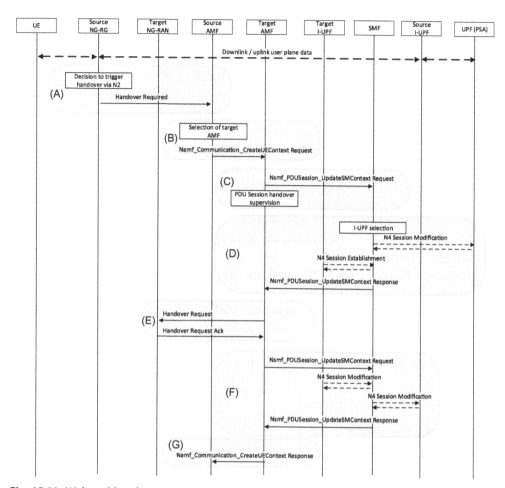

Fig. 15.11 N2-based handover—preparation phase.

A. The source NG-RAN decides to trigger an N2 handover and sends an N2 Handover Required message to AMF. The message contains the ID of the target NG-RAN and also information about the PDU Sessions to be handed over.

B. The AMF determines whether it can serve the new UE location (target NG-RAN ID). If it cannot, it selects a target AMF and sends a Create UE Context Request message to target AMF, including the information received from source NG-RAN.

C. The target AMF now informs the SMF(s) corresponding to each PDU Session about the handover. At the same time AMF starts to supervise the replies. The AMF need to get a reply from each SMF before it can send a message to target NG-RAN (step D) but it can at the same time not wait too long since the handover need to be executed to ensure that the UE can receive service from NG-RAN.

D. When receiving the update messages from AMF, the SMF(s) need to decide whether the existing UPF(s) can serve the target NG-RAN or if a new I-UPF is needed. In this call flow we assume that a source I-UPF needs to be replaced by a target I-UPF. SMF prepares the target I-UPF and then replies to AMF.

E. When the AMF has received responses from all SMF(s) or when it cannot wait any longer, the AMF sends a Handover Request to target NG-RAN, including information about the PDU Sessions to be handed over. The target NG-RAN replies with NG-RAN N3 tunnel information for each PDU Session that can be handed over. The reply also contains information about PDU Sessions that cannot be accepted by the target NG-RAN (if any).

F. The AMF forwards the PDU Session information from RAN to each SMF. If indirect forwarding is to be used, i.e., data forwarding from source NG-RAN to target NG-RAN via UPF, the SMF also establishes the required forwarding tunnels.

G. Finally, the target AMF replies to source AMF that the preparation phase is completed. This message also includes additional information that should be forwarded to source NG-RAN, e.g., about indirect forwarding tunnels and PDU Sessions that failed to be setup in target NG-RAN. But that will be done during the execution phase…

15.6.3.3 Execution phase

The procedure is briefly described in the following steps (Fig. 15.12):

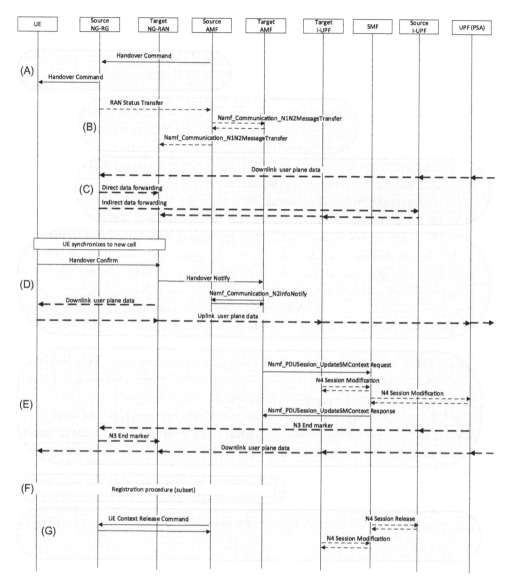

Fig. 15.12 N2-based handover—Execution phase.

A. The source AMF sends a Handover Command to source NG-RAN. This message includes, e.g., information received in step G during the preparation phase. The source NG-RAN now sends a Handover Command to the UE. After receiving the Handover Command, the UE will leave the source cell and start the connection towards the target cell.

B. The source NG-RAN may send status information to the target NG-RAN (via AMF). This is done if some of the radio bearers of the UE shall be treated with PDCP

The source AMF sends a Handover Command to source NG-RAN.

status preservation. If there is an AMF relocation, the source AMF sends the status information to the target AMF.

C. Down-link packets that reach the source NG-RAN after step A will now be forwarded to the target NG-RAN, either directly using Direct data forwarding or via a UPF using Indirect data forwarding.

D. After receiving the Handover Command, the UE moves to the target cell and sends a Handover Confirm message to the target NG-RAN. By sending this message the UE considers the handover successful.

The target NG-RAN then sends a Handover Notify to target AMF, and by this the target NG-RAN considers the handover successful.

The target AMF then notifies source AMF about the N2 Handover Notify received from the T-RAN. The source AMF will now start a timer and at expiry of that timer the AMF will instruct source NG-RAN to release the resources related to this UE.

Up-link User Plane data and the forwarded down-link User Plane data is now sent via the target NG-RAN. What remains is to indicate to the SMF/UPF (s) that the down-link data path can be switched towards NG-RAN.

E. The AMF will now notify all SMF(s) that have PDU Sessions that are handed over. This message includes NG-RAN N3 tunneling information that the SMF(s) will provide to the UPF(s) for setting up the down-link forwarding towards the target NG-RAN. In order to assist the reordering function in the target NG-RAN, the UPF (PSA) sends one or more GTP-U "end marker" packets for each N3 tunnel on the old path immediately after switching the down-link path.

F. The UE initiates a mobility Registration procedure. However, not all steps are needed now after the handover procedure. For example, the context transfer between source and target AMF is skipped.

G. Finally, the network cleans up resources that are no longer needed. The source AMF (based on the timer started in step D) tells source NG-RAN to release the UE context. The SMF(s) (based on its timers) release User Plane resources that are not needed anymore, including indirect data forwarding tunnels (if any) as well as the source I-UPF.

15.7 EPS interworking with N26

15.7.1 Introduction

As described in Chapters 3 and 7, the 5GS supports interworking with EPS and session continuity for a UE moving between EPC and 5GC. In this section we describe handover and mobility procedures for the case where N26 interface is supported between AMF and MME. The N26 interface allows the UE context to be transferred between the systems and ensures that a handover can be prepared also in case the UE only supports single-registration, i.e., the UE is only connected to either EPC or 5GC at each instant.

The N26 interface enables seamless mobility for single-registered UEs, e.g., fulfilling requirements for real-time services like voice.

15.7.2 Handover from 5GS to EPS

The handover from 5GS to EPS is triggered by the source NG-RAN by informing (source) AMF that a handover is required (Fig. 15.13).

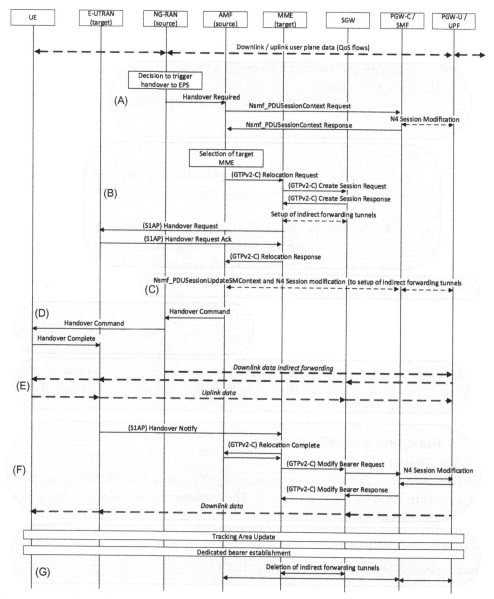

Fig. 15.13 Handover from 5GS to EPS.

The procedure is briefly described in the following steps:

A. The NG-RAN decides to trigger a handover to EPS and sends a Handover Required message to AMF. Based on the target "RAN" identifier the AMF determines it is a handover to E-UTRAN, and the AMF then request PGW-C+SMF to provide SM Context that is needed for AMF to transfer SM context to MME.

B. The AMF then selects a target MME (based on information about target E-UTRAN provided by NG-RAN). The AMF then acts as a source MME and sends a Relocation Request to the MME. The MME will initiate session setup and handover in target E-UTRAN access, basically following the procedure for S1-based handover. If indirect forwarding applies (i.e., user data forwarding from source NG-RAN to target E-UTRAN via the core network), the MME initiates setup of forwarding tunnels on the EPC side.

C. If indirect forwarding applies, the AMF triggers SMF to setup forwarding tunnels also on 5GC side.

D. The source side now issues the Handover Command to the UE, and UE moves to target cell.

E. The data path for indirect forwarding (if applicable) is established and down-link data can be forwarded. Also, the up-link data path is available over the target access.

F. The target E-UTRAN then sends a Handover Notify to target MME, and by this the target E-UTRAN considers the handover successful. The MME now notifies the source side (AMF) that the handover is complete. MME also provides the E-UTRAN tunneling information to the SGW and PGW-C+SMF to establish the down-link data path and switch the down-link User Plane path to target E-UTRAN access.

G. Finally, the UE initiates a Tracking Area Update procedure (as done according to normal S1-based handover). The network also triggers establishment of dedicated bearers that were not established during the actual handover procedure. The network also releases the indirect forwarding tunnels when they are no longer needed.

15.7.3 Handover from EPS to 5GS

15.7.3.1 General

The handover from EPS to 5GS is triggered by the source E-UTRAN by informing (source) MME that a handover is required. The handover consists of a preparation phase and an execution phase.

15.7.3.2 Preparation phase

The preparation phase is described in Fig. 15.14.

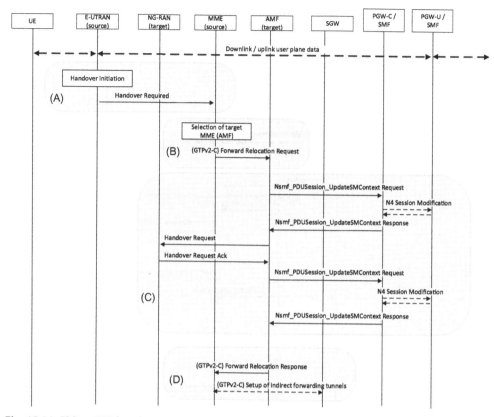

Fig. 15.14 EPS to 5GS handover—preparation phase.

The procedure is briefly described in the following steps:

A. The handover is initiated by the source E-UTRAN that sends a Handover Request to MME.

B. The MME selects a target node (AMF) and sends a Forward Relocation Request (GTPv2-C) message.

C. The handover is now prepared in the target side, similar to the preparation phase during N2-based handover. The AMF notifies each affected SMF. The SMF(s) provide QoS information and UPF tunnel information towards RAN. AMF provides this to target NG-RAN in a Handover Request. The NG-RAN then provides information about accepted (and rejected) PDU Sessions, including NG-RAN tunnel information. This information is forwarded to the respective SMF(s).

D. Finally, the AMF sends a reply to the Forward Relocation Request. This completes the preparation phase.

15.7.3.3 Execution phase

The execution phase is shown in Fig. 15.15.

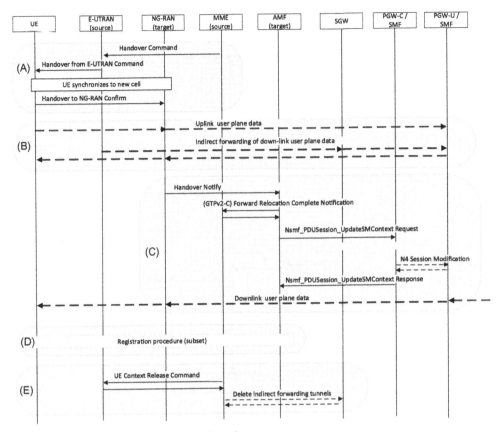

Fig. 15.15 EPS to 5GS handover—execution phase.

The procedure is briefly described in the following steps:

A. The MME sends a Handover Command to the UE via E-UTRAN, and the UE moves to the new cell.

B. The up-link data path is now available in the target 5G system, as well as indirect forwarding of down-link data.

C. The target NG-RAN notifies the AMF that the UE has arrived. The AMF now notifies each impacted SMF about the handover, and the SMF(s) can switch the down-link data path to the target NG-RAN.

D. As in the N2-based handover, a subset of the mobility registration procedure is performed.

E. Finally, the source EPC side cleans up resources in E-UTRAN as well as indirect forwarding tunnels (if any).

15.7.4 Idle mode mobility from 5GS to EPS

The procedure involves a Tracking Area Update to EPC and setup of default and dedicated EPS bearers and is shown in Fig. 15.16.

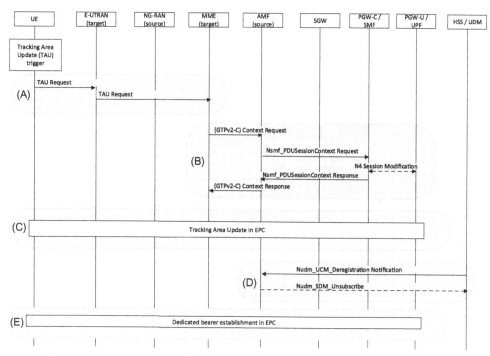

Fig. 15.16 5GS to EPS Idle mode mobility using N26 interface.

The procedure is briefly described in the following steps:

A. The UE initiates mobility to EPC by a Tracking Area Update (TAU) Request sent via E-UTRAN to the MME.

B. The MME contacts the source AMF to request the UE context. The AMF need to go to the corresponding SMFs to get the SM context for the affected PDU Sessions. The AMF then provides the UE context (including SM context) to the MME.

C. The TAU procedure continues, as specified for EPC.

D. When the UE is registered in EPC side, the UDM cancels the AMF registration for serving the UE in 3GPP access in 5GS. The AMF may now also unsubscribe to subscription data updates.

E. Finally, dedicated bearer establishment takes place in EPC side, as needed.

15.7.5 Idle mode mobility from EPS to 5GS

The procedure for idle mode mobility from EPS to 5GS is shown in Fig 15.17.

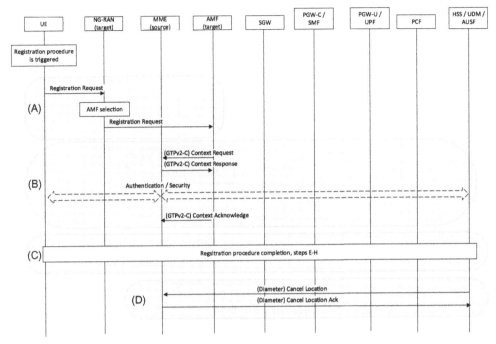

Fig. 15.17 Idle mode mobility from EPS to 5GS.

The procedure is briefly described in the following steps:

A. The UE triggers the Registration procedure towards 5GC and sends a NAS Registration Request. The NG-RAN selects an AMF (as described in the general Registration procedure) and forwards the Registration request message.

B. If the target AMF can identify the source MME, the AMF requests the UE context. If the UE provided an additional GUTI, then the AMF may request the UE context from an old AMF rather from the source MME. The AMF may also decide to authenticate the UE. If the target AMF accepts to serve the UE, the target AMF sends Context Acknowledge to the MME.

C. The Registration procedure continues with steps E–H of Fig. 15.1.

D. When the AMF has been registered in HSS/UDM as serving the UE in 3GPP access (as part of step C), the HSS/UDM sends a Cancel Location to the MME, to inform the MME that is no longer registered as serving the UE.

15.8 EPS fallback

In order to support various deployment scenarios for obtaining IMS voice service, the UE and NG-RAN may support a mechanism to direct or redirect the UE from NG-RAN connected to 5GC towards E-UTRAN connected towards EPC.

The solution is based on that the AMF indicates towards the UE during the Registration procedure that IMS voice over PS session is supported. The UE thus assumes that it can get IMS services over 5GS. If then a request from the core network for establishing a QoS flow for IMS voice (with 5QI = 1) arrives at the NG-RAN, the NG-RAN rejects this request and initiates either a handover or redirection to EPS. The procedure is briefly described in Fig. 15.18 (adapted from 3GPP TS 23.502).

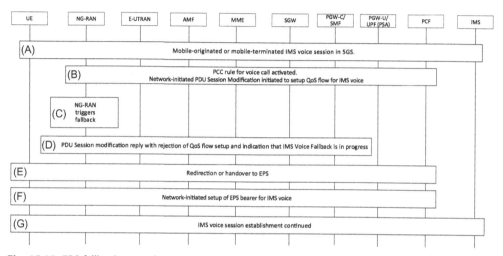

Fig. 15.18 EPS fallback procedure.

The procedure is briefly described in the following steps:

A. UE camps on NG-RAN in the 5GS and a Mobile Originated or Mobile Terminated IMS voice session establishment has been initiated.

B. The PCF installs the required PCC rules for the IMS voice call, and the SMF triggers a PDU Session Modification procedure to request the activation of a corresponding QoS flow.

C. If the NG-RAN is configured to support EPS fallback for IMS voice, it can decide to trigger fallback to EPS.

D. NG-RAN responds to the PDU Session Modification request and indicates that the setup of the QoS flow for IMS voice is rejected. NG-RAN also indicates that mobility due to fallback for IMS voice is ongoing.

E. NG-RAN now initiates either handover or AN Release with inter-system redirection to EPS, taking into account UE capabilities. In EPS, the UE initiates Tracking Area Update or PDN Connection Establishment depending, e.g., on whether N26 is used or not.

F. The PGW-C + SMF now reinitiates the setup of the QoS resources for IMS voice, mapping the 5G QoS to EPC QoS parameters. Since this is EPS, the PGW-C + SMF initiates dedicated bearer establishment.

G. The IMS voice session establishment is continued, now over E-UTRAN in EPS.

15.9 Procedures for untrusted non-3GPP access

15.9.1 Introduction

Since non-3GPP accesses and N3IWF are connected to 5GC using N2/N3, and the UE uses NAS (N1) also over non-3GPP access, the procedures for a UE accessing to 5GC via untrusted non-3GPP access follows the general procedures. There are however differences when it comes to the interface between the UE and the Access Network. When 3GPP access is used (NG-RAN), the 3GPP radio protocols such as RRC are used between UE and NG-RAN. With untrusted non-3GPP access and N3IWF, instead IKEv2 and IPSec is used between UE and N3IWF. In this section we will describe some of the interactions that are specific to untrusted non-3GPP access, but referring to the general procedures for steps that or common to all access technologies.

15.9.2 Registration over untrusted non-3GPP access

The Registration procedure for untrusted non-3GPP access follows the general Registration procedure described in Section 15.2. There are however access-specific aspects for how the 3GPP signaling is carried between UE and N3IWF. As mentioned in Chapter 8, IKEv2 and IPSec are used between UE and N3IWF to provide access control, ciphering, and integrity protection. Authentication in IKEv2 is however based on shared keys or EAP while authentication in 5GC is carried in NAS signaling (with EAP-AKA' or 5G AKA inside NAS). There was however a challenge in how to follow IKEv2 specifications and at the same time authenticate the UE using NAS. The solution defined by 3GPP was to define a new EAP method (called EAP-5G) that is basically only used to carry NAS. The EAP-5G method is thus not providing any authentication on its own, which makes it different from other EAP methods such as EAP-AKA or EAP-TLS that have authentication built in. When the NAS level authentication is completed and successful, the network informs the EAP Authenticator (N3IWF) that NAS authentication is completed and thus also EAP-5G can complete.

Fig. 15.19 shows the Registration procedure for untrusted non-3GPP access on high level, focusing on the differences to the general Registration procedure, i.e., on the interactions between UE and N3IWF.

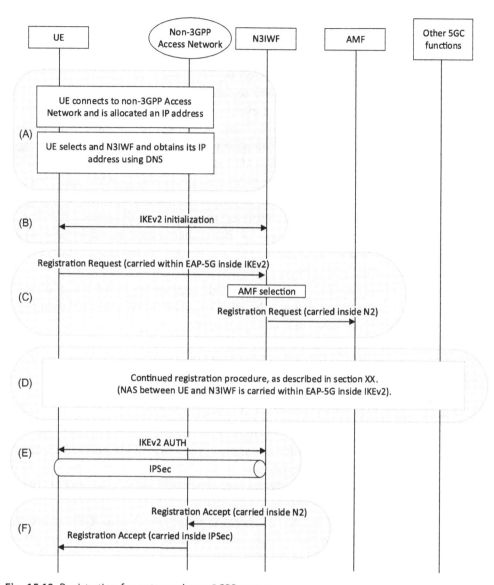

Fig. 15.19 Registration for untrusted non-3GPP access.

The procedure is briefly described in the following steps:

A. The UE discovers a non-3GPP access (e.g., WiFi), connects to it, and receives a local IP address from the non-3GPP access. The UE then discovers available N3IWF (s) using DNS and selects one to connect to.

B. The UE initiates the IKEv2 procedure towards the N3IWF. This includes IKE_SA_INIT and IKE_SA_AUTH exchanges between UE and N3IWF.

C. The UE then sends a NAS Registration Request to the AMF, just like in the general Registration procedure. This NAS message is carried inside an EAP-5G message, which in turn is carried inside an IKE_AUTH message. When this message reaches the N3IWF, the N3IWF extracts the NAS message, selects an AMF, and forwards the NAS message together with other information over N2 to AMF.

D. The Registration procedure then continues, as described in the general call flow in Section 15.2.1. NAS messages, e.g., for the authentication signaling are carried between UE and N3IWF as in step C, i.e., inside EAP-5G messages, which in turn are carried inside IKE_AUTH messages.

E. When the authentication is complete, the AMF notifies the N3IWF that authentication is successful, together with a N3IWF key. The N3IWF and UE then complete the IKEv2/IPSec setup and key agreement.

F. The AMF then sends a NAS Registration Accept to the UE. Once the IKEv2 has completed and the IPSec tunnel has been established in step E, any following NAS messages are carried inside IPSec between UE and N3IWF.

15.9.3 PDU Session Establishment over untrusted non-3GPP access

The PDU Session Establishment procedure for untrusted non-3GPP access (Fig. 15.20) follows the general PDU Session Establishment procedure described in Section 15.5. There are however access-specific aspects for how the 3GPP signaling is carried between UE and N3IWF.

As described in Section 15.9.2 for Registration in untrusted non-3GPP access, the Registration procedure including authentication results in that IKEv2 SA and IPSec SA are established between UE and N3IWF. That IPSec SA is used for sending NAS signaling between UE and N3IWF, as well as for carrying User Plane data for the active PDU Sessions.

To support QoS, separate Child IPSec SAs may be generated between UE and N3IWF. The reason is that IPSec has an anti-replay protection feature that can drop packets that arrive too much out of order. If multiple QoS classes would be carried over a single IPSec SA, discarding of lower priority packets due to this feature may occur. Therefore, separate QoS classes are preferably sent via different IPSec SAs.

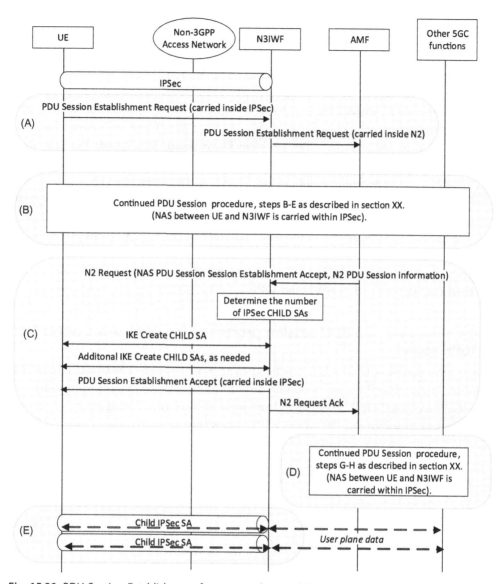

Fig. 15.20 PDU Session Establishment for untrusted non-3GPP access.

The procedure is briefly described in the following steps:

A. The UE sends a NAS PDU Session Establishment Request message, carried in IPSec between UE and N3IWF.

B. The procedure continues as in the general PDU Session Establishment procedure in Section 15.5, steps B-E. NAS between UE and N3IWF is carried within IPSec.

C. When the SMF provides the PDU Session information (QoS information and UPF N3 tunnel information) it is forwarded by AMF to N3IWF in an N2 message. The N3IWF then initiates the setup of User Plane resources towards the UE. This includes establishing at least one IPSec SA for the PDU Session, and optionally additional IPSec SAs depending on the QoS information received from SMF/AMF. The N3IWF also forwards the NAS PDU Session Establishment Accept message to the UE.

D. The procedure continues as in the general PDU Session Establishment procedure in Section 15.5, steps G-H.

E. User Plane is now carried between UE and N3IWF in the IPSec Child SA(s), and then via N3 between N3IWF and UPF.

15.9.4 Handover of a PDU Session procedure from untrusted non-3GPP to 3GPP access

This section describes how a UE can hand over from a source Untrusted non-3GPP access to a target 3GPP access, including how to move a PDU Session from untrusted non-3GPP access to 3GPP access. Such handovers are performed using the PDU Session Establishment procedure in the target 3GPP access (Fig. 15.21).

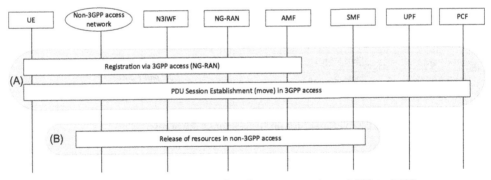

Fig. 15.21 Handover of a PDU Session procedure from untrusted non-3GPP to 3GPP access.

The procedure is briefly described in the following steps:

A. If the UE is not yet registered in 3GPP access (NG-RAN) it first needs to register. Then, in order to move a PDU Session from non-3GPP access to 3GPP access the PDU Session Establishment procedure is executed over 3GPP access. The UE includes an "Existing PDU Session" indication in the PDU Session Establishment Request message which indicates to AMF that the request refers to an already existing PDU Session. This ensures that AMF will look up which SMF is serving the PDU Session and forwards the PDU Session Establishment Request message to that SMF.

B. After the PDU Session has been moved, the SMF will initiate resource release in the source non-3GPP access. This releases the User Plane resources and PDU Session context in N3IWF.

15.9.5 Handover of a PDU Session procedure from 3GPP to untrusted non-3GPP access

Handover from 3GPP access to untrusted non-3GPP access follows the same principles (Fig. 15.22).

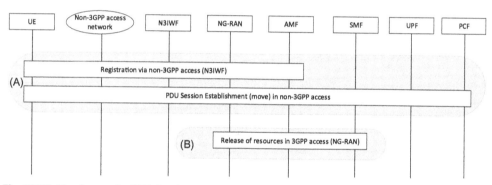

Fig. 15.22 Handover of a PDU Session procedure from 3GPP to untrusted non-3GPP access.

The procedure is briefly described in the following steps:

A. If the UE is not yet registered in non-3GPP access (N3IWF) it first needs to register. Then, in order to move a PDU Session from 3GPP access to non-3GPP access, the PDU Session Establishment procedure is executed over non-3GPP access. The UE includes an "Existing PDU Session" indication in the PDU Session Establishment Request message which indicates to AMF that the request refers to an already existing PDU Session. This ensures that AMF will look up which SMF is serving the PDU Session and forwards the PDU Session Establishment Request message to that SMF.

B. After the PDU Session has been moved, the SMF will initiate resource release in the source 3GPP access. This releases the User Plane resources and PDU Session context in AMF and NG-RAN.

CHAPTER 16

Architecture extensions and vertical industries

16.1 Overview

The 3GPP Release-15 specifications define the first generation of 5G network architecture and functionality. It is the result of a major undertaking by the telecom industry.

These specifications include the fundamental capabilities for providing data connectivity over NR and LTE access technologies, with advanced functionality for mobility, network slicing, Quality of Service and traffic steering. Release-15 also defines the Service-Based Architecture for the 5G Core Network and support for Untrusted non-3GPP access connectivity.

3GPP Release-16 extends the 5G specifications in two main areas—enhancing the 5G architecture, and providing support for some selected new service capabilities, specifically targeting use cases assumed to be important for enterprise and industry deployments. In this chapter we describe some of the key Release-16 enhancements.

16.2 Architecture enhancements and extensions

Some of the most important components of Release-16 architecture enhancements are described below. They target improvements of the 5G Core architecture to achieve a higher operational efficiency and enable further deployment optimizations.

16.2.1 Enhancing the service based architecture

There are three main aspects of the architecture that was addressed based on the eSBA study in 3GPP:
- Addition of indirect communication and delegated discovery
- Introduction of NF Sets and NF Service Sets
- Support for Context Transfer between NFs

16.2.1.1 Indirect communication and delegated discovery

Motivated, e.g. by separation of service business logic from discovery and selection functionality 3GPP embarked on defining indirect communication models. In addition to the so-called direct communication mode of 3GPP Release-15 (where NFs interact directly with each other), an indirect communication mode is introduced as an option for the NFs to communicate via a 'Service Communication Proxy' (SCP). In this indirect communication mode, the discovery is made either by consumer NFs themselves or the consumer NF transfers the responsibility for discovery to the SCP. The latter case is referred to as 'delegated discovery'.

16.2.1.2 NF Sets and NF Service Sets

An NF Set is a grouping of NF instances. All NF instance(s) in the NF Set shall have access to the UE related data. Thus, they are in principle interchangeable. The NF instance(s) in an NF Set come from the same vendor.

An NF Service Set is a grouping of NF service instances. The NF service instance(s) in the NF Service Set are interchangeable. An NF Service Set is only present within a single NF instance; thus, it does not span across NF instances.

In a procedure, a service producer may indicate stickiness to a resource via a binding indicator, i.e. if a resource is created, the producer will indicate in the response to what level a certain resource is bound; service instance, NF instance, NF Service Set or NF Set.

The NF Set and NF Service Set concepts will be introduced and described by 3GPP and it is expected that the additional flexibilities, impact on resiliency and individual procedures will be described while the overall architecture illustrations remain the same.

16.2.1.3 Context Transfer

To improve support for use cases like swapping one vendor's NF for another vendor's, NF context transfer solutions between different vendors' NFs can be employed. The ambition in 3GPP is to create context transfer solutions for several different NF types.

16.2.2 Network automation

The functionality of NWDAF has been extended significantly in 3GPP Release-16. New data collection and new analytics events have been defined to support a range of additional use cases.

The NWDAF provides analytics services to 5GC NFs, AFs, and OAM. As described further in 3GPP TS 23.288, analytics are either statistical information about the past, or predictive information of the future. Different NWDAF instances may specialize in different categories of analytics. The capabilities of a NWDAF instance are described in the NWDAF profile stored in the NRF and can be used by NF consumers to discover NWDAFs with the correct profile.

16.2.3 Enhanced network slicing

16.2.3.1 Introduction

Network slicing is an integral part of 5GS, i.e. in principle every feature introduced somewhat need to consider network slicing.

In 3GPP Release-16, the Network Slice selection was enhanced when interworking with EPS as described in Chapter 11.

As described in Chapter 8, 5GS supports primary authentication, and as further described in Chapter 6 5GS also supports the option to use secondary authentication between the UE and an external AAA on the Data Network. Such secondary authentication can be performed per PDU Session using EAP messages between the UE and the external AAA (see Chapter 6 for further details). In addition, it is possible to perform a Network Slice-Specific Authentication and Authorization during registration time, which can avoid the need for performing the secondary authentication at the time of establishing the PDU Sessions.

16.2.3.2 *Network Slice-Specific Authentication and Authorization*

The Network Slice-Specific Authentication and Authorization (SSAA) is controlled by the subscription having additional indication per Subscribed S-NSSAI whether it is subject to SSAA. Fig. 16.1 provides a high-level description of the SSAA procedure.

A. The subscription in UDM is enabled for SSAA by an indication per Subscribed S-NSSAI whether it is subject to SSAA. Normally at least one default Subscribed S-NSSAI is not subject to SSAA.

B. The UE indicates during Registration procedure whether the UE supports SSAA.

The AMF/NSSF, as described in Chapter 11, determines an Allowed NSSAI, but the S-NSSAI that are subject to SSAA are not included in the Allowed NSSAI.

The S-NSSAIs for which SSAA needs to be performed are included in the list of Rejected S-NSSAIs with a rejection cause value indicating pending Network Slice-Specific Authentication and Authorization.

C. The AMF triggers SSAA for each of the S-NSSAIs that are subject for SSAA.

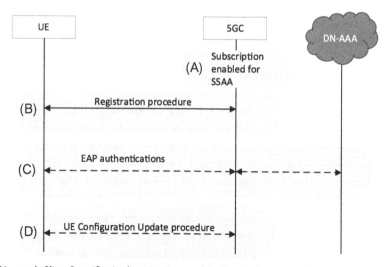

Fig. 16.1 Network Slice-Specific Authentication and Authorization procedure.

The SSAA is performed by EAP messages exchanged between the UE and the DN-AAA, relayed by the 5GC. The DN-AAA may be hosted by the HPLMN operator or a third party which has a business relationship with the HPLMN.

D. After the SSAA, the AMF initiates the UE Configuration Update procedure to provide the UE with an updated Allowed NSSAI.

The 5GC or DN-AAA may revoke the authorization or trigger re-authentication and re-authorization of S-NSSAIs at any time.

16.2.4 Enhanced SMF/UPF deployment flexibility
16.2.4.1 Background

One of the main assumptions in Release-15 5GC is that a PDU Session is served by a single SMF in non-roaming cases. A PDU Session can however be served by one or more UPFs. At least one UPF is required (i.e. a PDU Session Anchor UPF–PSA) but additional UPFs may be inserted when needed. A reason for inserting an Intermediate UPF (I-UPF) between RAN and the PDU Session Anchor is if the UE has moved to a location where there is no N3 connectivity between the serving RAN node and the PSA. An I-UPF is then inserted and this I-UPF will have the N3 interface towards RAN and an N9 interface towards the PSA UPF. See Chapter 6 for further details on I-UPF and PSA UPF.

If the UE moves far away from its initial location, and SMF needs to insert a UPF acting as I-UPF, it needs to be assumed in Release-15 that the SMF can find and control UPFs in that area. To provide PDU Session continuity in the whole PLMN, the SMF needs to have access to UPFs that can cover the whole PLMN. However, in large networks, e.g. in large countries, there may be a desire to deploy SMFs and UPFs using a more regionalized topology. In that case an SMF will only be controlling UPFs that cover one region, i.e. a subset of the PLMN. The assumption made in Release-15 that a single SMF can serve the PDU Session in non-roaming cases, even as the UE moves within a PLMN, may therefore not be valid in all deployments. To accommodate such deployments, Release-16 work was done to support SMFs with a limited service area.

As mentioned above, a specific UPF may depending on deployment only cover a limited area, i.e. a UPF may only have N3 connectivity to a certain set of RAN nodes. The area covered by a UPF is referred to as a *UPF Service Area (UPF SA)*. Also, as mentioned above in certain deployments each SMF may only be able to interface and control (via N4) a subset of the UPFs in a PLMN. That subset of UPFs together cover a certain area (the union of all UPF Service Area). This area, i.e. the collection of UPF Service Areas of all UPFs which can be controlled by one SMF, is referred to as *SMF Service Area (SMF SA)*. Fig. 16.2 illustrates a schematic example of UPF SAs and SMF SAs. It can be noted that SMF SAs may or may not overlap, this depends on deployment.

If the UE happens to move out of the area/region served by the SMF, the SMF is no longer able to select and control a UPF for the PDU Session that can act as I-UPF with

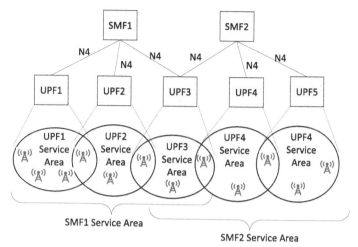

Fig. 16.2 Example of deployment topology with limited SMF Service Areas.

N3 towards the new RAN node. In this case (in Release-15) the PDU Session must be released.

In order to enable PDU Session continuity also across such borders, a solution has been developed in Release-16 to address the Release-15 limitations.

16.2.4.2 Architectures with I-SMF

The solution defined in Release-16 is that an Intermediate SMF (I-SMF) is added, when needed, to the non-roaming architecture between the AMF and the SMF. The purpose of the I-SMF is to select and control UPFs that acts as I-UPF and has the N3 interface to RAN. Fig. 16.3 shows a scenario where a UE creates a PDU Session (A) and later moves out of SMF SA (B) causing an I-SMF to be inserted to maintain the PDU Session. A reader may recognize that this architecture is very similar to the home-routed roaming architecture. We will discuss the relation to roaming more in Section 16.2.4.3.

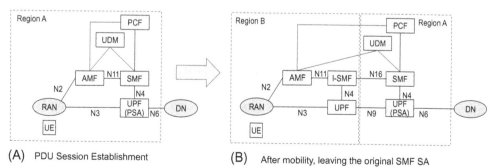

(A) PDU Session Establishment   **(B)** After mobility, leaving the original SMF SA

Fig. 16.3 (A) PDU Session Establishment with single SMF. (B) Architecture with I-SMF inserted.

In Fig. 16.3, an example with I-SMF insertion is shown. However, several use cases are supported:

— I-SMF insertion: This case applies to a PDU Session that is served by a single SMF only, and UE moves out of the SMF SA (as in Fig. 16.3).

— I-SMF removal: This case applies to a PDU Session that is served by I-SMF and SMF, and the UE moves out of the I-SMF SA and into SMF SA.

— I-SMF change: This case applies to a PDU Session that is served by I-SMF and SMF, and the UE moves out of the I-SMF SA but still outside the SMF SA. In this case a different I-SMF has to be selected serving the new location.

In order to support the three cases above, basically all procedures defined in Release-15 for handover, Service Request, PDU Session Establishment, etc. are enhanced in Release-16 to also incorporate the I-SMF insertion, change and removal.

16.2.4.3 Inter-PLMN mobility

As mentioned above, the architecture with I-SMF is similar to the home-routed roaming architecture, but with V-SMF replaced by I-SMF and H-SMF replaced by SMF. Another difference is that I-SMF and SMF are located in the same PLMN, while V-SMF and H-SMF are located in different PLMNs. This similarity was actually one motivation why 3GPP selected the I-SMF architecture shown in Fig. 16.3 to support the new use cases. By doing that, the enhancements to Handover procedures for insertion, removal and change of I-SMF can be applied to home-routed roaming cases as well for V-SMF insertion, removal and change. Having support for V-SMF insertion, removal and change enables PDU Session continuity in several use cases not supported in Release-15:

— Inter-PLMN mobility between two VPLMNs (V-SMF change)

— Inter-PLMN mobility from a VPLMN to a HPLMN (V-SMF removal)

— Inter-PLMN mobility from a HPLMN to a VPLMN (V-SMF insertion)

— Mobility within a VPLMN where a V-SMF may have a limited SMF Service Area (V-SMF change)

Support of inter-PLMN mobility use cases does however have specific requirements on roaming agreement. For example, inter-PLMN handover requires direct communication between AMFs in the two PLMNs, etc.

16.2.4.4 Traffic breakout in architectures with I-SMF

One aspect that was also included in Release-16 in architectures with I-SMF is to support selective routing of User Plane traffic to a local DN from a UPF controlled by I-SMF. This feature builds on the Release-15 features for selective traffic routing to a DN, as described in Chapter 6. The only difference is that the UPFs doing the breakout of traffic is controlled by an I-SMF. The architecture is shown in Fig. 16.4. See Chapter 6 for more information on the selecting routing of traffic to a local DN.

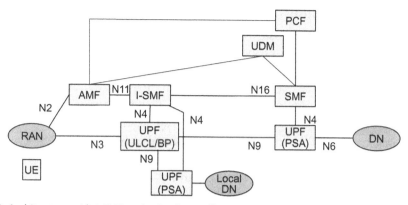

Fig. 16.4 Architecture with I-SMF and selective traffic routing to Local DN.

When the architecture shown in Fig. 16.4 applies, there needs to be a way to control what traffic is routed to the Local DN by the UPF(s) controlled by I-SMF. There must also be support for charging, QoS enforcement, etc. of this traffic. The solution defined for that case is that I-SMF will inform SMF about its capability to break out traffic as well as the applicable DNAI(s) that can be used. The DNAI(s) represent specific access locations into the DN (see Chapter 6 for more information on DNAIs). The SMF will then, based on the PCC rules received, charging requirements, etc. generate N4 rules and send to I-SMF. These N4 rules sent from SMF to I-SMF are basically the same N4 rules sent to UPFs (PDR, FAR, etc.) but the content differs somewhat from the N4 rules sent from an SMF to UPF. The differences are, e.g.:

— The FAR may contain a DNAI to indicate where traffic should be routed.
— Some parameters of the N4 rules are left out by SMF since they need to be added by
   I-SMF. This includes, e.g. N3 tunnel identifiers as these are only known to I-SMF.
When the I-SMF receives the N4 rules from SMF, the I-SMF will need to add some parameters (e.g. N3 tunnel identifiers) before I-SMF sends the rules to the UPF(s) controlled by I-SMF.

It can be noted that this breakout of traffic in UPFs controlled by I-SMF is not applicable to the home-routed roaming scenario, i.e. 3GPP has not defined a solution where traffic can be broken out in a UPF controlled by a V-SMF.

16.3 New feature capabilities

With Release-16, 3GPP takes an important step towards enabling new feature capabilities. The main target is industrial use cases, but also the integration of fixed access support is included in Release-16.

16.3.1 Support for industrial IOT applications

16.3.1.1 Background/drivers for IIoT in 5GS

New verticals for industry groups and industrial partners are posing emerging and lucrative business opportunities for 5G. 3GPP has been developing various tools for facilitating use of 5G System for these use cases, e.g. rail-bound mass transit, building automation, factory of the future, eHealth, smart city, electric-power distribution, central power generation, smart agriculture as well as mission critical applications. Some of these areas have been investigated in 3GPP and has been documented in 3GPP TR 22.804 .

3GPP has taken technical leadership in developing features discussed in this chapter to address the needs and requirements for these areas.

In the following four sections we will describe four topics related to support for industrial IOT applications that are done in 3GPP Release-16:

- 5G LAN-type services
- Support for Non-Public Networks
- Ultra-Reliable Low-Latency Communication
- Time Sensitive Networks

16.3.2 5G LAN-type services

16.3.2.1 Introduction

There are multiple market segments in the realm of residential, office, enterprise and factory, where Local Area Network (LAN) and Virtual Private Network (VPN) technologies are deployed today. This is an important area where the 5G System will need to provide services with similar functionalities as LANs and VPNs but improved with 5G capabilities (e.g. high performance, long distance access, mobility and security). One feature defined in Rel-16 for this type of deployment is the "5G-LAN type services" where the 5G System is evolved to offer private communication for UEs that are members of a 5G Virtual Network (5G VN) group. A 5G VN in this context is a virtual network based on 5GS. Access to a 5G VN is provided with a PDU Session that is established for a specific 5G VN. UEs that are members of a specific 5G VN group are authorized to establish PDU Sessions for that 5G VN group, and can communicate with other UEs in the group and can also, if applicable, access services on the DN. A 5G VN supports private communication, i.e. it is not possible to use a PDU Session to one 5G VN group to communicate with a UE belonging to another 5G VN group. A 5G VN group may be configured to use either IP PDU Session types or Ethernet PDU Session type.

Support for 5G LAN-type services is based on the Rel-15 5G System, i.e. the regular architecture and procedures are re-used. There are however two main additional aspects that have been enhanced in Rel-16 to better support 5G LAN-type services:

- Group management to enable the NEF to expose an API for 5G VN group management. This allows a third party AF to create, modify and delete 5G VNs and to add and remove 5G VN group members.

— Enhanced User Plane traffic handling, where new features have been added to SMF and UPF for additional capabilities for UE-to-UE communication within a 5G VN group.

Below we will describe these two aspects in some more detail.

16.3.2.2 5G VN group management

The northbound API exposed by the NEF towards third parties have been enhanced in Release-16 to support functionalities to create, modify and delete 5G VN groups. By making use of the API, a third party, such as a corporate, can manage 5G VN groups, including adding and removing group members. The AF may provide the following information to the NEF:

— Group Identifier
— Group membership information (GPSIs of the 5G VN group members)
— Group data (DNN, S-NSSAI, PDU Session type, etc. of the 5G VN group)

The NEF provides the received information to UDM that in turn stores it in UDR under the relevant Data Types. When a UE requests establishment of a PDU Session to a DNN that corresponds to a 5G VN group, the UDM will fetch the subscription data from UDR in the normal way and will also fetch the 5G VN group data (DNN, etc.) if the UE is subscribed to that 5G VN group. The subscription data is then provided to AMF and SMF in the normal way. PCF will also request the 5G VN group data from UDR, in order to generate URSP rules with the corresponding DNN, S-NSSAI, etc. Fig. 16.5 illustrates the overall procedure.

16.3.2.3 5G VN User Plane handling

One target of the work on 5G VNs was to enable efficient support of UE-to-UE communication. To accomplish that, two User Plane forwarding enhancements have been added as part of Release-16:

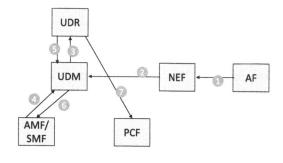

At Group Management by AF
1. AF request to manage the 5G VN group
2. NEF provides to UDM:
   - Group ID
   - Group membership
   - Group data
3. UDM stores 5G VN group info in UDR

At Registration and PDU Session Establishment
4. AMF/SMF request SubsData from UDM
5. UDM fetches relevant information from UDR.
6. UDM provides AM/SM SubsData to AMF/SMF
7. PCF fetches Group Data from UDR.

Fig. 16.5 5G VN group management.

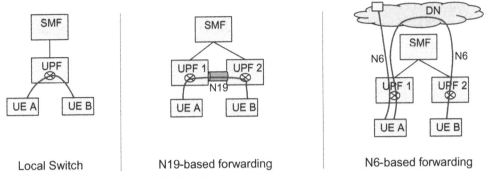

Fig. 16.6 User plane forwarding for 5G VN groups.

— Local switch, where up-link traffic from one UE is locally forwarded by a UPF as down-link traffic to another UE. This option requires that this UPF is the common anchor point (PSA UPF) of the different PDU Sessions for the UEs in the 5G VN group.
— N19-based forwarding, where direct UPF-to-UPF forwarding via an N19 tunnel is done. With this mechanism the traffic for the 5G VN communication is forwarded between PSA UPFs of different PDU Sessions via a shared (group-level) tunnel connecting PSA UPFs of a single 5G VN group.

These mechanisms are added as extensions to the Release-15 mechanisms for User Plane handling. The regular UPF handling of traffic forwarding, QoS enforcement, measurements, etc., including N6-based forwarding, still apply to 5G VN groups. Fig. 16.6 summarizes the different data forwarding options.

The different forwarding options are not mutually exclusive, they may all be applied in a 5G VN group for different PDU Sessions depending on which UPFs are serving the different PDU Sessions. It is the SMF that is in charge of configuring the UPFs with appropriate forwarding rules for the local switch, N19 based forwarding and N6 based forwarding. The SMF does this be including relevant N4 rules in each UE's N4 session. In addition, the SMF may establish a group level N4 session with each UPF that has PDU Sessions in the group, in order to manage the N19 tunnels. In Release-16, it is assumed that a single SMF manages all PDU Sessions in a 5G VN group. This allows the SMF to have visibility of all PDU Sessions and corresponding UPFs and can thus generate forwarding rules for the N19 tunnels.

16.3.3 Support for Non-Public Networks

16.3.3.1 Introduction

A Non-Public Network (NPN) is intended for the sole use of a private entity such as an enterprise. The NPNs may be deployed as a completely standalone network, or they may be integrated by a PLMN (i.e. public network), e.g. they may be offered as a Network Slice of a PLMN.

When an NPN is deployed as a Standalone NPN (SNPN), the NPN does not rely on network functions provided by a PLMN.

When an NPN is deployed as a Public Network Integrated NPN (PNI-NPN), the NPN is made available via the PLMN. There are different options how an PNI-NPN can be provided, e.g. access to the NPN can be made available using dedicated DNNs, or a Network Slice can be dedicated to an NPN with various levels of shared resources and Network Functions between the NPN and the PLMN.

Fig. 16.7 shows a high-level description of example deployment options for NPN.

16.3.3.2 Stand-alone Non-Public Networks

A PLMN is identified by a PLMN ID consisting of Mobile Country Code (MCC) and Mobile Network Code (MNC). The MCCs is three digits in length and each value is allocated to a country, while the MCCs for Stand-alone Non-Public Networks are in the 90× range and are non-geographic MCCs (country-agnostic). The MNC is two or three digits in length and is administered by the respective national numbering plan administrator, i.e. a country, except for the MNCs under MCC ranges 90× that are administered by the ITU Director of Telecommunication Standardization Bureau. The MNC, in combination with the MCC, traditionally have provided enough

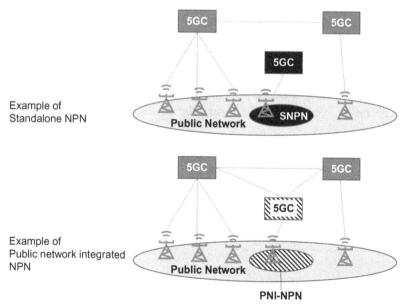

Fig. 16.7 Examples of NPN deployment options.

information to identify a network. However, to support the deployments of many SNPNs, the network identifier used needs to be extended.

To identify an SNPN, a Network Identifier (NID) has been added to be used with the PLMN ID, i.e. the combination of a PLMN ID and Network identifier identifies an SNPN.

In principle, a NID can be used in combination with any PLMN ID. However, the ITU has, in ITU OB 1156 (2018), allocated the MCC equals to 999 for internal use within a private network, and with no restrictions to the MNC used with MCC equals to 999. Therefore, such MCC is a natural option for usage by an SNPN. Several regions/countries have allocated specific MNC numbers for closed networks or networks for private use. 3GPP allows any PLMN ID to be used together with a NID.

Therefore, to enable support for SNPNs many of the procedures that includes a PLMN ID have been extended with an optional NID. An interested reader is referred to 3GPP specifications for further details of the enhancements added to support SNPN, e.g. 3GPP TS 23.501.

16.3.3.3 Access to PLMN services via an SNPN, and access to SNPN services via a PLMN

It is possible for a UE that has successfully registered with an SNPN to access PLMN services as depicted in Fig. 16.8. The UE first registers in the SNPN and establishes a PDU Session for obtaining IP connectivity via the SNPN to discover and establish connectivity to an N3IWF provided by the PLMN. The connectivity to the N3IWF in the PLMN re-uses the same functionality as specified for untrusted Non-3GPP access via NWu. The UE, using the credentials of the PLMN, then registers to the AMF in the PLMN via the 'NWu-PLMN' and 'N1-PLMN' to be able to access the services provided by the PLMN.

In a similar way, a UE that has successfully registered with a PLMN may perform another registration with an SNPN, using the credentials of that SNPN, following

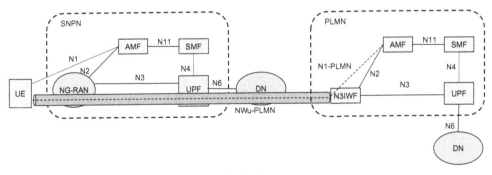

Fig. 16.8 Access to PLMN services via a Non-Public Network.

the same principles as described above, and in Fig. 16.8, but with the SNPN exchanged with a PLMN, and the PLMN exchanged with an SNPN.

When the UE moves between access networks from an SNPN to a PLMN, service continuity for PDU Sessions established in the PLMN via the SNPN can be achieved by re-using the procedure 'Handover of a PDU Session procedure from untrusted non-3GPP to 3GPP access' described in Chapter 15. The procedure maintains IP address preservation, and a seamless experience can be achieved if the UE is able to keep simultaneous access to both the NPN and the PLMN access networks. Again, similar service continuity for PDU Sessions established in the SNPN via the PLMN can be achieved using the same procedure.

16.3.3.4 Public network integrated NPN
5GS supports ways for a PLMN to enable access for specific purposes, e.g. special DNN or dedicated Network Slices. However, in case there is a need to prevent UEs that are not authorized to access the NPN from even trying to access the network, some further mechanism is required as the available mechanisms either implies a rejection of the UE access attempts or it requires to enable some barring of the cell, e.g. using UAC. The mechanism enabling such control of UE access attempts is called Closed Access Group (CAG).

A Closed Access Group identifies a group of subscribers who are permitted to access one or more CAG cells associated to the CAG. That is, CAG is used to prevent UE(s), which are not allowed to access the NPN via the associated cell(s), from automatically selecting and accessing the associated cell(s).

A CAG is identified by a CAG Identifier which is unique within the scope of a PLMN ID. A CAG cell broadcasts one or multiple CAG Identifiers per PLMN.

To support CAG, the UE is configured, with an Allowed CAG list, i.e. a list of CAG Identifiers the UE can access, and optionally, an indication whether the UE is only allowed to access 5GS via CAG cells. The 5GC also provides the same CAG information to NG-RAN for NG-RAN to apply during connected mode mobility, i.e. to avoid selecting target cells that the UE is not authorized to access.

16.3.4 Ultra-Reliable Low-Latency Communication (URLLC)
16.3.4.1 Overall architectural aspects
3GPP requirements in 3GPP TS 22.261 and 3GPP TS 22.104 have defined cyber-physical systems as *"Cyber-physical systems are to be understood as systems that include engineered, interacting networks of physical and computational components. Cyber-physical control applications are to be understood as applications that control physical processes. Cyber-physical control applications in automation follow certain activity patterns, which are open-loop control, closed-loop control, sequence control, and batch control."*

Applications such as robotic surgery performed at a distance over a network would have requirements of ultra reliable, highly available and low to very low end to end latency performance (such as less than 10 ms to 1 ms range; and may be able to achieve even better performance with improvements in NR technologies) and deterministic and periodic communication patterns.

3GPP has addressed the ultra-reliable and low latency requirements via updated architectures and solutions end to end, which is under development when writing this book. It has done this in several ways.

3GPP defined new standardized 5QIs with QoS characteristics (see Chapter 9) dedicated to serving low latency requirements. These address application needs such as Low Latency eMBB applications like Augmented Reality, Discrete Automation, Intelligent transport systems, Electricity Distribution- high voltage and their characteristics are further defined in 3GPP TS 22.261.

To address low latency, 3GPP also developed means to control the end to end Packet Delay Budget (RAN PDB and CN PDB) within the network between UPF (the edge of the network) and RAN (gNB) for the User Plane traffic. As explained in Chapter 9, the CN PDB portion is assumed to be static for 5QIs and this assumption restricts fulfillment of actual latency requirements by RAN as it lacks the actual values. Now an SMF can provide specific CN PDB based on the 5QI being used, and RAN may be configured to have different CN PDB values for different deployment of UPF entities.

3GPP also defined a dedicated standardized SST value for URLLC specific Network Slice an operator may choose to deploy. Chapter 11 provides details of the use of such Network Slices for networks dedicated to serving very specific requirements end to end.

The UE should request an Always on PDU session for URLLC sessions to ensure that the PDU sessions are always connected. If an UE does not request this, then the SMF should establish the PDU session as Always on.

To address high reliability requirements, 3GPP has been developing solutions addressing end to end redundancy as well as certain portions of the network's User Plane redundancy and transport network redundancy. Following sections describe on a high level some of these options that an operator may deploy addressing specific business requirements.

In addition to the redundant data transmission, 3GPP is also developing mechanism to monitor the packet delay associated with a URLLC traffic using the 5QI associated with the URLLC service between the UE and the UPF or within network entities like the NG-RAN and UPF. Per QoS Flow approach and GTP-U path approach may be used based on operator's configuration and network and device support. This can allow operators or actual URLLC service (i.e. the AF) to make changes to better provide packet delay improvement.

16.3.4.2 Dual connectivity based end to end redundant user plane paths

This solution provides end to end redundancy for User Plane paths from the devices (UE) to the applications/devices to the other end. This solution utilizes existing mechanism

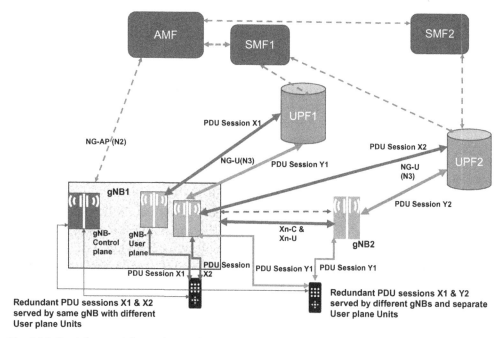

Fig. 16.9 Dual Connectivity end to end redundant PDU Sessions.

such as dual connectivity to provide redundant paths using two PDU connections for two redundant sessions for the same application. Fig. 16.9 is an example that illustrates how this is realized without any changes to the actual architecture itself.

The device is configured using UE provisioning by the PCF so that for certain applications requiring high reliability, the UE is to set up redundant PDU Sessions for this application and transmit to both PDU Sessions thus creating redundant data. The receiving end discards redundant data received.

When the SMF detects that the DNN is marked for redundant connectivity, it informs NG-RAN. Once the second PDU Session is established, based on input from SMF, the NG-RAN ensures that the User Plane for the two PDU Sessions use distinctly different User Plane nodes creating redundant paths from NG-RAN. SMF(s) ensure that the UPFs selected are distinct ensuring redundant paths in the CN.

Operators should ensure that underlying transport network also provides redundancy to have the full benefit. In addition, overall network topology, geographical location of network entities and power supply should be sufficiently distributed to ensure proper redundancy. Even though not all Control Plane entities can achieve redundancy, SMFs can support redundancy for the session management functions when the operator network is configured with SMF capabilities (e.g. in NRF) such that two different SMFs are selected for the redundant PDU Sessions.

16.3.4.3 Redundant user plane paths based on multiple UEs per device

This option relies on separate paths end to end, including on the device side where duplicate entities transmit/receive data independently. RAN deployment where redundant coverage by multiple gNBs (in case of NR) is expected and mechanisms available for example such as the IEEE TSN (Time Sensitive Networking), can make use of the multiple user plane paths.

The UEs belonging to the same terminal device request the establishment of PDU Sessions that use independent RAN and CN network resources.

Fig. 16.10 illustrates how this option may work. No 3GPP specification work is planned to be performed for this option, but operators may consider such deployment based on agreements with the vendors.

This option relies on certain configuration in the device where two UE entities reside, to create end to end redundancy. Each UE are connected to gNB1 and gNB2, respectively and sets up a PDU Session via gNB1 to UPF1, and a PDU Session via gNB2 to UPF2. UPF1 and UPF2 connect to the same Data Network (DN), though the traffic via UPF1 and UPF2 may be routed via different User Plane elements within the DN. UPF1 and UPF2 are controlled by different SMFs, SMF1 and SMF2 respectively.

Specialized applications and devices may make use of concepts like reliability groups (using, e.g. SUPI, PEI, S-NSSAI, RFSP) to coordinate and ensure that distinct NG-RAN nodes are selected and then followed up by selecting distinct CN entities and thus creating end to end redundant paths. To achieve this, certain specific configurations would be required, including, but not limited to:

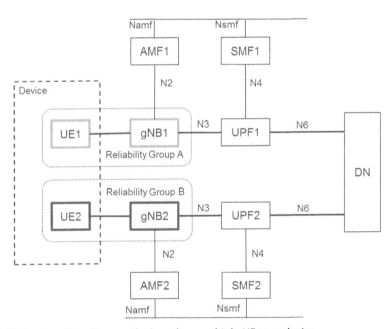

Fig. 16.10 Redundant User Plane paths based on multiple UEs per device.

- UEs belonging to the same device request establishment of PDU Sessions that use independent RAN and CN network resources.
- NG-RAN selects different AMFs by proper configuration of parameters such as reliability group, for AMF selection. NG-RAN uses information such as reliability group to ensure that UEs from same group are not handed over outside of the group served by NG-RANs (this can be configured via OAM in NG-RAN).
- UPF selection mechanism may be enhanced via configuration to ensure distinct UPFs are selected for the UEs belonging to the group.
- Different DNN setting can direct AMF to select different SMFs.

16.3.4.4 Support of redundant transmission on N3/N9 interfaces

This option provides redundant paths between NG-RAN and UPF via creating two redundant N3 (or N9) tunnels for the same traffic. Figs. 16.11 and 16.12 illustrates how multiple tunnels are established between two end points.

For the specific ULRRC 5QI(s), once the duplicate tunnels have been established, NG-RAN needs to identify and then duplicate the packets received from uplink direction and send them to both N3 tunnels. Similarly, UPF creates the duplicate packets and transfers via the two tunnels for downlink traffic. Each node at the end of the tunnel is responsible for discarding any duplicate packets received. SMF establishes the duplicate tunnels via Control Plane signaling and operator configuration ensures that there is distinct transport and IP routing for redundant paths.

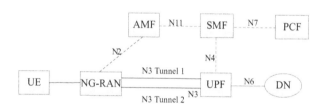

Redundant transmission: two N3 tunnels between the single UPF single NG-RAN node

Fig. 16.11 Redundant transmission on N3/N9 interfaces, single UPF.

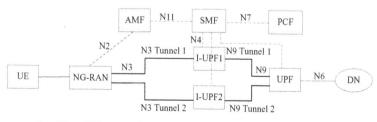

Two N3 and N9 tunnels between NG-RAN and UPFs for redundant transmission

Fig. 16.12 Redundant transmission on N3/N9 interfaces, multiple UPFs.

16.3.4.5 Support for redundant transmission at transport layer

This option relies on the operator providing redundant backhaul which provides two transport paths between UPF and NG-RAN. The Redundant functionality within NG-RAN and UPF make use of the independent paths at transport layer and maintains single User Plane tunnel between them. Redundant backhaul duplicates the traffic and the redundant data is eliminated at the receiving end (i.e. NG-RAN and UPF).

16.3.5 Time Sensitive Networks in 5GS

3GPP is developing Time Sensitive Networking support for applications like industrial IIoT services and factory automation applications, based on service and performance requirements defined in 3GPP TS 22.104. The requirements for clock synchronization include: support mechanism to synchronize the user-specific time clock of UEs with a global clock and mechanism to synchronize the user-specific time clock of UEs with a working clock. The 5G system needs to also support two types of synchronization clocks, the global time domain and the working clock domains (up to 32). These requirements would follow (IEEE 802.1AS-Rev/D7.3) sync domains.

The initial phase of 3GPP development focuses on integration of the 5G System as part of the IEEE Time Sensitive communications which is a set of standards to define mechanisms for the time-sensitive (i.e. deterministic) transmission of data over Ethernet networks. 3GPP 5G system already supports PDU Session type Ethernet, together with these 5GS can provide the necessary communications tools to be able to deliver Time-Sensitive Networking (TSN) of the IEEE P802.1Qcc.

The key functions 3GPP have been working on include an overall architecture for Time Sensitive Networks. Fig. 16.13 (3GPP TS 23.501), which describes 5GS integration using centralized model for supporting TSN and defines the overall 3GPP architecture. Additional functions have been introduced to interoperate with TSN related functions that do not interact or influence 3GPP functions and procedures not related to TSN, and these functions are confined at the edges of the 5GS. These additional functions for User Plane integration are Device-side TSN translator (DS-TT) at the UE and Network-side TSN translator (NW-TT) at the UPF. For Control Plane integration, a TSN AF is introduced which is responsible for interacting with the TSN Control Plane entity for bridge configuration, QoS mapping and any management requirements for 5GS acting as a TSN bridge.

URLLC capabilities enable 5G system prepared for wireless deterministic and time-sensitive communication, which is essential for integration with TSN. To support synchronization and multiple working time domains, 3GPP has been developing an architecture where TSN sync information and working domain clock synchronization for multiple working domains are handled transparently through the User Plane. Current agreement on an architecture that has been developed is described in Fig. 16.14 (from 3GPP TS 23.501). The main principles for this architecture are as follows:

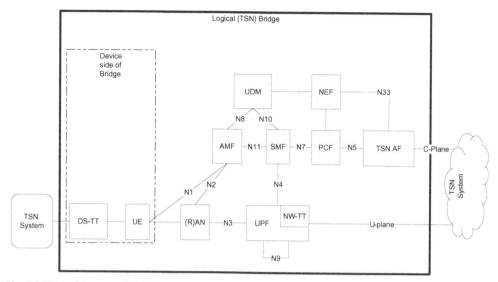

Fig. 16.13 Architecture for 5GS appearing as TSN bridge.

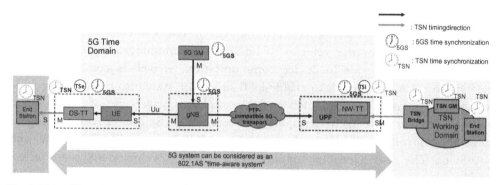

Fig. 16.14 5GS IEEE 802.1AS-Rev/D7.3 compliant time aware system for TSN time synchronization.

- Entire 5GS acts as an IEEE 802.1AS-Rev/D7.3 compliant entity performing as an 802.1AS "time-aware system".
- 5GS TSN Translators (DS-TT, NW-TT) at the edges of the 5G system support the IEEE 802.1AS-Rev/D7.3 operations.
- UE, gNB, UPF, NW-TT and DS-TTs are synchronized with the 5G internal system clock (5G GM).
- Two types of synchronization process: the 5GS synchronization (uses NG-RAN synchronization) and the TSN domain synchronization, as well as the Master (M) and Slave (S) ports are used when the working clock (TSN GM) is located at TSN working domain.

- The two synchronization processes are independent from each other and the gNB is synchronized to the 5G GM clock only.
- 5G internal clock is distributed to all User Plane entities like UPF NW–TT and provided via the transport network.
- Time stamping is performed at the edge on the User Plane TT functions.

All PDU Sessions connected to TSN via a specific UPF are grouped into a single virtual bridge.

To support multiple working clock domains, each working clock domain clock executes the time stamping independently from the other working clock domain and existing TSN parameters are used to determine the domain associated with each working clock.

3GPP work is still not stable nor complete in this area. As the technical work is currently in progress, the above description is for informational purposes only.

16.3.6 Automotive use cases

16.3.6.1 Background: V2X in EPS

3GPP started enabling use of LTE and EPS for supporting automotive industries requirements to support use cases for vehicular communications using 3GPP connectivity, known also as Vehicle-to-Everything (V2X). The Vehicular communication requires the ability for the vehicles to be able to communicate with each other directly, known as device to device communication, as well as communicate via the infrastructure. The goal is to enable LTE and 3GPP system to be able to provide automotive, infrastructure providers and other industry players use of 3GPP system for their services. Intelligent transportation services (ITS) can provide intelligent messaging and other services to the users using 3GPP system like LTE/EPC and NR/5GC. The system enables support for the devices/vehicles/RSUs to collect intelligent information about their surroundings and provide this information to each other or to an application server. This allows the V2X services to provide further intelligent enhancements and allow better safety and improved services. Three basic set of applications for providing ITS services have been considered initially: road safety, traffic efficiency, and other applications. Though 3GPP focused on providing the network and device capabilities to enable industries like ITS to provide their services. Some of the key performance requirements such as latency, range, reliability, priority, message size, frequency of transmission puts requirements for the 3GPP infrastructure and design of the PC5 and Uu communication services and are fulfilled by dedicated QoS parameters and design principles.

Fig. 16.15 provides a high-level connectivity illustration for V2X for LTE and EPC network.

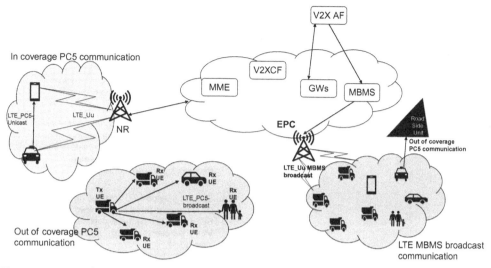

Fig. 16.15 LTE and EPC based V2X architecture.

The concept of V2X includes communication between vehicle to vehicle, vehicle to pedestrian, vehicle to infrastructure (that is a Road Side Unit (RSU) stationary entity providing V2X applications support to other vehicles/UEs same V2X services) and vehicle to network. Communication between vehicles and vehicles and pedestrian directly without traversing the network (with or without assistance from the eNB) require developing what is known as PC5 based communication (using what is known as sidelink communication capabilities of LTE and in 5G for NR). The concept of "in coverage" and "out of coverage" for the sidelink operation involves whether the vehicles/UEs are under the supervision/control of the RAN node (eNBs for LTE and gNB for NR) or outside of the coverage (also the terminology used to describe if served by the RAN or not).

Whereas using Uu communication, and standard 3GPP EPS connectivity, vehicles may connect to any application servers or other vehicles and also support efficiently reaching a larger group of UEs/vehicles within and outside its current geographic locations to relay messages and other communications using 3GPP system's multicast and broadcast services (MBMS).

Fig. 16.16 illustrates the various mode of operation that applies to different V2X connection.

Three main components of V2X communication are:

1. Proper configuration of the devices which includes, but not limited to, PC5 and Uu specific configurations, radio frequencies allowed for operating within operator's coverage area and outside of the coverage/radio frequencies, service related configurations as well as necessary radio parameters to operate V2X services.

V2X Type	V2V	V2P	V2I	V2N	Configuration
PC5(out of coverage of the specific RAT)	X	X			Devices pre-configured (in ME or UICC) to use the allocated frequency/RAT and other information to be able to use this mode (known as autonomous mode)
PC5 (in coverage)	X	X			Devices maybe pre-configured (in ME or UICC) or configured by the network when connected via Uu and eNB assists in the PC5 transmission resources setup (network scheduled mode)
Uu			X	X	Devices may be preconfigured (in ME or UICC) or configured by the network. Specific QCIs have been defined to fulfil V2X characteristics for communication over Uu.

Fig. 16.16 Connectivity and configuration for V2X.

2. PC5 specific communication support, which includes
   - Connectionless, and there is no signaling over PC5 Control Plane for connection establishment.
   - Broadcast only operation.
   - V2X messages are exchanged between UEs over PC5 User Plane.
   - Both IP based and non-IP based V2X messages are supported.
   - For IP based V2X messages, only IPv6 only.
   - Per Packet Priority and transmission (Tx) profile to select the PC5 radio and priority, a simplified method enabling some level of differentiation and priority among different PC5 data.
   - Subscription enabling PC5 support when network scheduled mode is used allowing eNB to enforce UE usage of resources and prioritize different PC5 traffic transmission using PPPP.
3. Uu specific communication support, which includes, in addition to normal Uu operation for connectivity via EPC/LTE:
   - Connectivity over Uu towards V2X Application Server and other vehicles/UEs.
   - Dedicated QCI values for V2X messages and other applications (i.e. QCI 3 (GBR bearer) and QCI 79 (Non-GBR bearer) for the unicast delivery of V2X messages and QCI 75 (GBR bearer) for V2X messages delivery over MBMS bearers only).
   - Subscription control for V2X and support for V2X Control Function.
   - Support for broadcast of V2X messages (using MBMS) and other services to group of UEs/vehicles on a specific geographic location.

— Provide eNB with necessary V2X user subscription information to enable V2X service authorization and enforcement of policies for PC5.

For EPS, to manage the UE configuration with V2X related parameters, 3GPP introduced a dedicated control function called V2X Control Function. The UE in the vehicle needs to connect to the EPS network and establish PDN connection in order to enable V2X Control Function to configure the devices (via V3 interface). The preconfiguration of the ME and UICC are other two mechanism for device configuration as described in the above table. In cases where a dedicated V2X Application Server may want to configure the vehicle with necessary parameters, this is enabled via the UE to V2X AS connectivity over Uu interface (V1 in Fig. 16.17). Fig. 16.17 as defined in 3GPP specification 3GPP TS 23.285.

For vehicle communication support, it is important that the vehicles/UEs have been configured to work across PLMNs when the vehicle is under PLMN/operator managed connectivity. Even when communicating directly using PC5 link, operator managed radio resources usage needs to be provided so that the resources are fairly used and prioritized. A device can connect using PC5 and Uu links independently and in case of regulatory services being used over Uu, like emergency services, Uu communication is given priority.

Security for applications over PC5 are provided by application specific security mechanism and not specified by 3GPP. Network level security provided by 3GPP is used on the network interfaces defined for V2X. In case of Uu, all 3GPP defined security

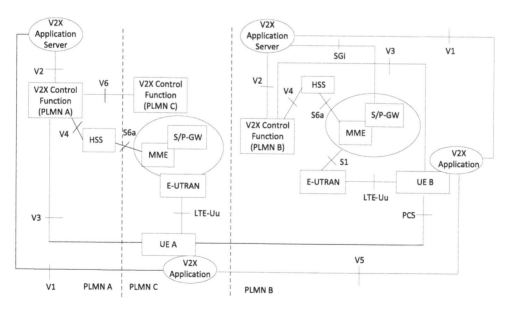

Fig. 16.17 Overall architecture in EPS for V2X.

mechanism apply for V2X traffic as well. Privacy of the users are ensured by changing the trackable identities (such as IP address, Layer 2 identities used by V2X communication link over PC5) frequently as per configuration, in addition to mechanism that may be available on the application itself.

16.3.6.2 V2X in 5GS for NR

In Release-16, 3GPP initiated work on supporting V2X over NR using 5GC as well as NR as a secondary RAT using EPC (i.e. EN-DC or 5G in EPC as described in Chapter 12). The 5G V2X architecture follows same principles as described in Section 16.3.6.1, with additional enhancements on NR-PC5 communication and UE configuration mechanisms. To date, for 5GS the support for MBMS has not yet been developed and as such the Uu based broadcast services are not available.

Fig. 16.18 illustrates the key aspects of NR connected 5GC V2X architecture.

The key difference for NR-PC5 (independent of whether NR-PC5 is used in connection with 5GS or EPS) compared to LTE-PC5 are as follows:

- NR PC5 supports unicast, groupcast and broadcast communication support.
- NR PC5 supports PC5 signaling in order to establish the PC5 unicast communication.
- Broadcast and Groupcast uses configured information about the service (e.g. source and destination layer 2 ID, Application ID). Groupcast group management is provided by the application layer.
- Unicast user discovery for PC5 communication may use broadcast and response mechanism to discover other UEs in the vicinity and then establish direct peer to peer unicast communication.

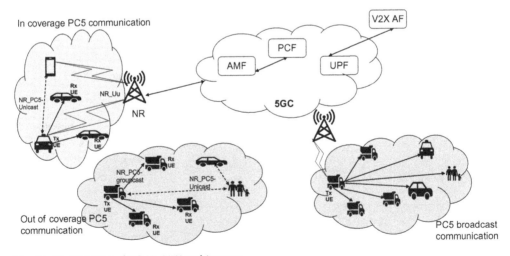

Fig. 16.18 NR-PC5 and NR Uu V2X architecture.

- For Unicast communication between two UEs using NR PC5, PC5 signaling is used to further negotiate the details about the specific application/service of interest.
- Enhanced PC5 QoS support aligned with Uu QoS using 5QI. This allows different application to use different PC5 QoS profile for different QoS Flows according to the service requirements by the application or when none available then use pre-configured information.
- UE has the choice of using LTE-PC5 or NR-PC5 based on availability, subscription/authorization and application capability.
- Range is an added optional parameter taken into account for PC5 communication. Range indicates the distance applied for the PC5 for the service to be meaningful to be part of the transmission.

3GPP is currently developing the architecture of NR based PC5 V2X communication and the development and finalization of the details are not yet complete. As such, we refrain from elaborating further.

For the UE configuration of parameters related to PC5 communication for 5GC, 3GPP chose an architecture more in line with 5GC capabilities. Using non-session based Policy management mechanism as described in Chapter 10, the V2X related UE configuration utilizes PCF based policy provisioning. UE provides the 5GC with UE capability related to V2X and based on such information the PCF triggers UE configuration of V2X policies. These include parameters related to authorization (e.g. NR or LTE PC5), related frequencies, other radio related parameters, application related data, PC5 QoS related information, etc.

For Uu based V2X, as per 5GS:

- The following additional parameters may be configured when the NR-Uu is used: PDU Session Type (i.e. IP type or Unstructured type), Transport layer protocol (i.e. UDP or TCP, only applicable for IP PDU Session type), SSC Mode, S-NSSAI(s), DNN(s).
- Dedicated SST value for a V2X specific slice has been introduced with the intent that such dedicated slice across operators PLMNs would facilitate vehicles/UEs moving through possibly various PLMNs.
- The 5QIs with same characteristics as for EPS are also available in 5GS.
- Enhanced service exposure and support for edge computing features available in 5GS like User Plane relocation/(re)selection and breakout near UE location, Local Area Data Network and traffic steering and routing for optimal location based on AF influence allows for better User Plane performance.

For Uu based V2X, the transport protocol is no longer restricted to UDP only and the transport protocol selection is therefore dependent on the application in question.

Support for Uu based broadcast services using mechanism like MBMS are not yet available in 5GS.

For Uu based V2X, further enhancements have been made to improve the V2X service experience. Two areas have been included for enhancement (note that these are under development in 3GPP at the time of the writing of this book, as such may develop differently when final solutions are specified):

- Providing gNB with additional QoS profile(s) in combination with QoS Notification control function, where the additional QoS profile(s) are provided by the V2X application server through the PCF. This is to allow RAN flexibility to adjust to varying radio condition and attempt to fulfill one of the candidates QoS profile for delivering the service.
- Enabling V2X AS triggered QoS monitoring over a period and over a geographical location from the OAM system. NWDAF uses the historical data over a period of time and provides analytics to the V2X AF that may allow a more predictable adjustment of the possible varying network conditions over that location for moving vehicles.

16.3.7 Integration of fixed access

16.3.7.1 Background and drivers

Current legacy wireline networks use a different "core network" than 3GPP mobile networks. In the wireline networks defined by the Broadband Forum (BBF), the "core network" includes a Broadband Network Gateway (BNG) acting as access router towards the end-user and providing IP services. The end-user is typically a residential customer where a Residential Gateway (RG)/Customer Premises Equipment (CPE) has been deployed. The RG/CPE requests connectivity from the wireline "core network", e.g. via DSL or Fiber access, in order to access Internet, voice, and IPTV services. The wireline core network may also include a AAA Server holding subscription data and taking part in authorization of connections, and DHCP server for providing IP addresses. A simple illustration is provided in Fig. 16.19.

In 2016–2017 the Broadband Forum and 3GPP started to discuss the possibility for 5G network convergence, whereby an RG would get connectivity service from the 5G Core, basically replacing the legacy wireline core network (BNG, AAA, DHCP, etc.) with the 5G Core (SMF, UPF, UDM, etc.). This was driven by numerous operators that provide both wireline and wireless services, currently deploying separate network

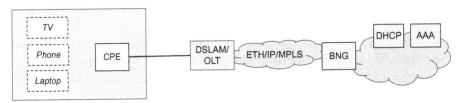

Fig. 16.19 Typical legacy wireline access.

infrastructure for each access. They saw opportunities to have a common 5G core network for both wireline access and wireless access, enabling, e.g. service convergence and CAPEX/OPEX savings.

Based on the initial discussions between 3GPP and BBF, the work on Wireline Wireless Convergence (WWC) was started in Release-16. The idea was to split the work based on each group's expertise and scope. 3GPP would define the required extensions to the 5GC, while BBF would define the access-specific aspects related to the RG and wireline access network. Later CableLabs, specifying wireline cable access, also contributed to the 3GPP work. The WWC work would thus define support for wireline integration with 5GC, for both BBF and CableLabs wireline access.

16.3.7.2 Migration considerations

In Release-15, the 5G Core was defined with the target to be a "common core network" for all accesses. As mentioned in Chapter 3, the N1, N2 and N3 interfaces are specified primarily for NG-RAN but re-used for Untrusted non-3GPP access. It was therefore very natural to integrate wireline access as another type of non-3GPP access in a similar way, i.e. using N1, N2 and N3. There were, however, a few important aspects to consider when it comes to wireline access specifically:

— The impact on the Access network: Many wireline operators obtain access to copper and fiber via third party entities which means an intervening network between the 5GC and the RG. Therefore, the new protocols and procedures for 5G support needed to be specified in such a way that they could transit existing access networks and integrate into existing operational procedures.
— The impact on existing services: Not all services are delivered by a BNG in an access network, and not all services will be reimplemented in 5G versions by all operators. Linear IPTV being an example of an access network integrated service likely to migrate to OTT and an "on-demand" paradigm for 5G. At the same time there is also the configuration and management of the RG to consider in a 5G context. Therefore, a design decision was made to adapt BBF TR-69/369 to the 5GC in order to provide an enhanced RG management platform.
— The impact on the RG. Many wireline operators have a large installed base with legacy RGs and it was desired to have a solution for convergence that did not require all these RGs to be replaced either all at once, or outside of normal business cycles. A migration strategy where the core network could be upgraded to 5GC before replacing all RGs was needed. Therefore, 3GPP and BBF agreed on supporting two scenarios:
  ○ 5G-capable RGs (called 5G-RG), where the RG is acting as a UE and requests access to 5GC using N1. This scenario requires RGs to support 3GPP specific functionality such as NAS (N1), i.e. RGs with new functionality.

○ Legacy RGs (called FN-RG), which do not have any 5G or 3GPP specific functionality. These FN-RGs use legacy mechanisms to access a legacy wireline core
network (e.g. PPPoE or IPoE protocol methods).

16.3.7.3 Network architecture

The network architecture for the two classes of RG support is shown in Fig. 16.20.

The Wireline Access Gateway Function (WAGF) is a function in the wireline access
network further specified by BBF and CableLabs. It acts in a sense like a RAN node,
supporting N2 and N3 towards 5GC and relays data traffic between the RG and the
UPF. As will be further described more below, the W–AGF has somewhat different functionality depending on whether an FN-RG or 5G-RG is to be served. The Y4 interface
in Figure 16.20 is defined by BBF and CableLabs and is a new interface supporting 5G
capabilities, e.g. NAS transport and based upon Ethernet and IP protocols. The Y5 interface in Fig. 16.20 is on the other hand a legacy interface with no specific 5G capabilities
and using existing wireline session models and protocols.

As mentioned above, a 5G-RG acts as a UE, including the support of NAS. The
5G-RG will thus Register with the network and request establishment of PDU Sessions.
The W-AGF supports an access-specific interface towards the 5G-RG (Y4) and relays
the NAS signaling between 5G-RG and AMF in 5GC (N1 interface in Fig. 16.20).
For FN-RGs the situation is different. The FN-RGs do not support NAS and instead
it will be the W-AGF that acts as UE towards 5GC on behalf of the FN-RG. This
can be thought of as similar to tethering, but in a slightly more sophisticated form.
The W-AGF will thus Register with the 5GC on behalf of the RG, request establishment
of PDU Sessions, etc. in response to the combination of subscription information in the
5GC and FN-RG behavior using existing protocols. The N1 interface is in this case
terminated on the W-AGF (as can be seen in Fig. 16.20).

The main purpose of an RG is to give devices that connect to the RG (e.g. laptops,
tablets, set-top boxes, etc.) access to services in the network (e.g. Internet, voice or TV).

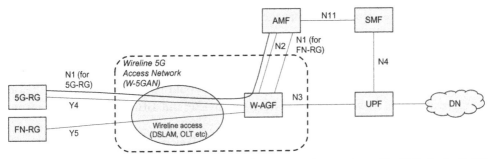

Fig. 16.20 Overall network architecture for wireline access integration with 5GC. (not all NFs
are shown).

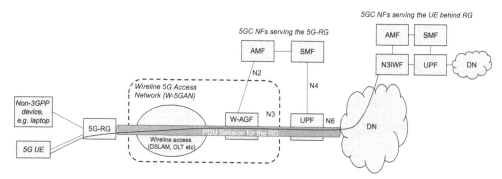

Fig. 16.21 Devices behind the RG.

These devices that are behind the RG (seen from a 5G Core perspective) typically have no 3GPP functionality and also no 5GC subscription or credentials. They just use the connectivity that has been setup by the RG (PDU Session). However, in case the device behind the RG is a 3GPP 5G capable UE, such UE could connect to 5GC via the RG using, e.g. Untrusted non-3GPP access procedures towards a N3IWF in the network. Fig. 16.21 illustrates how devices behind the RG (both 5G capable UEs and other devices) connect via the RG PDU Session.

16.3.7.4 Fixed wireless access and hybrid access
A 5G-RG may also support 3GPP radio access (e.g. NG-RAN) in which case it can connect via radio access towards the 5G Core network. This scenario is often referred to as Fixed Wireless Access (FWA). The 3GPP radio access may, e.g. be used as a fallback if the wireline access does not exist or if it for some reason fails. The 3GPP radio access may however also be used simultaneously with wireline access, e.g. as capacity booster or to load-balance between wireline and wireless access. These scenarios where an RG can use wireline and wireless accesses, either simultaneously or sequentially, are sometimes referred to as Hybrid Access (HA). Fig. 16.22 illustrates a network architecture with Hybrid Access support. Release 16 supports Hybrid Access for 5G-RGs in two ways; either by moving PDU Sessions between wireline and wireless or by using the ATSSS solution for simultaneous multi-access PDU Session connectivity (ATSSS is further described in Section 16.3.8).

16.3.7.5 Conclusions
The resulting architecture and specifications provide a roadmap to convergence and a streamlining of the operator's network, OSS/BSS and inventory of network functions. Most of the artifacts of legacy access network and FN-RG support are confined to the W-AGF in the model such that the end result can be considered a true convergence architecture, and not simply an accumulation of additional functionality unrelated to

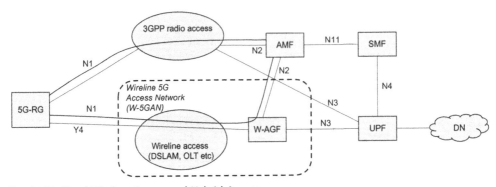

Fig. 16.22 Fixed Wireless Access and Hybrid Access.

the basic functionality of the 5GC. The additional features that have been needed are primarily for home network support and can also be repurposed to industrial, transportation and IoT applications.

16.3.8 Multi-access PDU Sessions
16.3.8.1 Introduction
5GS supports mobility between 3GPP and untrusted non-3GPP access in Rel-15. This is done by moving PDU Sessions between 3GPP access and non-3GPP access. Each PDU Session is thus at a given time only active in one access; either 3GPP access or non-3GPP access. This means that all traffic of that PDU Session is carried over a single common access type and all traffic is moved between accesses at the same time. It can however be beneficial to have a more general solution where, e.g.:
— Steering: Access type (3GPP and non-3GPP) can be selected for each packet flow (e.g. an IP 5-tuple or application) separately, e.g. when a new application or packet flow is started.
— Switching: A packet flow can be moved between access types independently from other packet flows.
— Splitting: A packet flow (e.g. IP 5-tupl) can even be split across both 3GPP access and non-3GPP access simultaneously, on a per-packet basis.
— Steering, Switching and Splitting allows a single PDU Session to use resources from both 3GPP access and non-3GPP access simultaneously to improve the total throughput for a PDU Session.
The Release-16 work on Access Traffic Steering, Switching and Splitting (ATSSS) introduces such possibilities. It should be noted that the mechanisms introduced for ATSSS, and described in this section, only apply to cases where one access is a 3GPP access and the other access is a non-3GPP access. Simultaneous connectivity over different 3GPP access is handled, e.g. using Dual Connectivity (see Chapters 3 and 12) and is not part of ATSSS.

For readers familiar with EPC, it can be noted that the use cases covered by ATSSS are similar to those addressed by NBIFOM in EPC. The solution defined for 5GS is however quite different from NBIFOM. The aim has been to make a better solution in 5G which does not suffer from complexity and Control Plane overhead that is present in the NBIFOM solution. The 5G solution relies more on delegating the user data steering/switching/splitting decisions to UE and UPF and avoids frequent signaling over Control Plane.

16.3.8.2 Multi-access PDU session

A key concept introduced to support ATSSS is the Multi-access PDU Session (MA PDU Session). This is a generalization of the regular "single-access" PDU Session (that we have discussed throughout the other parts of this book), where the MA PDU Session can have simultaneous User Plane resources on 3GPP and non-3GPP access. From a UPF point of view, this means that there are two N3 tunnels available for the PDU Session (and two N9 tunnels in case intermediate UPFs are on the path). From the UE point of view there are two access resources associated to the PDU Session. Fig. 16.23 illustrates the MA PDU Session.

A MA PDU Session is established using the regular PDU Session Establishment procedure but with extra information elements to negotiate Multi-Access aspects. There is also an additional User Plane establishment in order to create the second "leg" of the MA PDU Session. In addition, the URSP rules are extended to indicate to the UE that a MA PDU Session shall be requested instead of a normal "single-access" PDU Session.

16.3.8.3 Steering functionality and performance measurements

Once MA PDU Session has been established and the two User Plane paths are available, there is a need for selecting which path to use. In the down-link direction, the UPF needs to determine which N3 (or N9) tunnel to use for each packet. In the up-link direction the

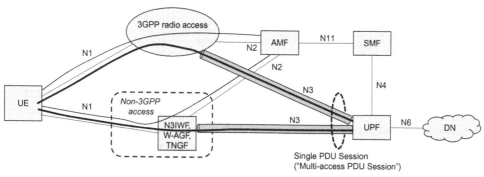

Fig. 16.23 Multi-access PDU Session.

UE needs to determine which access (3GPP or non-3GPP) to use for each packet. This whole procedure involves two basic parts:

— The control of which access to select and when to switch between accesses for each application, based on operator policies and rules as well as performance aspects of each User Plane path.
— The actual handling of User Plane packets between UE and UPF and how the transmitter (UE for up-link and UPF for down-link) splits traffic and then how the receiver (UPF for up-link and UE for down-link) recombines traffic.

We will first have a look at the second aspect, referred to in 3GPP specifications as "steering functionality". There are two "steering functionalities" supported for handling packets between UE and UPF across the multiple accesses:

— Multi-path TCP (MPTCP) with MPTCP client in the UE and MPTCP proxy in the UPF. This option operates on layer 4 and applies to TCP traffic only.
— A lower layer steering functionality referred to as "ATSSS Lower Layer" (ATSSS-LL). This steering functionality operates below the PDU layer and can be applied to steer, switch and split all types of traffic, including TCP traffic, UDP traffic, Ethernet traffic, etc.

A UE may support MPTCP, ATSSS-LL or both.

The MPTCP option is based on IETF RFC 6824, or actually the version that is being worked in IETF targeting to replace RFC 6824 (at the time of writing this book it is still an Internet-Draft; draft-ietf-mptcp-rfc6824bis). When MPTCP is used, separate TCP sub-flows are established between UE and UPF over the different accesses, and TCP traffic is steered, switched and split over these sub-flows.

ATSSS-LL on the other hand does not use any additional protocol between UE and UPF other than what is used for single-access PDU Sessions. Down-link packets in UPF are thus sent over one of the two N3 GTP-U tunnels, and up-link packets in the UE are sent over either 3GPP access or non-3GPP access as normal. Since there is no specific function in the receiving side that recombines and re-orders packets over the two paths and ensures, e.g. that packets arrive in order, ATSSS-LL should only be used for packet flow switching, not for packet flow splitting.

We now take a look at the first aspect, the control of the ATSSS feature. The multi-access capability is controlled by PCF, by inserting multi-access steering information in the PCC rule. The PCC rule can contain information whether multi-access communication shall be used, which steering functionality to apply (ATSSS-LL or MPTCP), and also what "steering mode" to apply. The "steering mode" determines how the traffic matching the PCC rule should be distributed across 3GPP and non-3GPP accesses. The following steering modes are supported:

— Active-Standby: Used to steer traffic on one access (the active access) when this access is available, and to switch the traffic to the other access (the standby access) when the active access becomes unavailable.

- Smallest Delay: Used to steer traffic to the access that is determined to have the smallest Round–Trip Time (RTT). UE and UPF measure the RTT in order to determine which access has the lowest RTT.
- Load–Balancing: Used to split traffic across both accesses according to a percentage for how much traffic that should be sent over 3GPP access and over non-3GPP access.
- Priority-based: Used to steer all the traffic matching a PCC rule to the high priority access, until this access is determined to be congested. In this case, the traffic is sent also to the low priority access, i.e. the traffic is split over the two accesses.

The SMF will take the information in the PCC rule into account and generate an "ATSSS rule" that is sent to the UE, so that the UE can determine how to steer/switch/split the up-link traffic. The ATSSS rule contains information on steering functionality and steering mode corresponding to a packet filter or application identity. The SMF also generates corresponding N4 rules to UPF so that UPF can determine how to steer/switch/split the down-link traffic.

CHAPTER 17

Future outlook

With the extensive work resulting in the specification of the 5G Core, 3GPP has laid a foundation for a network architecture that can be expected to play a major role for communication services for years to come. One of the fundamental principles behind the creation of the 5G Core Network Architecture has been to apply a long-term perspective and to design a future-proof architecture that can support future generations of radio access networks beyond 5G NR. Initial 5G network deployments have focused on supporting mobile broadband services or fixed wireless access services as an alternative to fixed broadband connections, but many more use cases are being implemented and will be rolled out over the next few years. Various studies, such as the joint study conducted by Arthur D. Little and Ericsson (A.D. Little, 2017), points at the significant values 5G technologies can bring to various industry segments within for example the transportation, manufacturing and energy sectors.

Work on 3GPP specifications following Release-15 are focusing on adding capabilities for more advanced use cases, and to further enhance the new service-based architecture. 5G technology in general – and the 5G Core Network architecture specifically – are well positioned to play a significant role for a wide range of applications across all parts of the society.

The authors are convinced that for 5G to achieve wide-scale adoption, it is very important that the industry align the evolution of 5G technology towards clear commercial values, both for industrial applications as well as more traditional consumer or business services. These services may rely on a set of different access technologies, not only 5G/NR, and on efficient support in the 5G Core Network for simultaneously serving devices that both use access networks with quite different capabilities and vary in numbers quite dramatically. Requirements for a network collecting data from millions of small cheap sensor devices will naturally be different compared to a network that serves a few devices built-into robots that form one part of a business-critical manufacturing process.

Given the wide range of use cases that create different requirements on network configurations, it is also important to ensure a very cost-efficient way of deploying and operating networks. This calls for network infrastructure vendors to design innovative solutions with a very high degree of flexibility and automation support. Network slicing capabilities will be key to tailor network configurations to different services, and machine learning features can be expected to play a major role both for tracking the performance as well as for automatically tuning the network configurations to optimize capacity, end user experiences and overall performance of the networks.

In parallel with the evolution of technology itself, business models and the overall economic landscape will also evolve. As the range of target markets and customers is significantly broadened, the role of the service provider may change over time, adding new business-to-business service offerings to the portfolio. New actors could also emerge on the market, as they attempt to exploit the disruptive nature of some of the potential use cases. Operators will benefit from providing services beyond pure connectivity through building higher value services and solutions, and tightly integrating with their customers' networks.

Summing up, with the creation of the 5G specifications, the industry has taken an important step toward creating a technology that can truly realize the vision of a society where everything that benefits from being connected is connected. For this to happen, the business values and market drivers need to be clearly understood and applied to direct the future evolution of 5G technology.

We look forward to the exciting times that lie ahead.

References

3GPP SP-160455, 3GPP Tdoc: SP-160455, "5G Architecture Options", Deutsche Telekom, 2016.

3GPP TR 22.804, 3GPP Technical Report 22.804, "Study on Communication for Automation in Vertical domains (CAV)".

3GPP TR 23.714, 3GPP Technical Report 23.714, "Study on control and user plane separation of EPC nodes".

3GPP TR 23.799, 3GPP Technical Report 23.799, "Study on Architecture for Next Generation System".

3GPP TR 38.913, 3GPP Technical Report 38.913, "Study on Scenarios and Requirements for Next Generation Access Technologies".

3GPP TS 22.104, 3GPP Technical Specification 22.104, "Service requirements for cyber-physical control applications in vertical domains".

3GPP TS 22.261, 3GPP Technical Specification 22.261, "Service requirements for next generation new services and markets".

3GPP TS 23.003, 3GPP Technical Specification 23.003, "Numbering, addressing and identification".

3GPP TS 23.041, 3GPP Technical Specification 23.041, "Technical realization of Cell Broadcast Service (CBS)".

3GPP TS 23.122, 3GPP Technical Specification 23.122, "Technical Specification Group Core Network; NAS Functions related to Mobile Station (MS) in idle mode".

3GPP TS 23.203, 3GPP Technical Specification 23.203, "Policy and charging control architecture".

3GPP TS 23.214, 3GPP Technical Specification 23.214, "Architecture for Control and User Plane Separation for EPC nodes".

3GPP TS 23.287, 3GPP Technical Specification 23.287, "Architecture enhancements for V2X services".

3GPP TS 23.288, 3GPP Technical Specification 23.288, "Architecture enhancements for 5G System (5GS) to support network data analytics services".

3GPP TS 23.401, 3GPP Technical Specification 23.401, "General Packet Radio Service (GPRS) enhancements for Evolved Universal Terrestrial Radio Access Network (E-UTRAN) access".

3GPP TS 23.501, 3GPP Technical Specification 23.501, "System architecture for the 5G System (5GS)".

3GPP TS 23.502, 3GPP Technical Specification 23.502, "Procedures for the 5G System (5GS)".

3GPP TS 23.503, 3GPP Technical Specification 23.503, "Policy and charging control framework for the 5G System (5GS)".

3GPP TS 24.007, 3GPP Technical Specification 24.007, "Mobile radio interface signalling layer 3; General Aspects".

3GPP TS 24.501, 3GPP Technical Specification 24.501, "Non-Access-Stratum (NAS) protocol for 5G System (5GS); Stage 3".

3GPP TS 24.502, 3GPP Technical Specification 24.502, "Access to the 3GPP 5G Core Network (5GCN) via non-20GPP access networks".

3GPP TS 28.530, 3GPP Technical Specification 28.530, "Management and orchestration; Concepts, use cases and requirements".

3GPP TS 29.244, 3GPP Technical Specification 29.244, "Interface between the Control Plane and the User Plane nodes".

3GPP TS 29.281, 3GPP Technical Specification 29.281, "General Packet Radio System (GPRS) Tunnelling Protocol User Plane (GTPv1-U)".

3GPP TS 29.303, 3GPP Technical Specification 29.303, "DNS Procedures for UP Function Selection".

3GPP TS 29.500, 3GPP Technical Specification 29.500, "5G System; Technical Realization of Service Based Architecture; Stage 3".

3GPP TS 29.501, 3GPP Technical Specification 29.501, "5G System; Principles and Guidelines for Services Definition; Stage 3".

3GPP TS 29.518, 3GPP Technical Specification 29.518, "5G System; Access and Mobility Management Services; Stage 3".

3GPP TS 29.571, 3GPP Technical Specification 29.571, "5G System; Common Data Types for Service Based Interfaces; Stage 3".

3GPP TS 29.891, 3GPP Technical Specification 29.891, "5G System—Phase 1 CT WG4 Aspects".

3GPP TS 33.126, 3GPP Technical Specification 33.126, "Lawful Interception requirements".

3GPP TS 33.210, 3GPP Technical Specification 33.210, "3G security; Network Domain Security (NDS); IP network layer security".

3GPP TS 33.401, 3GPP Technical Specification 33.401, "3GPP System Architecture Evolution (SAE); Security architecture".

3GPP TS 33.402, 3GPP Technical Specification 33.402, "3GPP System Architecture Evolution (SAE); Security aspects of non-3GPP accesses".

3GPP TS 33.501, 3GPP Technical Specification 33.501, "Security architecture and procedures for 5G System".

3GPP TS 36.300, 3GPP Technical Specification 36.300, "Evolved Universal Terrestrial Radio Access (E-UTRA) and Evolved Universal Terrestrial Radio Access Network (E-UTRAN); Overall description; Stage 2".

3GPP TS 37.324, 3GPP Technical Specification 37.324, "Service Data Adaptation Protocol (SDAP) specification".

3GPP TS 37.340, 3GPP Technical Specification 37.340, "NR; Multi-connectivity; Overall description; Stage-2".

3GPP TS 38.101-1, 3GPP Technical Specification 38.101-1, "NR; User Equipment (UE) radio transmission and reception; Part 1: Range 1 Standalone".

3GPP TS 38.101-2, 3GPP Technical Specification 38.101-2, "NR; User Equipment (UE) radio transmission and reception; Part 2: Range 2 Standalone".

3GPP TS 38.300, 3GPP Technical Specification 38.300, "NR; Overall description; Stage-2".

3GPP TS 38.304, 3GPP Technical Specification 38.304 "NR; User Equipment (UE) procedures in Idle mode and RRC Inactive state".

3GPP TS 38.321, 3GPP Technical Specification 38.321, "NR; Medium Access Control (MAC); Protocol specification".

3GPP TS 38.401, 3GPP Technical Specification 38.401, "NG-RAN; Architecture description".

3GPP TS 38.413, 3GPP Technical Specification 38.413, "NG-RAN; NG Application Protocol (NGAP)".

3GPP TS 38.423, 3GPP Technical Specification 38.423, "NG-RAN, Xn application protocol (XnAP)".

Dahlman, et al., 2018. 5G NR: The Next Generation Wireless Access Technology. Elsevier.

Fielding, R., 2000. Architectural Styles and the Design of Network-Based Software Architectures (PhD thesis). University of California, Irvine.

IEEE 802.1AS-Rev/D7.3, IEEE Std 802.1AS-Rev/D7.3, August 2018, "IEEE Standard for Local and metropolitan area networks—Timing and Synchronization for Time-Sensitive Applications".

IEEE P802.1, IEEE P802.1Qcc, "Standard for Local and metropolitan area networks—Bridges and Bridged Networks—Amendment: Stream Reservation Protocol (SRP) Enhancements and Performance Improvements".

ITU OB 1156, ITU Operational Bulletin No. 1156, International Telecommunication Union (ITU), Standardization Bureau (TSB), "Operational Bulletin No. 1156".

ITU-R Recommendation M, ITU-R Recommendation M-2083, "IMT Vision—Framework and overall objectives of the future development of IMT for 2020 and beyond".

ITU-R TR M.2410-0, ITU-R Technical Report M.2410-0, "Minimum requirements related to technical performance for IMT-2020 radio interface(s)".

Little, A.D., 2017. The 5G Business Potential, Ericsson Report 2017.

MEF 6.4, Metro Ethernet Forum Specification 6.4.

Olsson, et al., 2014. EPC and 4G Packet Networks—Driving the Mobile Broadband Revolution. Elsevier.

RFC 2784, IETF RFC 2784, "Generic Routing Encapsulation (GRE)".

RFC 2890, IETF RFC 2890, "Key and Sequence Number Extensions to GRE".

RFC 3748, IETF RFC 3748, "Extensible Authentication Protocol (EAP)".

RFC 3758, IETF RFC 3758, "Stream Control Transmission Protocol (SCTP) Partial Reliability Extension".

RFC 4187, IETF RFC 4187, "Extensible Authentication Protocol Method for 3rd Generation Authentication and Key Agreement (EAP-AKA)".

RFC 4191, IETF RFC 4191, "Default Router Preferences and More-Specific Routes".

RFC 4301, IETF RFC 4301, "Security Architecture for the Internet Protocol".

RFC 4303, IETF RFC 4303, "IP Encapsulating Security Payload (ESP)".

RFC 4304, IETF RFC 4304, "IP Authentication Header".

RFC 4555, IETF RFC 4555, "IKEv2 Mobility and Multihoming Protocol (MOBIKE)".

RFC 4861, IETF RFC 4861, "Neighbor Discovery for IP version 6 (IPv6)".

RFC 4960, IETF RFC 4960, " Stream Control Transmission Protocol".

RFC 5216, IETF RFC 5216, "The EAP-TLS Authentication Protocol".

RFC 5246, IETF RFC 5246, "The Transport Layer Security (TLS) Protocol Version 1.2".

RFC 5448, IETF RFC 5448, "Improved Extensible Authentication Protocol Method for 3rd Generation Authentication and Key Agreement (EAP-AKA')".

RFC 6347, IETF RFC 6347, "Datagram Transport Layer Security Version 1.2".

RFC 6749, IETF RFC 6749, "The OAuth 2.0 Authorization Framework".

RFC 7296, IETF RFC 7296, "Internet Key Exchange Protocol Version 2 (IKEv2)".

RFC 7515, IETF RFC 7515, "JSON Web Signature (JWS)".

RFC 7516, IETF RFC 7516, "JSON Web Encryption (JWE)".

RFC 7540, IETF RFC 7540, "Hypertext Transfer Protocol Version 2 (HTTP/2)".

RFC 8259, IETF RFC 8259, "The JavaScript Object Notation (JSON) Data Interchange Format".

RFC 8446, IETF RFC 8446, "The Transport Layer Security (TLS) Protocol Version 1.3".

SNS Telecom and IT, 2018, SON (Self-Organizing Networks) in the 5G Era: 2019—2030—Opportunities, Challenges, Strategies & Forecasts.

Abbreviations

5GC	5G Core Network
5GLAN	5G Local Area Network
5GS	5G System
5G-AN	5G Access Network
5G-EIR	5G-Equipment Identity Register
5G-GUTI	5G Globally Unique Temporary Identifier
5G-BRG	5G Broadband Residential Gateway
5G-CRG	5G Cable Residential Gateway
5G-RG	5G Residential Gateway
5G-S-TMSI	5G S-Temporary Mobile Subscription Identifier
5QI	5G QoS Identifier
AF	Application Function
AMBR	Aggregate Maximum Bit Rate
AMF	Access and Mobility Management Function
ANDSF	Access Network Discovery and Selection Functionality
APN	Access Point Name
ARP	Allocation and Retention Priority
AS	Access Stratum
ATM	Asynchronous Transfer Mode
ATSSS	Access Traffic Steering, Switching, Splitting
ATSSS-LL	ATSSS Low-Layer
AUSF	Authentication Server Function
BSF	Binding Support Function
CAG	Closed Access Group
CAPIF	Common API Framework for 3GPP northbound APIs
CBC	Cell Broadcast Center
CBE	Cell Broadcast Entity
CHF	Charging Function
CP	Control Plane
CSCF	Call Session Control Function
DL	Downlink
DN	Data Network
DNAI	DN Access Identifier
DNN	Data Network Name
DRB	Data Radio Bearer
DRX	Discontinuous Reception
ePDG	evolved Packet Data Gateway
EBI	EPS Bearer Identity
eMBB	enhanced Mobile Broadband
EN-DC	E-UTRAN New Radio-Dual Connectivity
EPC	Evolved Packet Core
EPS	Evolved Packet System
E-UTRAN	Evolved Universal Terrestrial Radio Access Network
FAR	Forwarding Action Rule

FDD	Frequency Division Duplex
FN-BRG	Fixed Network Broadband RG
FN-CRG	Fixed Network Cable RG
FN-RG	Fixed Network RG
FQDN	Fully Qualified Domain Name
GBR	Guaranteed Bit Rate
GFBR	Guaranteed Flow Bit Rate
GMLC	Gateway Mobile Location Centre
GPRS	General Packet Radio Services
GPS	Global Positioning System
GPSI	Generic Public Subscription Identifier
GSM	Global System for Mobile Communications (2G)
GTP-U	GPRS Tunneling Protocol for User Plane
GUAMI	Globally Unique AMF Identifier
HPLMN	Home PLMN
HR	Home Routed (roaming)
HSS	Home Subscriber Server
HTTP	Hypertext Transfer Protocol
IETF	Internet Engineering Task Force
I-SMF	Intermediate SMF
IMS	IP Multimedia Subsystem
KPI	Key Performance Indicator
LADN	Local Area Data Network
LBO	Local Break Out (roaming)
LMF	Location Management Function
LRF	Location Retrieval Function
LTE	Long Term Evolution (4G)
MCC	Mobile Country Code
MCX	Mission Critical Service
MCPTT	Mission Critical Push To Talk
MDBV	Maximum Data Burst Volume
MFBR	Maximum Flow Bit Rate
MICO	Mobile Initiated Connection Only
MIMO	Multiple-Input-Multiple-Output
mIoT	Massive Internet of Things
MME	Mobility Management Entity
MNC	Mobile Network Code
MPS	Multimedia Priority Service
MPTCP	Multi-Path TCP Protocol
MR-DC	Multi RAT Dual Connectivity
MRU	Mobility Registration Update
N3IWF	Non-3GPP InterWorking Function
NaaS	Network as a Service
NAI	Network Access Identifier
NAS	Non Access Stratum
NAT	Network Address Translation
NEF	Network Exposure Function
NF	Network Function
NGAP	Next Generation Application Protocol
NG-RAN	Next Generation Radio Access Network

NID	Network identifier
NPN	Non-Public Network
NR	New Radio
NRF	Network Repository Function
NSI	Network Slice Instance
NSI ID	Network Slice Instance Identifier
NSSAI	Network Slice Selection Assistance Information
NSSF	Network Slice Selection Function
NSSP	Network Slice Selection Policy
NWDAF	Network Data Analytics Function
O&M	Operation and Maintenance
OFDM	Orthogonal Frequency-Division Multiplexing
PCF	Policy Control Function
PDB	Packet Delay Budget
PDCP	Packet Data Convergence Protocol
PDP	Packet Data Protocol
PDR	Packet Detection Rule
PDU	Protocol Data Unit
PEI	Permanent Equipment Identifier
PER	Packet Error Rate
PFD	Packet Flow Description
PGW	Packet Data Network Gateway
PGW-C	PDN Gateway CP
PLMN	Public Land Mobile Network
PPD	Paging Policy Differentiation
PPF	Paging Proceed Flag
PPI	Paging Policy Indicator
PSA	PDU Session Anchor
QCI	QoS Class Identifier
QFI	QoS Flow Identifier
QoE	Quality of Experience
QoS	Quality of Service
RA	Registration Area
(R)AN	(Radio) Access Network
RFSP	RAT/Frequency Selection Priority
RG	Residential Gateway
RQA	Reflective QoS Attribute
RQI	Reflective QoS Indication
RRC	Radio Resource Control
RSN	Redundancy Sequence Number
SA NR	Standalone New Radio
SBA	Service Based Architecture
SBG	Session Border Gateway
SBI	Service Based Interface
SCP	Service Communication Proxy
SCTP	Stream Control Transmission Protocol
SD	Slice Differentiator
SDAP	Service Data Adaptation Protocol
SDN	Software Defined Networking
SEAF	Security Anchor Functionality

SEPP	Security Edge Protection Proxy
SGSN	Serving GPRS Support Node
SGW	Serving Gateway
SIP	Session Initiation Protocol
SLA	Service Level Agreement
SM	Session Management
SMF	Session Management Function
SMS	Short Message Service
SMSF	Short Message Service Function
SN	Sequence Number
SNPN	Stand-alone Non-Public Network
S-NSSAI	Single Network Slice Selection Assistance Information
SSC	Session and Service Continuity
SSCMSP	Session and Service Continuity Mode Selection Policy
SST	Slice/Service Type
SUCI	Subscription Concealed Identifier
SUPI	Subscription Permanent Identifier
TA	Tracking Area
TAI	Tracking Area Identity
TCP	Transmission Control Protocol
TDD	Time Division Duplex
TMSI	Temporary Mobile Subscription Identifier
TNAN	Trusted Non-3GPP Access Network
TNAP	Trusted Non-3GPP Access Point
TNGF	Trusted Non-3GPP Gateway Function
TNL	Transport Network Layer
TNLA	Transport Network Layer Association
TSC	Time Sensitive Communication
TSN	Time Sensitive Networking
TSP	Traffic Steering Policy
UDM	Unified Data Management
UDP	User Datagram Protocol
UDR	Unified Data Repository
UDSF	Unstructured Data Storage Function
UL	Uplink
UL CL	Uplink Classifier
UPF	User Plane Function
URLLC	Ultra Reliable Low Latency Communication
URRP-AMF	UE Reachability Request Parameter for AMF
URSP	UE Route Selection Policy
V2X	Vehicle-to-Everything
VID	VLAN Identifier
VLAN	Virtual Local Area Network
VPLMN	Visited PLMN
W-5GAN	Wireline 5G Access Network
W-5GBAN	Wireline BBF Access Network
W-5GCAN	Wireline 5G Cable Access Network
W-AGF	Wireline Access Gateway Function
WCDMA	Wideband Code Division Multiple Access (3G)
WLAN	Wireless Local Area Network

Index

Note: Page numbers followed by *f* indicate figures and *t* indicate tables.

A

Access and mobility management function (AMF), 287–288, 293–300, 295*f*, 406–407, 412–414
  charging and policy control, 224
  HTTP Request
    SMF, 356–357
    UDM, 356
  Namf_Communication service, 293–298
  Namf_EventExposure service, 298–299
  Namf_Location service, 300
  Namf_MT service, 299–300
  N2 management, 152
Access Network Discovery and Selection Policy (ANDSP), 224–227
Access Traffic Steering, Switching and Splitting (ATSSS), 460
Application Function (AF), 43, 83, 131–132, 198, 292
Application Service Provider (ASP), 227, 241, 309
Application Specific Policy (ASP) identifier, 229
Architecture
  automotive use cases
    EPS, V2X in, 450–454
    5GS for NR, V2X in, 454–456
  core components, 26–28
  data storage, 59
  device positioning services, 48–49
  enhanced network slicing
    3GPP Release-16, 432
    5GS, 432
    Network Slice-Specific Authentication and Authorization (SSAA), 433–434
  enhanced SMF/UPF deployment flexibility
    deployment topology, 434, 435*f*
    Intermediate SMF (I-SMF), 435–436
    Intermediate UPF (I-UPF), 434
    Inter-PLMN mobility, 436
    PDU Session, 434–435
    PLMN, 434
    Release-15 5GC, 434
    selective traffic routing, 436–437, 437*f*
  enhancing service based architecture
    context transfer, 432
    indirect communication and delegated discovery, 431–432
    NF Sets and NF Service Sets, 432
  evolution and disruption, 15–16
  fixed access, integration of
    drivers, 456–457
    fixed wireless access and hybrid access, 459
    migration considerations, 457–458
    network architecture, 458–459
  5GC interworking, EPC
    with N26 interface, 38–40
    without N26 interface, 40
  5G core perspectives, 19–22
  5G LAN-type services
    group management, 5G Virtual Network (5G VN), 439
    Rel-15 5G System, 438–439
    User Plane handling, 5G Virtual Network (5G VN), 439–440
  3GPP, 16–19
  5G radio networks
    advanced antenna techniques, 66–67
    base station internal architecture, 71–72
    3GPP specifications Release-15, 59
    5G targets, 61–63
    mobile network fundamentals, 60–61
    New Radio (NR), 59, 64–71
  industrial IOT applications, 438
  messaging services, 44
    IP, 44–45
    NAS solution, 45–46
  mobile devices and radio networks, core network to, 28–30
  mobility and data connectivity, 30–35
  modeling, 105
  multi-access PDU Sessions
    Access Traffic Steering, Switching and Splitting (ATSSS), 460
    3GPP access and non-3GPP access, 460–461
    steering functionality and performance measurements, 461–463
    URSP rules, 461

Architecture *(Continued)*
  network automation, 432
  Network Data Analytics Function (NWDAF),
      49–50
  network information, exposure of, 46–48
  network slicing, 54–55
  non-3GPP access networks, 52–54
  Non-Public Network (NPN)
      PLMN services, SNPN, 442–443
      public network integrated NPN, 443
      stand-alone Non-Public Networks, 441–442
  policy control and charging, 35–37
  Public Warning System (PWS), 50–52
  roaming, 55–59
  service-based architecture (SBA)
      concept of, 22
      HTTP REST interfaces, 22–23
      registration and discovery, 23–26
  Time Sensitive Networks, in 5GS, 448–450
  ultra-reliable low-latency communication
      (URLLC)
      cyber-physical systems, 443
      end to end redundant user plane paths, dual
          connectivity based, 444–445
      low latency eMBB applications, 444
      N3/N9 interfaces, redundant transmission on,
          447
      redundant user plane paths, multiple UEs,
          446–447
      3GPP, 444
      transport layer, redundant transmission at,
          448
  voice services
      Evolved Packet System (EPS) fallback,
          41–42
      voice-over-LTE, 41
      voice-over-NR, 42–43
Artificial Intelligence, 12–13
AUSF. *See* Authentication server function (AUSF)
Authentication, 186–188
Authentication credential Repository and
      Processing Function (ARPF), 177
Authentication header (AH), 383–385
Authentication Server Function (AUSF), 177, 290,
      322–324
  Nausf_SoRProtection service, 323–324
  Nausf_UEAuthentication service, 323
  Nausf_UPUProtection service, 324

Authentication Vector (AV), 187
Automation
  network, 432
  new technologies, 5G Drivers, 12–13

B
BAR. *See* Buffering action rule (BAR)
Billing domain (BD), 244
Binding Support Function (BSF), 43
Buffering action rule (BAR), 366, 374, 375*t*

C
Call flows
  deregistration, 399–400
  inter-NG-RAN handover, 409
      N2-based, 412–416
      Xn-based, 409–411
  PDU session establishment, 407–409
  registration, 396–398
  service request, 400
      network triggered, 402–404
      UE triggered, 401–402
  UE configuration update
      access and mobility management related
          parameters, 404–406
      for transparent UE policy delivery, 406
  untrusted non-3GPP accesses procedures, 424
      PDU session establishment, 426–428
      registration procedure, 424–426
Canary testing (small-scale testing), 11
Charging and policy control
  access and mobility management related policies,
      222–223
  AMF, 224
  charging data records (CDR), 243
  converged charging system (CCS), 243–244
  direct debiting, 243
  end-users/subscribers, 242
  future background data transfer, negotiation for,
      228–229
  home routed User Plane anchor, 220, 221*f*
  network status analytics, 228
  non-Session Management related policy control,
      218
  offline charging data, 242–243
  packet flow descriptions, management of,
      227–228

PCC roaming architecture, local User Plane
    anchor, 220, 220f
PCF deployment, 222
PLMNs, 220
Policy Control Request Triggers, 224
service based architecture model, 243–244
session management related policy control,
    218–219, 219f
    application detection, 237
    concepts, 229–231
    event reporting, PCF, 241–242
    policy decisions and the PCC rule, 231–232
    QoS flow binding, 235–236
    service data flow detection, 236
    SMF related policy authorization request
        triggers, 236
    spending limits, policy decisions based on,
        239–240
    sponsored connectivity, 241
    traffic steering control, 237–238
    usage monitoring control, 238–239
    use case, application authorization, 232–235
UE policy control
    Access Network Discovery and Selection
        Policy (ANDSP), 224–227
    UE Route Selection Policy (URSP), 224–227
Unit Reservation, 243
VPLMN, 221
Charging Data Function (CDF), 244
Charging Gateway Function (CGF), 244
Charging Trigger Function (CTF), 244
Ciphering, 173, 179
Cloud deployment, 8
Cloud-native strategy, 10–11
Container as a Service (CaaS), 11
Containers, 11–12
Control and user plane separation (CUPS), 89–101
    5GC Session Management, 126
    and N4 interface
        Packet Forwarding Control Protocol (PFCP),
            124
        selective activation and deactivation, UP
            connections, 125–126
        UPF discovery and selection, 124–125
    selective traffic routing, DN
        IPv6 multi-homing, 130–131
        PSA UPF, 129
        Up-link Classifier (UL CL), 129–130

service and session continuity (SSC) modes
    SSC mode 1, 127
    SSC mode 2, 128
    SSC mode 3, 128
traffic routing, application function on, 131–132
UPF-reselection, 127

D
Datagram Transport Layer Security (DTLS) version
    1.2 protocol, 362
Data network (DN), connectivity service to
    Session Management
        basic PDU Session connectivity, 111
        multiple PDU Sessions, 113
        PDU Session properties, 114
        transport network, PDU Session and
            application traffic relation, 112–113
Deep Packet Inspection (DPI) filters, 232
DevOps, 10–11
Drivers, 5G
    new technologies
        automation, 12–13
        cloud-native strategy, 10–11
        containers, 11–12
        microservices, 12
        virtualization, 9–10
    use cases, 7–9
Dual connectivity (DC)
    E-UTRA Cell groups, 265–266
    Master Cell Group (MCG), 266–267
    Master-Node-terminated (MN-terminated), 266
    multiple Receive/Transmit (Rx/Tx), 265
    Multi-Radio DC (MR-DC), 265–266, 267f
        E-RABs, 274–277
        MR-DC bearers, 274–277
        QoS flows, 274–277
        RAN perspective, 272–274
        subscription, 274–277
        UE perspective, 272–274
    Multi-RAT Dual Connectivity, 268–272
    "Options 1 to 8", 266
    RAN nodes, 265
    Secondary Cell Group (SCG), 265
    secondary RAN node handling, mobility and
        session management, 278–282
    security, 282–283
    User Data Volume traversing via SN, 283–285
Duplicate Address Detection (DAD), 117

E

EAP. *See* Extensible authentication protocol (EAP)
Encapsulated security payload (ESP), 196, 383–385
Encapsulation/tunnel protocol, 392, 393*f*
Enhanced Dedicated Core Networks ((e)DECOR), 84–89
Equipment Identity Register (EIR), 358–360, 398, 399*f*
ESP. *See* Encapsulated security payload (ESP)
Evolved Packet Core (EPC), 5G
   control and user plane separation (CUPS), 89–101
   core EPS architecture, LTE, 73, 74*f*
   DEdicated CORE networks (DECOR), 75
   (enhanced) Dedicated Core Networks ((e) DECOR), 84–89
   functions
      control-plane aspects, 80–81
      default and dedicated bearers, 82
      EPS bearer, *E*-UTRAN access, 77, 81–82
      mobility management, 78–80
      PDN GW (P-GW), 78
      policy control and charging, 83–84
      QoS, 81
      Serving GW (S-GW), 78
      session management, 80
      simplified EPS architecture, 77, 77*f*
      subscription, 78–79
      3GPP radio access networks, 77
      user-plane aspects, 82–83
   MBB, 75
   MME and Serving/PDN GW selection path, 75, 76*f*
   non-Stand-Alone Architecture (NSA), 73
   radio access network (RAN), 75
   simplified EPC, 73, 74*f*
Evolved Packet System (EPS), 41–42, 181–182, 203
   fallback procedure, 422–424
   interworking with N26, 416–422
      EPS to 5GS handover, 418–420
      EPS to 5GS idle mode mobility, 421–422
      5GS to EPS handover, 417–418
      5GS to EPS idle mode mobility, 420–421
Extensible authentication protocol (EAP), 181–183, 379–381

F

FAR. *See* Forwarding action rule (FAR)
5G core (5GC) network, 1, 294*f*
   access and mobility management function, 287–288
   application function, 292
   authentication server function, 290
   core requirements, 4
   equipment identity registry, 290
   location management function, 292–293
   network data analytics function, 291–292
   network exposure function, 291
   network repository function, 289
   network slice selection function, 291
   non-3GPP inter working function, 292
   policy control function, 290–291
   security edge protection proxy, 292
   session management function, 288
   short message service function, 292
   unified data management function, 289–290
   unified data repository, 290
   unstructured data storage function, 290
   user plane function, 288–289
5G equipment identity registry (5G-EIR), 290, 327–328, 328*f*
5G Globally Unique Temporary Identifier (5G-GUTI), 108
5G mobility management (5GMM), 138, 337–340
5G networks
   network access security
      flexibility, 5GS, 176–177
      5G AKA based primary authentication, 183–186
      in 5GS, 178–179
      key derivation and key hierarchy, 188–190
      logical architecture for, 177–178, 177*f*
      NAS security, 190
      security entities, 177–178
      USIM content, Steering of Roaming, 191
   network deployments, 2
   operators, 2
   radio networks
      advanced antenna techniques, 66–67
      architecture
         advanced antenna techniques, 66–67
         base station internal architecture, 71–72
         3GPP specifications Release-15, 59
         5G targets, 61–63

mobile network fundamentals, 60–61
    new radio (NR), 59, 64–71
    base station internal architecture, 71–72
    5G targets, 61–63
    mobile network fundamentals, 60–61
    new radio (NR), 59, 64–71
security domains
    application domain security, 175
    network access security, 174–175
    Network Functions (NFs), 175
    SBA domain security, 176
    user domain security, 175
    visibility and configurability, 176
security requirements, 172
services, 172–174
3GPP release 15 and 16, 2–4
wireless communications, 1
5G non-access stratum (5G NAS), 337–338
    5G mobility management, 338–340
    5G session management, 340
    message structure, 341–343
Fixed-mobile convergence, 8
Forwarding action rule (FAR), 366, 369, 370–371*t*

G
Generic Public Subscription Identifier (GPSI), 109
Generic routing encapsulation (GRE), 392
    delivery protocol, 393, 393*f*
    packet format, 394
    payload packet and payload protocol, 392, 393*f*
    tunnel protocol, 392, 393*f*
Globally Unique AMF ID (GUAMI), 109
GPRS tunneling protocol control-plane (GTP-C), 378
GPRS tunneling protocol user plane (GTP-U), 378
GRE. *See* Generic routing encapsulation (GRE)
5G session management (5GSM), 337, 340–341
GSM (2G), 1, 15, 61
GTP-U. *See* GPRS tunneling protocol user plane (GTP-U)

H
Handshake protocol, TLS, 360–362
HTTP. *See* Hypertext transfer protocol (HTTP)
Hyperscalers, 10
Hypertext transfer protocol (HTTP), 347–348
    documented versions, 348, 408
    exchange, 349*f*

interface definition language, 357–360
    methods, 350–351
    principles, 348–349
    protocol format, 353–355
    RESTful design, 351–353
    serialization protocol, 355–357
    uniform resource identifier, 349–350

I
Identifiers, 107–109
IDL. *See* Interface definition language (IDL)
IKE. *See* Internet key exchange (IKE)
Industry digitalization, 9
Integrity protection, 173, 179
Interface definition language (IDL), 357–360
Internet key exchange (IKE), 196, 385–387
Internet Security Association and Key Management
        Protocol (ISAKMP) framework, 386
IP Multimedia Subsystem (IMS), 41, 230
IP security (IPSec), 382–383
    encapsulated security payload and authentication
        header, 383–385
    IKEv2 mobility and multi-homing, 386–387
    internet key exchange, 385–386
ISAKMP. *See* Internet Security Association and
        Key Management Protocol (ISAKMP)
        framework

J
JavaScript Object Notation (JSON), 356
JSON Web Encryption (JWE), 195–196
JSON Web Signatures (JWS), 195–196

K
Key derivation function (KDF), 191–192

L
Law Enforcement Agencies (LEA), 198–199
LMF. *See* Location management function (LMF)
Local Area Network (LAN), 438–440
    5G Virtual Network (5G VN)
        group management, 439
        User Plane handling, 439–440
    Rel-15 5G System, 438–439
Location management function (LMF), 292–293,
        331–332, 331*f*
LTE (4G), 1, 15

M

Machine Learning, 12–13
Machine-to-machine communications services, 4
Master Cell Group (MCG), 265–267, 272–273
Mean Time Between Failures (MTBF), 11
Microservices, 12
MOBIKE protocol, 387
Mobile Broad Band (MBB), 1
Mobile Country Code (MCC), 179–180, 183–185, 441–442
Mobile devices, 28–30
Mobile Network Code (MNC), 179–180, 183–185, 441–442
Mobility management
  connectivity establishment
    cellular connected mode mobility, 143
    network discovery and selection, 138–140
    registration and mobility, 140–142
  5GS related functions, 137–138
  interworking with EPC
    and 5GC, 162, 163f
    using 3GPP access, 163–169
  NAS message types for, 339–340t
  N2 management
    AMF management, 152
    RAN optimizations, 5GC assistance for, 152–153
    service area and mobility restrictions, 153–157
  non-3GPP aspects, 161–162
  overload, control of
    control channel resources, congestion in, 158
    random access channel (RACH) resources, congestion in, 158
    release/reject UE RRC connection, 158–159
    severe and uncontrollable congestion, 159
    unified access control (UAC), 159–160
  principles, 137
  procedures, 137–138
  reachability
    mobile Initiated Connection Only (MICO) mode, 144
    paging, 144
    UE's reachability and location, 144–146
  RRC Inactive, 146–149

N

NEF. See Network exposure function (NEF)
Network access security, 5G system
  flexibility, 176–177

  5G AKA based primary authentication, 183–186
  in 5GS, 178–179
  key derivation and key hierarchy, 188–190
  logical architecture for, 177–178, 177f
  NAS security, 190
  security entities, 177–178
  USIM content, Steering of Roaming, 191
Network Address Translators (NATs), 115
"Network as a Service" (NaaS) business model, 251
Network Data Analytics Function (NWDAF), 49–50, 328, 328f
  Nnwdaf_Analytics_Info service, 329
  Nnwdaf_EventsSubscription service, 328–329
Network Exposure Function (NEF), 46–48, 291, 332–336, 332f
  Nnef_AFsessionWithQoS service, 336
  Nnef_BDTPNegotiation service, 335
  Nnef_ChargeableParty service, 336
  Nnef_EventExposure service, 332–333
  Nnef_ParameterProvision service, 334
  Nnef_ParameterProvision_Update service operation, 334
  Nnef_PFDManagement service, 333–334
  Nnef_TrafficInfluence service, 335–336
  Nnef_Trigger service, 334
Network Functions (NFs), 105–106, 106f, 175. See also 5G core (5GC) network
Network Function Virtualisation (NFV), 10
Network Identifier (NID), 442
Network Provided Location Information (NPLI), 300
Network Repository Function (NRF), 106, 289, 318–319, 319f
  Nnrf_AccessToken service, 322
  Nnrf_NFDiscovery service, 321–322
  Nnrf_NFManagement service, 319–320
Network Slice Selection Function (NSSF), 291, 330, 330f
  Nnssf_NSSAIAvailability service, 331
  Nnssf_NSSelection service, 330–331
Network slicing, 54–55
  architecture, 54–55
  benefits, 248
  definition, 247, 247f
  examples, 248, 249f
  logical networks, 247
  management and orchestration
    commissioning phase, 250
    decommissioning phase, 251

network slice instance (NSI), lifecycle
    management of, 249
  operation phase, 250–251
  preparation, 249–250
Mobile Network Code (MNC), 248–249
selection mechanism, 255–262
  availability, 252–255
  identifiers, 251–252, 253–254t
  interworking with EPS, 262–264
  registration procedure, 258–262, 258f
Network-triggered Service Request, 400, 402–404
New Radio (NR), 1, 59, 64–71
NG application protocol (NGAP), 338, 343
  elementary procedures, 344–347
  non UE-associated services, 343
  UE-associated services, 344
N3IWF. *See* Non-3GPP inter working function
  (N3IWF)
Non-access stratum (NAS)
  message types, 341–343
  for mobility management, 339–340t
Non-3GPP inter working function (N3IWF), 292
Non-Public Network (NPN)
  PLMN services, SNPN, 442–443
  public network integrated NPN, 443
  stand-alone Non-Public Networks, 441–442
Non stand-alone (NSA) architecture, 17–19
NPLI. *See* Network Provided Location Information
  (NPLI)
NSSF. *See* Network slice selection function (NSSF)
NWDAF. *See* Network data analytics function
  (NWDAF)

O

OpenAPI specification (OAS), 358
Operating System (OS), 11

P

Packet Data Network (PDN), 78
Packet detection rule (PDR), 366–368
Packet Flow Description (PFD), 227–228
Packet forwarding control protocol (PFCP), 363
  buffering action rule, 366, 374
  control plane protocol stack, 364f
  forwarding action rule, 366, 369, 370–371t
  message format, 365f
  node related procedures, 364–366
  packet detection rule, 366–368
  QoS enforcement rule, 366, 369–371
  rules, 367f

  session related procedures, 364–365
  SMF *vs.* UPF, data forwarding, 377, 378f
  UPF to SMF, reporting from, 375–376
  usage reporting rule, 366, 371–373
PCF. *See* Policy control function (PCF)
PDR. *See* Packet detection rule (PDR)
Permanent equipment identifier (PEI), 107–108
Permanent subscription identifier (SUPI), 178–180
PFCP. *See* Packet forwarding control protocol
  (PFCP)
Point of Interception (POI), 200
Point-to-Point Protocol (PPP), 379
Policy and Charging Enforcement Function
  (PCEF), 78, 83–84
Policy and Charging Rules Function (PCRF), 39,
  75, 77–78, 83
Policy control function (PCF), 290–291, 304–312
  Npcf_AMPolicyControl service, 305–306
  Npcf_BDTPolicyControl service, 308–309
  Npcf_EventExposure service, 311–312
  Npcf_PolicyAuthorization service, 306–307
  Npcf_SMPolicyControl service, 308
  Npcf_UEPolicyControl service, 309–311
Policy Control Request Triggers (PCRT), 236
Policy Section Identifier(s) (PSI(s)), 226
PPP. *See* Point-to-Point Protocol (PPP)
Protocol Configurations Options (PCO) field, 116
Public Warning System (PWS), 50–52

Q

QoS Class Identifiers (QCIs), 203
QoS enforcement rule (QER), 366, 369–371,
  372–374t
QoS Notification Control (QNC), 231–232
Quality of Service (QoS)
  characteristics, 213–214
  Evolved Packet System (EPS), 203
  5G QoS frameworks, 203
  flow based QoS framework, 205–207
  NG-RAN, 204
  parameters, 213
  QoS Flow ID (QFI), 204
  reflective, 210–213
  signaling of, 207–210
  standardized 5QI to QoS characteristics mapping,
    214–216

R

Radio access network (RAN), 15, 75, 203
Record protocol, TLS, 361–363

Representational State Transfer (REST), 351–353
Roaming, 55–59
Router Advertisement (RA), 117, 377

S
SCTP. *See* Stream control transmission protocol
        (SCTP)
Secondary Cell Group (SCG), 265, 272–273
Security
    application layer security, 171
    5G system, 171
        security domains, 174–176
        security requirements, 172
        services, 172–174
    interworking with EPS/4G
        dual registration mode, 192
        single registration mode with N26, 191–192
        single registration mode without N26, 192
    lawful interception (LI), 172, 198–201
    network domain security
        IP based communication, 196–197
        N2 and N3 interfaces, 197–198
        Network Exposure/NEF, 198
        service based interfaces, 193–196
        2G (GSM/GERAN), 193
    primary authentication and key derivation,
        180–183
    privacy protection, 174
    telecommunications traffic, 172
    user domain security, 198
    web-browsing, 171
    wireless communication, 171
Security Anchor Function (SEAF), 177
Security Associations (SAs), 383, 386
Security edge protection proxy (SEPP), 195, 292
Security parameter index (SPI), 383
Security Policy Database (SPD), 383
SEID. *See* Session Endpoint Identifier (SEID)
Selective traffic routing, 237
Self-Organising Networks (SON), 12–13
SEPP. *See* Security edge protection proxy (SEPP)
Serialization protocol, 355–357
Service-based architecture (SBA), 105–107
    concept of, 22
    HTTP REST interfaces, 22–23
    registration and discovery, 23–26
Service based interface (SBI), 105–106, 193–195
Service data flow (SDF) template, 231–232

Service level agreements (SLAs), 81, 251
Service request procedure, 400–404
Serving network identity (SN ID), 183–185
Session endpoint identifier (SEID), 365
Session Initiation Protocol (SIP), 234
Session management
    data network (DN), connectivity service to
        basic PDU Session connectivity, 111
        multiple PDU Sessions, 113
        PDU Session properties, 114
        transport network, PDU Session and
            application traffic relation, 112–113
    edge computing, 132–134
    Ethernet PDU Session type
        broadcast, handling of, 120
        MAC address, 118–119
        QoS and charging aspects, 119–120
        Residential Gateway (RG), 117–118
        SSC modes 1 or 2, 118
        UE, 117–118
        Virtual LANs (VLANs), 119
    IP based PDU Session types
        allocation, 116–117
        5GC, 115
        IPv4, IPv6 and IPv4v6, 115–116
        Network Address Translators (NATs), 115
        QoS features, 115
    Local Area Data Networks (LADN), 135–136
    NAS message types for, 341*t*
    PDU Session concepts, 111–114
    session authentication and authorization, 134
    unstructured PDU Session type, 120
    user plane handling
        control-plane and user-plane separation
            (CUPS) and N4 interface, 124–126
        CP-UP split, 121–122
        Data Radio Bearers (DRBs), 121
        GTP-U tunnels, 121
        UPF roles, 122–124
        user plane protocol stack, 121, 121*f*
Session management function (SMF), 43, 288, 297,
        300–301, 301*f*
    Nsmf_EventExposure, 303–304
    Nsmf_PDUSession_ContextRequest, 303
    Nsmf_PDUSession_Create, 302
    Nsmf_PDUSession_CreateSMContext, 301
    Nsmf_PDUSession_Release, 303
    Nsmf_PDUSession_ReleaseSMContext, 302

Nsmf_PDUSession_SMContextStatusNotify, 302

Nsmf_PDUSession_StatusNotify, 303

Nsmf_PDUSession_Update, 302

Nsmf_PDUSession_UpdateSMContext, 301–302

vs. UPF, data forwarding, 377, 378*f*

Short Message Service (SMS), 44

 IP, 44–45

 NAS solution, 45–46

Short message service function (SMSF), 292, 324–325, 324*f*

Single Network Slice Selection Assistance Information (S-NSSAI), 251–252, 256*f*

SMF. *See* Session management function (SMF)

SPD. *See* Security Policy Database (SPD)

SPI. *See* Security parameter index (SPI)

Stateless IPv6 Address Auto Configuration (SLAAC), 117

Steering of Roaming (SOR), 191, 323–324

Stream control transmission protocol (SCTP), 387

 features, 387–388

 multi-homing, 390–391, 391*f*

 multi-streaming, 389–390

 packet structure, 391–392

 vs. transmission control protocol, 388, 389*t*

 vs. user datagram protocol, 388, 389*t*

Subscription Concealed Identifier (SUCI), 107, 179–180

Subscription Permanent Identifier (SUPI), 107

T

TAU. *See* Tracking Area Update (TAU)

TEID. *See* Tunnel Endpoint Identifier (TEID)

3rd Generation Partnership Project (3GPP), 2–4, 8, 12

 Evolved Packet Core (EPC), 5G, 77

 interworking with EPC

  dual-registration mode, 165

  NAS protocols, 163–164

  network selection, 163–164

  N26 interface, 167–168

  with N26 interface, 167–168

  single-registration mode, 164

  UE selection, 166–167

  without N26 interface, 168–169

 ultra-reliable low-latency communication (URLLC), 444

Time Sensitive Networks (TSN), 448–450

Tracking area update (TAU), 396, 421

Transport layer security (TLS), 348, 360

 handshake protocol, 360–362

 record protocol, 361–363

Tunnel endpoint identifier (TEID), 378

U

UDM. *See* Unified data management function (UDM)

UDR. *See* Unified data repository (UDR)

UDSF. *See* Unstructured data storage function (UDSF)

UE Route Selection Policy (URSP), 224–227

Ultra-reliable low-latency communication (URLLC)

 cyber-physical systems, 443

 end to end redundant user plane paths, dual connectivity based, 444–445

 low latency eMBB applications, 444

 N3/N9 interfaces, redundant transmission on, 447

 redundant user plane paths, multiple UEs, 446–447

 3GPP, 444

 transport layer, redundant transmission at, 448

Unified access control (UAC), 159–160

Unified data management function (UDM), 289–290, 312–318

 Nudm_EventExposure service, 317–318

 Nudm_ParameterProvision service, 318

 Nudm_SubscriberDataManagement (SDM) service, 315–317

 Nudm_UEAuthentication service, 317

 Nudm_UECM (Nudm_UEContextManagement) service, 313–315

Unified data repository (UDR), 290, 325–327

Uniform Resource Identifier (URI), 409

Uniform Resource Locator (URL), 409

Unstructured data storage function (UDSF), 290, 329

UPF. *See* User plane function (UPF)

Up-link Classifier (UL CL), 129–130

Usage reporting rule (URR), 366, 371–373

User plane function (UPF), 288–289

V

Vertical industries. *See* Architecture
Virtualization, 9–10
Virtual machines (VMs), 10, 250

Voice services
    Evolved Packet System (EPS) fallback, 41–42
    Voice-over-LTE (VoLTE) services, 41
    voice-over-NR, 42–43

Printed in the United States
By Bookmasters